82 6605277 0
TELEPEN
DILEWYD O STOC
WITHDRAWN FROM STOCK
£50.20
UCW
CPC
LLYFRGELL
LIBRARY
ABERYSTWYTH

AF606552

Marmosets in Experimental Medicine

Primates in Medicine

Vol. 10

Series Editors
E.I. Goldsmith, New York, N.Y.
J. Moor-Jankowski, New York, N.Y.

S. Karger · Basel · München · Paris · London · New York · Sydney

Proceedings of the Conference on Marmosets in Experimental Medicine
Oak Ridge, Tenn., March 16–18, 1977

Marmosets in Experimental Medicine

Volume Editors
NAZARETH GENGOZIAN, Oak Ridge Associated Universities, and
FRIEDRICH DEINHARDT, Max von Pettenkofer-Institut

44 figures and 65 tables, 1978

Proceedings of a Symposium sponsored by
Oak Ridge Associated Universities
Oak Ridge, Tenn., March 1977

The Symposium was supported by the U.S. National Institutes of Health,
U.S. Energy Research and Development Administration,
Abbott Laboratories, and Litton Bionetics

S. Karger · Basel · München · Paris · London · New York · Sydney

Primates in Medicine

Vol. 5: Conservation of Nonhuman Primates in 1970. Editor: B. HARRISSON (Ithaca, N.Y.). VI + 102 p., 3 tab., 1971. ISBN 3-8055-1243-0

Vol. 6: Chimpanzee: Immunological Specifities of Blood. Editor: C.H. KRATOCHVIL (Kalmazoo, Mich.). VI + 146 p., 6 fig., 27 tab., 1972. ISBN 3-8055-1389-5

Vol. 7: Transplantation in Primates. Editor: G.P. MURPHY (Buffalo, N.Y.). VIII + 140 p., 36 fig., 27 tab., 1972. ISBN 3-8055-1408-5

Vol. 8: The Baboon. Microbiology, Clinical Chemistry and Some Hematological Aspects. Editor: S.S. KALTER (San Antonio, Tex.). VIII + 172 p., 8 fig., 42 tab., 1973. ISBN 3-8055-1442-5

Vol. 9: Atherosclerosis in Primates. Editor: J.P. STRONG (New Orleans, La.). VIII + 404 p., 111 fig., 68 tab., 1976. ISBN 3-8055-2195-2

Cataloging in Publication

Conference on Marmosets in Experimental Medicine, Oak Ridge, Tenn., 1977
Marmosets in experimental medicine: proceedings of the
Conference on Marmosets in Experimental Medicine, Oak Ridge, Tenn.,
March 16–18, 1977 / Volume editors, F. Deinhardt, N. Gengozian. –
Basel; New York: Karger, 1978. (Primates in medicine; v. 10)
1. Marmosets – congresses 2. Research – congresses
3. Animals, Laboratory – congresses
I. Deinhardt, Friedrich, ed. II. Gengozian, Nazareth, ed. III. Title IV. Series
W1 PR522E v. 10 QY 60.P7 C746m 1977
ISBN 3–8055–2750–0

Printed in Switzerland by Werner Druck AG, Basel
ISBN 3-8055-2750-0

Contents

Introduction

Scientific investigators have been using marmosets, small primates of the family *Callithricidae,* in experimental studies only during the past 50 years. Until recently, most species were abundant in their wild habitat in Central and South America; they were not hunted extensively because they have little food value, although some were captured for zoological collections and others as pets. A marmoset as a pet is most charmingly portrayed in the 1567 painting by Hans Eworth of the Family Cobham, which can be seen in the Longleat collection in England. Now, however, the growing demand by scientists for these extraordinarily useful primates has resulted in a threat to their continued availability. To analyze this problem and to gain a better understanding of the scope of the research in which marmosets are being used, particularly its implications for future demand for the animals, we convened a conference at Oak Ridge, Tennessee in March 1977.

Behavioral, psychological and toxicological researchers were the first to make experimental use of marmosets, although in small numbers. Since then, studies have been made of their natural chimerism, their immunological systems, their dentition, and they have been used in studies of dental diseases. Marmosets have also been used in larger numbers for investigations of infectious diseases and virus-induced malignancies. In particular, the demonstration that some *Saguinus* ssp. were susceptible to human hepatitis A rapidly increased the number of animals used for biomedical research, an increase that accelerated when their susceptibility to oncogenic viruses was demonstrated; in these fields of endeavor alone the marmosets have been shown to be the most valuable of all nonhuman primates tested.

The relative ease with which marmosets can be bred in captivity – at least some species – has made possible the use of larger numbers of newborn animals for these investigations. It was subsequently shown that several C-type RNA (Rous sarcoma, feline sarcoma and simian-sarcoma viruses) and DNA viruses (*Herpesvirus saimiri, Herpesvirus ateles* and the human Epstein-Barr virus) could induce sarcomas, gliomas or lymphoproliferative diseases in marmosets. The first oncogenic primate C-type virus (simian sarcoma virus type I [Lagothrix]) was isolated in marmosets and oncogenicity of a human virus in a primate host (Epstein-Barr virus) was demonstrated also in these animals.

Breeding colonies of some species of marmosets were established in the late 1950s and early 1960s due to (i) the realization that the research use of marmosets might endanger the species' survival in the wild, and (ii) the need for newborn animals in certain investigations. Of importance also was that colony-born animals are better characterized and free of the many infections and parasitic diseases often present in feral animals. Nevertheless, the breeding colonies could not keep pace with rapidly growing needs and larger numbers of animals were captured every year. In addition and still much more important is the danger to the survival of the animals through the destruction of large areas of their natural habitats. Severe restrictions on the capture and export of marmosets from South America have been imposed leading to a severe impediment to some crucial medical research of major and immediate public health importance; e.g., studies of human viral hepatitis, oncogenic and slow viruses, as well as basic biologic research of potential long-range importance, such as behavioral studies and investigations of the normal physiology and immune mechanisms of marmosets. For these reasons it seemed appropriate to convene researchers from different fields, all interested in marmosets, to pool our knowledge, define goals and discuss the means by which the animals can be protected in the wild and, at the same time, insure that enough animals can be procured for essential studies in which other species cannot be used. It is particularly important that the meeting brought together animal conservationists, taxonomists, primatologists, and the medical scientists using the animals. The progress and appreciation of the problems confronting each professional interest were discussed in group sessions and informally among individual members.

It is hoped that this conference and the published proceedings will lead to progress in medicine, as well as to a better understanding of primate biology and the protection of those marmoset species endangered in the wild.

FRIEDRICH DEINHARDT
NAZARETH GENGOZIAN

Acknowledgments

The co-organizers of the conference wish to express their appreciation for the continuing interest and support of the Interagency Primate Steering Committee of the National Institutes of Health, and particularly for the assistance of the committee's executive director, Dr. Benjamin Blood. They would also like to acknowledge the encouragement of Dr. Philip L. Johnson, executive director of Oak Ridge Associated Universities.

The U.S. National Institutes of Health, the U.S. Energy Research and Development Administration, Abbott Laboratories and Litton Bionetics are thanked for their financial support of the Symposium.

Session I: Taxonomy, Distribution and Conservation

Chairman: J.R. HELD, Bethesda, Md.

Prim. Med., vol. 10, pp. 1–11 (Karger, Basel 1978)

Some Problems Relevant to the Conservation of the Callitrichidae

R.W. THORINGTON, jr.

Department of Vertebrate Zoology, National Museum of Natural History, Smithsonian Institution, Washington, D.C.

Introduction

The major cause of the impending conservation crisis in South America is the destruction of the neotropical forests. The cutting and burning of the forest is increasing at such a rate that it seems only a matter of time before all the tropical forests of the continent will have been removed. The specter of a treeless Amazon valley is usually envisioned for the misty future of 2000 A.D. [19]. As I review satellite photographs and hear of the ambitious plans and accomplishments of various South American countries, I am inclined to change my estimate to 1990. For example, Brazil has almost completed its transamazonian highway, it has completed and paved two of the three north-south highways, and it plans to open a perimeter road north of the Amazon. Each of these is or will be accompanied by deforestation and colonization along a belt of tens of kilometers or up to 100 km wide [24]. Perhaps a greater factor in the destruction of Brazil's forests is the drive to produce more beef. The southern edge of the Amazon forest is being cut and burned at an impressive rate in order to increase the area available for the cattle industry. No matter how distressing this may seem to conservationists, we must realize that this is a very exciting phase in Brazil's history as viewed by most Brazilians. 'Opening up the Amazon' is not dissimilar to 'opening up' the American west. There are many parallels between the two.

Against this background, we must consider what will become of a select part of the South American fauna, the marmosets and tamarins. These primates, the Callitrichidae, seem to survive well in second growth forests, in small parks, etc., so the problem of conserving them seems much less difficult than it does for the larger platyrrhines, such as the spider monkeys

and woolly monkeys. However, the problems are by no means trivial, and we cannot expect the callitrichids to survive merely because they are small and shy. My phrase 'a treeless Amazon valley' is hyperbolic, but cattle ranchers do not have much affection for unproductive forested areas and in South America, as elsewhere, parks are more appreciated and used if they are cleared playing fields for soccer and baseball than if they are forest groves. However, the conservation of callitrichids is far more amenable to the strategy of opportunism, which VANZOLINI [24] considers to be the most practical course in Brazil, than is true for the larger primates. A small piece of forest can maintain a high population of tamarins or marmosets, and it need not be undisturbed rain forest. Thus it would seem possible to establish a feasible conservation scheme, based on a list of taxa to be conserved, knowledge of their geographic ranges, and decisions about the minimal sizes of reserves and the types of habitat to be sought.

Endangered Callitrichidae

As of March 1977, there are four species of Callitrichidae[1] on the US Endangered Species List. These are the lion-marmosets of Southeastern Brazil, *Leontopithecus rosalia,* the two tamarins of northern Colombia and Panama, *Saguinus oedipus* and *Saguinus leucopus,* and the pied tamarin, *Saguinus bicolor,* of Amazonian Brasil. Two other taxa were listed as near extinction by participants at the International Conference on the Biology and Conservation of Marmoset Monkeys, held in 1975. These are *Callithrix flaviceps* and *Callithrix aurita,* both considered subspecies of *Callithrix jacchus* by HERSHKOVITZ [11, 13]. HERSHKOVITZ [12] suggests that *Callithrix humeralifer* is also endangered, due to its small geographic range. This is a relatively short list, but it already presents us with some problems. Since we can expect many more taxa to be added to this list in the next decade or two, and because we will not be able to conserve the whole Callitrichid radiation, we should consider our priorities immediately.

There are four genera of marmosets and tamarins. Two of these are monotypic. *Leontopithecus* is now considered as a single extant species, *L. rosalia,* with three subspecies, all of which are endangered. *Cebuella* is the other monotypic genus. It is protected by law in Peru, but it is not considered

1 For the purposes of this conference, I have not included *Callimico* in the following discussions. This does not imply my agreement with the taxonomic conclusion that *Callimico* should be excluded from the *Callitrichidae.*

rare or endangered. From a taxonomic viewpoint, these two monotypic genera should be of greatest concern in a strategy of conservation. Since one of these, *Leontopithecus,* is endangered, it should be the focus of conservation efforts, and in fact it has been. The pygmy marmoset, *Cebuella,* may not be endangered, but we would be wise to concern ourselves with its status to ensure that it does not become endangered. If either *L. rosalia* or *C. pygmaea* becomes extinct, we will lose 25% of the generic diversity of the marmoset-tamarin radiation at one time. In these two cases we have all our eggs in one basket, so the basket should be handled gingerly and watched carefully. The conservation of these taxa should receive our highest priority.

The other two genera, *Saguinus* and *Callithrix,* include several species each and a larger number of subspecies. The combined geographic ranges are large and cover several countries for each genus. Therefore neither is particularly vulnerable and our focus should be below the generic level. Both genera can be divided into groups of closely related species. HERSHKOVITZ [13] has reviewed the evidence that *Callithrix* can be treated as two groups of species. These are the *C. argentata* group, comprising *C. argentata* and *C. humeralifer,* and the *C. jacchus* group, comprising five subspecies of *C. jacchus,* some of which are argued to be distinct species [4]. Within both these species groups there are taxa which I would consider endangered, but the species groups are not. The geographic range of *C. humeralifer* has recently been extended to the south, but I think HERSHKOVITZ' [12] concern is still well founded. This is a valid species with a small range in an area that may be greatly deforested in the near future. Fortunately *C. argentata* is wide spread and unendangered.

The status of the southern subspecies of *C. jacchus* appears to be more precarious. The taxa *flaviceps* and *aurita* occur in the coastal areas north and south of Rio de Janeiro, where substantial environmental change has occurred. It is further reported that the subspecies *jacchus* has been introduced into the vicinity of Rio and has displaced the local *aurita* [AVILA-PIRES, personal commun.]. To my knowledge, this has not been critically investigated. It is probably relevant that large numbers of *C. j. jacchus* have been exported from Paraguay, because it is most probable that these originated from Paraguay or neighboring Brazil, far south of the original range of this subspecies. This situation presents two important conservation problems. First, it keeps open the question whether the *C. jacchus* group contains more than one species. Even if these can be demonstrated to be subspecies, and especially if the two ends of the range constitute subspecies which will not interbreed, it will be important to conserve examples of the southern subspecies. Second, it suggests that careful management may be necessary, to exclude *C. j. jacchus* from reserves where *flaviceps* and *aurita* are being conserved. Presumably the endangered forms are well adapted to the Atlantic forest of southeastern Brazil, whereas *jacchus* is better adapted to a more open and drier habitat. As the forest habitat is degraded, as it has been extensively in this area, the situation favors the survival of *jacchus* over the endangered forms. There are few more efficient ways to cause an extinction than to degrade an animal's habitat and also introduce a competitor which does better in the degraded habitat.

The genus *Saguinus* can also be divided into species groups [10, 11]. The bare-faced tamarins include two groups that should be of concern to conservationists. These are the *S. bicolor* group, including only a single species which is already listed as endangered, and the *S. oedipus* group. The conservation status of *S. bicolor* is not well known [26], but it has a small range in the vicinity of Manaus, one of the cultural and economic centers of the Amazon. Since it is reputedly independently derived from the hairy-faced tamarins, *S. bicolor* deserves a higher priority for study and conservation than it has had.

The *S. oedipus* group is comprised of three species, *S. oedipus* and *S. leucopus* of Northern Colombia and Panama and *S. inustus*[2] of Colombia and Brazil. *S. inustus* is probably unthreatened, but as noted by HERNANDEZ and COOPER [9] this is a guess based on ignorance. For the other species there is less ignorance and more concern. For example, *S. oedipus* is listed as endangered by the US. As currently defined, the subspecies *S. o. oedipus* is clearly an endangered form, but this is moot for the Panamanian subspecies, *S. o. geoffroyi*. I do not begrudge the protection thereby given to the Panamanian subspecies, but I think these two subspecies are sufficiently distinct that their conservation statuses should be reviewed independently. (Several biologists continue to treat *geoffroyi* as a distinct species, e.g. MOYNIHAN [16] and HERNANDEZ and COOPER [9]). The Colombian subspecies should not be abandoned because the Panamanian subspecies is still reasonably safe, nor should all our efforts go toward preserving the Colombian subspecies. If either becomes extinct we will have lost half of the genetic heterogeneity of the species. If future taxonomists, in their greater wisdom, agree that these are two species rather than two subspecies, it may be too late to conserve the form that was previously ignored. We are vulnerable in this respect in our captive breeding programs. To my knowledge, all breeding

2 HERSHKOVITZ [personal commun.] is now treating *S. inustus* as a distinct species group within *Saguinus*, although he still considers it closely related to the *oedipus* group.

colonies of *S. oedipus* are comprised of the Colombian subspecies, none of the Panamanian form. The last species of this group is *S. leucopus*. It has a small geographic range and its habitat is being destroyed rapidly [9]. When patches of forest are left, this species (like other tamarins) can thrive [1]. Thus the future of *S. leucopus* is dependent on how thoroughly the forest is removed where it occurs. These assessments lead me to the conclusion that the *S. oedipus* group of tamarins is not as vulnerable as the *S. bicolor* group, but that it has some severe conservation problems which will be exacerbated in the future.

The other groups of species of *Saguinus* have extensive Amazonian distributions. Thus their futures are linked to the development of the Amazon, principally by Brazil, and their conservation problems appear to be of lower priority and to have less immediacy than those of species in northern Colombia or southeastern Brazil. I submit that this means only that we have more time to debate our strategies and establish our priorities. For these tamarins as for other taxa, it is tempting to say that we should be most concerned about the conservation of species. However, I consider this too glib. Some species are comprised of distinct subspecies, and it is not obvious that our current definitions of the species will remain unchanged. The history of taxonomy suggests the opposite. Further, as suggested by VANZOLINI [24], it is desirable that we conserve evolutionary processes, such as taxa in the process of speciating, as well as the clearly defined species. In considering these problems, we shall find it much more difficult to define our priorities.

This problem looms large when we deal with extremely polytypic species, such as *Saguinus fuscicollis*. As defined and illustrated by HERSHKOVITZ [11], this species has 13 distinctive subspecies. It would be desirable to conserve each subspecies in its own preserve, but it is doubtful that this can be done. If it is, it will be by accident not by design. But the issue must not be dodged. *S. fuscicollis* is more than any one subspecies. Perhaps it should be treated as groups of subspecies, in which case the subspecies *fuscus, lagonotus, fuscicollis,* and *weddelli* might be singled out for priority. On the other hand, another combination of subspecies might better preserve evolutionary processes, such as the *acrensis* and *melanoleucus* forms which are so pale in color. My own recommendation is that we place high priority on maintaining the genetic diversity of polytypic species, like *S. fuscicollis*. A first approximation is probably obtained by conserving geographically diverse subspecies, under the hypothesis that geographic distance and genetic difference are correlated. The resulting priorities should be tempered by our knowledge of the morphological differences between the subspecies and by our hypotheses about their origins. The details may be argued for any particular species, but I suggest that such a scheme would enable us to rank the relative importance of different populations of tamarins. I further submit that we need priorities of this sort, so that we will know where surveys should be conducted, where we should request reserves and parks, and when we should pull out all the stops for a conservation campaign.

Conservation in the Wild

Most marmosets seem to survive well in secondary forest, and given the right circumstances they can thrive in the vicinity of towns and cities. I have observed this for three species of *Saguinus,* including *S. bicolor,* and for *Callithrix* in the vicinity of Rio de Janeiro. MÜLLER [personal commun., 1964] noted this also for *Callithrix* near Belo Horizonte. MOYNIHAN [17] found *Cebuella* also in the vicinity of human activity and described them as commensals of man. Unfortunately this does not seem to be true for *Leontopithecus,* which seems to require tall forest.

The problems of conserving *Leontopithecus,* the golden marmoset, in the wild have been much discussed [e.g. 2, 3]. I have nothing to add, but I will reiterate Recommendation 1 of the Golden Lion Marmoset Conference held in 1972: we need a good field study on the ecology, social structure, breeding biology, and behavior of *Leontopithecus* [14]. It is evident that the main problems in the conservation of *Leontopithecus* are the politics and economics of conserving its habitat. These can only be solved in the context of modern Brazil, and I suspect they can only be solved by Brazilians.

Because the habitats of other tamarins and marmosets differ from those of *Leontopithecus,* the biological requirements for reserves should be less stringent. In fact, there is basis for optimism that small populations of Callitrichids will survive wherever they are given a decent chance, whether it be in protected watersheds around reservoirs, in forest parks, on protected private land, or in suburban neighborhoods where they are tolerated or even protected. I would term this 'conservation by accident'. Without denying its importance and relevance for marmosets, I would simply warn that it permits extinction by accident too. Thus I feel we must promote an active program of conservation of marmosets in South and Central America with special priorities and strategies.

One of the first questions that arises is the desirable size of reserves for Callitrichids. This question quickly degenerates into the question of what is the minimal effective size, because it is probably true that 'bigger is better'. Most of the estimates of tamarin populations lie between 10 and 50 animals/ km^2 [6–8, 15, 18, 23]. To support 100 breeding pairs we probably need an area large enough for 500 individuals, of which half will be juveniles. Thus we would need between 10 and 50 km^2 of good tamarin habitat to support a small population.

There is much recent debate about the relevance of island biogeography to the problem of the size of nature reserves. The contention is that populations have a high probability of extinction on small islands, therefore reserves should not be small. When they become isolated from similar habitat, the reserves will effectively be islands and will gradually become depauperate if they are small [5, 22, 25]. (Simberloff and Abele [20, 21] express some disagreements with this thesis.) Because of this I would expect many persons to think that 10 km^2 is too small an area. However, there are certain conditions under which the theory of island biogeography is irrelevant. One of these conditions is the active management of a reserve to prevent such extinctions or to reintroduce the taxon if it becomes extinct. This is feasible for marmosets if there are several small reserves within the range of each taxon we are conserving. I contend this should be a serious intention in any conservation program for the Callitrichids. Under such conditions, smaller reserves can be very useful tools in management, providing experimental opportunities. For example, Barro Colorado Island is only 1.5 km^2, yet it has supported a population of tamarins for 60 years. I think it could be managed to maintain a tamarin population indefinitely. The population seems to be decreasing, but this I believe is probably due to the regeneration of the forest. This hypothesis could be tested by cutting strips of forest on this or some other reserve of similar size. If the hypothesis is correct, the population of tamarins should increase. If it is wrong and the tamarins become extinct in the 'managed' reserve, it can be allowed to regenerate and tamarins could be reintroduced from another small reserve which had not been managed in this manner. The experimental management of large reserves would be much more difficult, and it would also be more difficult to evaluate the results. It is important that we learn how to manage for the Callitrichids, for every reserve will need to be managed. The example of Barro Colorado Island suggests the management strategy of allowing a forest to return to its primary condition is not the best for tamarins. Therefore I think that conservation of marmosets and tamarins would be well served by a series of small reserves, in size between 1,000 and 10,000 hectares, which could be managed in different manners. In forested reserves larger than 10,000 hectares, it is probable that native populations of Callitrichids will survive under almost any management program.

Captive Propagation

There are two major problems of concern to me in the maintenance of Callitrichids in captivity. First is the species diversity. Our serious breeding efforts are confined to very few taxa. In the research laboratories only four species have been bred in numbers: *C. jacchus, S. fuscicollis, S. nigricollis,* and *S. oedipus* [ILAR Report, 1975]. To this we may add a new colony of *S. mystax*. Zoos are breeding a much larger variety, but necessarily in small numbers. The program to breed golden marmosets, *Leontopithecus,* is a well-publicized example. In this case, the captive population is holding steady [Kleiman, personal commun., 1977], with recruitments equaling deaths. To my knowledge there are no significant colonies of other species of Callitrichidae. Of especial concern to me is the lack of these for *C. pygmaea, S. o. geoffroyi, S. leucopus, S. bicolor, C. humeralifer,* and either of the two southern subspecies, *aurita* or *flaviceps,* of *C. jacchus.*

The second problem is the maintenance, and measure where practical, of genetic heterogeneity in the captive populations. We are handicapped because we do not know the genetic heterogeneity of wild populations of Callitrichids. However, from what we know of other mammals, it would seem advisable to maintain a strategy of outbreeding. This means that colonies should be started with a large heterogeneous founding stock. I would not opt for intraspecific hybridization between subspecies, although this might be debated in some cases. It further means that there should be conscious efforts made to keep the coefficient of inbreeding as low as possible. If cases arise which require inbreeding or necessitate it, the same data on coefficients of inbreeding and measures of genetic heterogeneity would be valuable adjuncts to data on the reproductive success of the colony. Eventually we must be able to measure the genetic health of our captive colonies just as we presently monitor the physical health of the individuals.

Conclusions and Summary

I have reviewed the conservation status of the Callitrichidae from a taxonomic viewpoint. In doing so, I have established a series of priorities among biological problems which are relevant to the conservation of marmosets and tamarins. Below I list seven of these problems, roughly in the order of the priorities I would assign them.

(1) How should *L. rosalia,* the golden marmoset, be managed in the wild and in captivity? A detailed field study of this species could contribute significantly to our knowledge about it and hence to our attempts to conserve it in the wild and to breed it in captivity.

(2) Can *C. pygmaea,* the pygmy marmoset, be managed in the wild and can the species be bred in large numbers in captivity? Since this is the other monotypic genus of Callitrichids, we should critically review our conservation strategies now. Is *Cebuella* found at high densities in the wild in areas where its food trees are common? Could it be managed by planting these species in greater concentrations than they are usually found?

(3) In what habitats is *S. bicolor,* the pied tamarin, found at greatest densities? What is the ecological or behavioral interaction between it and *S. midas?* Can a single reserve be used for both species without reducing its effectiveness for *S. bicolor?*

(4) What is the ecology of *C. humeralifer?* Are there any special biological problems which should be reflected in the management of this species?

(5) What are the behavioral and ecological interactions between the *aurita* and *flaviceps* subspecies of *C. jacchus* and the introduced *C. j. jacchus?* Are the endangered subspecies restricted to tall forest?

(6) How much genetic heterogeneity is there in wild populations of callitrichids? Are some of the small populations or endangered species going through 'genetic bottlenecks' which could lead to problems in the long-term maintenance of these populations or species?

(7) Are there a series of areas in the Amazon where several species of tamarins could be or are protected by a single reserve or park? Is there an area in Colombia where *S. f. fuscus* and *S. nigricollis* coexist; an area in Peru which would protect *S. m. mystax* and *S. f. nigrifrons,* or in Brazil which would protect *S. m. mystax* and *S. f. fuscicollis?* Reserves between the Rios Madeira and Purus of Brazil could serve to protect *S. labiatus, S. m. pluto,* and possibly a form of *S. fuscicollis*. Do these three species coexist at reasonable densities? Is there an area between the Rios Purus and Jurua where *S. f. avilapiresi* and *S. m. pileatus* coexist?

In short, there is need for more surveys in the Amazon which will document the details of tamarin and marmoset distributions – the habitats in which they occur at reasonable densities and the places in which two or more species coexist – so that reserves and parks can be located wisely.

References

1 BERNSTEIN, I.S.; BALCAEN, P.; DRESDALE, L.; GOUZOULES, H.; KAVANAUGH, M.; PATTERSON, T., and NEYMAN-WARNER, P.: Differential effects of forest degradation of primate populations. Primates *17:* 401–411 (1976).

2 BRIDGEWATER, D.D.: Saving the lion marmoset (Wild Animal Propagation Trust, Wheeling 1972).

3 COIMBRA-FILHO, A.F. and MITTERMEIER, R.A.: Distribution and ecology of the genus *Leontopithecus* in Brazil. Primates *14:* 47–66 (1973).

4 COIMBRA-FILHO, A.F. and MITTERMEIER, R.A.: New data on the taxonomy of the Brazilian marmosets of the genus *Callithrix* Erxleben, 1777. Folia primatol. *20:* 241–264 (1974).

5 DIAMOND, J.M.: Island biogeography and conservation: strategy and limitations. Science *193:* 1027–1029 (1976).

6 EISENBERG, J.F. and THORINGTON, R.W., jr.: A preliminary analysis of a neotropical mammalian fauna. Biotropica *5:* 150–161 (1973).

7 FREESE, C.: A census of non-human primates in Peru; in Primate censusing studies in Peru and Colombia, pp. 17–41 (Pan American Health Organization, 1975).

8 FREESE, C.H.; FREESE, M.A., and CASTRO, R.: The status and conservation of *Callitricids* in Peru; in KLEIMAN, The biology and conservation of the *Callitrichidae* (Smithsonian Institution Press, Washington, in press).

9 HERNANDEZ-CAMACHO, J. and COOPER, R.W.: The nonhuman primates of Colombia; in THORINGTON and HELTNE Neotropical primates, pp. 35–69 (National Academy of Sciences, Washington 1976).

10 HERSHKOVITZ, P.: Taxonomic notes on tamarins, genus *Saguinus* (*Callithricidae*, primates), with descriptions of four new forms. Folia primatol. *4:* 381–395 (1966).

11 HERSHKOVITZ, P.: Metachromism or the principle of evolutionary change in mammalian tegumentary colors. Evolution *22:* 556–575 (1968).

12 HERSHKOVITZ, P.: Notes on New World monkeys. Int. Zoo Yb. *12:* 3–12 (1972).

13 HERSHKOVITZ, P.: Comments on the taxonomy of Brazilian marmosets *(Callithrix, Callitrichidae)*. Folia primatol. *24:* 137–172 (1975).

14 KLEIMAN, D.: Recommendations on research priorities for the lion marmoset; in BRIDGEWATER Saving the lion marmoset, pp. 137–139 (Wild Animal Propagation Trust, Wheeling 1972).

15 MITTERMEIER, R.A.; BAILEY, R.C., and COIMBRA-FILHO, A.F.: Conservation status of the *Callitrichidae* in the Brazilian Amazon, Suriname; in KLEIMAN The biology and conservation of the *Callitrichidae* (Smithsonian Institution Press, Washington, in press).

16 MOYNIHAN, M.: Some behavior patterns of platyrrhine monkeys. II. *Saguinus geoffroyi* and some other tamarins. Smithsonian Contributions to Zoology, No. 28 (1970).

17 MOYNIHAN, M.: Notes on the ecology and behavior of the pygmy marmoset *(Cebuella pygmaea)* in Amazonian Colombia; in THORINGTON and HELTNE Neotropical primates, pp. 79–84 (National Academy of Sciences, Washington 1976).

18 NEYMAN, P.F.: Preliminary report on a field study of the ecology of the cotton-topped tamarin *(Saguinus o. oedipus)* in Colombia; in KLEIMAN The biology and

conservation of the *Callitrichidae* (Smithsonian Institution Press, Washington, in press).
19 Richards, P.W.: The tropical rain forest. Sci. Am. *229:* 58–67 (1973).
20 Simberloff, D.S. and Abele, L.G.: Island biogeography theory and conservation practice. Science *191:* 285–286 (1976).
21 Simberloff, D.S. and Abele, L.G.: Island biogeography and conservation: strategy and limitations. Science *193:* 1032 (1976).
22 Terborgh, J.: Island biogeography and conservation, strategy and limitations. Science *193:* 1029–1030 (1976).
23 Thorington, R.W., jr.: Observations of the tamarin *Saguinus midas*. Folia primatol. *9:* 95–98 (1968).
24 Vanzolini, P.E.: Current problems of primate conservation in Brasil. Proc. 6th Cong. Int. Primatol. Soc. (in press, 1977).
25 Whitcomb, R.F.; Lynch, J.F.; Opler, P.A., and Robbins, C.S.: Island biogeography and conservation: strategy and limitations. Science *193:* 1030–1032 (1976).
26 Wolfheim, J.S.: The status of wild primates (manuscript).

R.W. Thorington, jr., Department of Vertebrate Zoology, National Museum of Natural History, Smithsonian Institution, *Washington, DC 20560* (USA)

Prim. Med., vol. 10, pp. 12–19 (Karger, Basel 1978)

Marmoset Evolution: The Molecular Evidence

J.E. Cronin and V.M. Sarich

Departments of Anthropology and Biochemistry, University of California, Berkeley, Calif.

Introduction

The New World monkeys form a group of primates whose origin and history are both complex and relatively obscure. Even the simplest of questions, as for example, defining non-molecular derived features common to the group, is presently very difficult to answer. To such a general query might be added a number of specific ones. What is the relationship of the Platyrrhini to the Old World anthropoids, the Catarrhini? What are the relationships among modern ceboids? What are the relationships among the marmosets? Is *Callimico* a marmoset? Where did anthropoids originate and how do ancestors of the modern Platyrrhini enter South America? Although there are a number of fossil New World monkeys, only one *(Neosaimiri)* seems to have anything to do with the modern forms [5]. We believe that such a situation is particularly well-suited to resolution using various lines of molecular evidence. We have reviewed in a previous paper [3] the origin of the Anthropoidea, the relationships within the modern New World monkeys, and the relationship of New World forms to Old World higher primates. One other immunological study of the ceboid serum proteins has also dealt with New World monkey evolution [1]. In this paper we review the molecular relationships within the Ceboidea with special emphasis on marmoset systematics.

Materials and Methods

Immunological Procedures

The serum proteins albumin and transferrin were purified according to methods detailed in CRONIN and SARICH [3]. The purified proteins were injected into rabbits and antisera produced according to published schedules [7]. The resulting antisera were titered using the microcomplement fixation (MC′F) procedure and pooled in reciprocal proportion to their MC′F titers [10]. Antisera were produced to the albumins and transferrins of numerous catarrhine primates for use in assessing intraplatyrrhine rates of evolution, and to 11 ceboid albumins and five ceboid transferrins for delineating platyrrhine relationships. All immunological differences are expressed in terms of immunological distance units, where for albumin and transferrin one unit is approximately equal to one amino acid substitution [2].

Electrophoretic Procedures

Electrophoresis of serum proteins was carried out in a vertical slab polyacrylamide gel (7% acrylamide, 0.12% bis) in a continuous system (0.1 M Tris, 0.05 M glycine) [4, 8, 9]. Under these conditions some 20–25 bonds can be resolved. As an indicator of genetic similarity in such comparisons, we simply use the ratio of the number of bands that have the same mobility to the total number of bands counted in both samples. Pairs of individuals within populations will generally give a coefficient of genic similarity (S) ≥ 0.95, and D (–ln S)≤ 0.05. For the serum proteins a D value of 1 is reached in approximately 2.5 million years.

Results

Figure 1 is the phylogeny of the New World monkeys based on all available platyrrhine albumin and transferrin immunological data. The tree is constructed according to procedures published in CRONIN and SARICH [3] and SARICH and CRONIN [9]. The immunological data matrices are from CRONIN and SARICH [3].

The immunological data, whether analyzed independently for each protein or for the summed albumin plus transferrin distances, lead to the following conclusions. Except for *Aotus,* all of the modern New World monkeys are clearly part of a single primary adaptive radiation that occurred some appreciable time after the catarrhine-platyrrhine divergence. There then followed three secondary radiations which led to modern forms. *Ateles, Alouatta,* and *Lagothrix* form a monophyletic unit, as do *Cacajao* and *Pithecia.* The callithricids are a monophyletic unit not more closely associated with any ceboid lineage to the exclusion of others. Within the marmosets *Callimico* appears to share a more recent common ancestry with

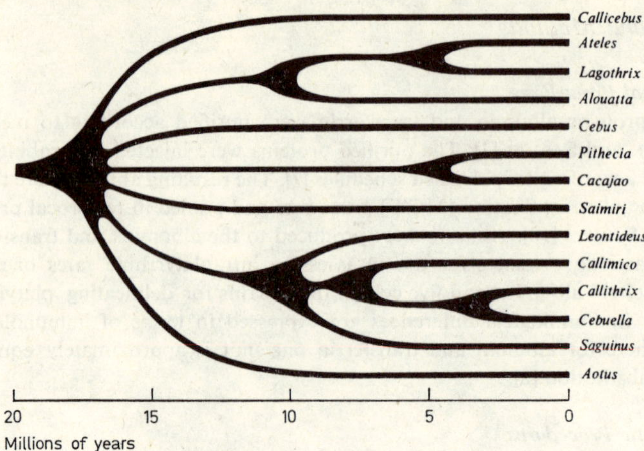

Fig. 1. A phylogeny based on all available platyrrhine albumin and transferrin immunological data. The shaded areas are meant to indicate the uncertainty as to the precise placement in time of that node or nodes. Thus we do not mean to rule out the possibility that one or another of the seven major lineages may have shared a brief period of common ancestry subsequent to the beginning of the adaptive radiation leading to the modern New World monkeys. We have here chosen to emphasize the transferrin placement of *Aotus* for reasons discussed in the text.

Callithrix and *Cebuella* than with either *Saguinus* or *Leontideus*. The other genera do not associate relative to one another. There is no indication in our data that the taxa Aotinae *(Aotus* plus *Callicebus)* and Cebinae *(Cebus* plus *Saimiri)* are of monophyletic origin.

The concordance between the albumin and transferrin data sets is extremely high. However, these are two discordant elements. In the first the albumin data suggest that the *Aotus* lineage represents the earliest separation among the extant New World monkeys; the transferrin data show the *Aotus* lineage as part of the general ceboid radiation. We tend to distrust the albumin placement in this case because of the peculiar, very conservative nature of its evolution in the *Aotus* lineage. Other immunological comparisons have been interpreted to suggest that *Aotus* is more closely aligned to the marmosets [1]. None of our data are consistent with this suggestion, and until DNA hybridization and further electrophoretic and immunological comparisons are carried out, we prefer the arrangement in figure 1. The second discordant result is more quantitative. This problem involves the marmosets and the striking difference in their rates of albumin and transferrin evolution. On the average, transferrin evolves about 50% faster than albumin [3]. Among the marmosets, however, the situation is reversed. The mean transferrin immunological distances among *Saguinus, Leontideus,* and *Callithrix* are about seven to ten units, while the corresponding albumin distances are about 20 units. Nonetheless, cladistic analysis of each data set shows the marmosets to be of monophyletic origin relative to the cebids. The question is the age of this monophyletic lineage. In our original publication, we emphasized the albumin data though we recognized that the Callithricidae may be of a more recent common ancestry as the transferrin data would suggest.

Recent allozyme electrophoretic studies [Bruce, thesis] give a *Saguinus-Callithrix* distance which is in line with that to be expected using the albumin plus transferrin immunological data. Specifically, take the series (1) *Saguinus – Callithrix,* (2) marmosets – other New World monkeys, (3) platyrrhines – catarrhines. Summed albumin plus transferrin immunological distances in these three comparisons are 27, 64, and 130, respectively. Genetic distance (D) values are 0.26, 0.60, and approximately 1.1. The ratios are extremely close and suggest that the marmoset radiation is extremely recent; that is, about 40% the age of the primary platyrrhine radiation.

Intramarmoset Relationships

The immunological and electrophoretic data do not as yet allow us to convincingly associate any pair of the three marmoset lineages to the exclusion of the third, though it may be that *Leontideus* split off slightly earlier (see below). The situation is very similar to that involving man, chimpanzee, and gorilla – where a much larger data base still cannot resolve that trichotomy – and may well simply be what remains to us to be seen some millions of years subsequent to the occurrence of an adaptive radiation. Those three lines are the monotypic genus *Leontideus, Saguinus,* and *Callimico-Callithrix-Cebuella* – with the latter two appearing, from the immunological data, to form well-defined clusters with no tested species being of ambiguous placement. That is, *oedipus, mystax, nigricollis, fuscicollis,* and *leucopus* are clearly members of one clade, and *Callimico, Callithrix,* and *Cebuella* are clearly members of a second. Intraclade systematics are not at all clear as yet due to a lack of effort and availability of representative samples of species (especially within *Callithrix*), but a few preliminary conclusions can be drawn using immunological and plasma protein electro-

Table I. Immunological (ID) and plasma protein electrophoretic (PPE) distances among the marmosets

	Albumin plus transferrin ID	PPE
S. fuscicollis-nigricollis	NA	0.4
(S.f.-S.n.)-S. mystax	NA	0.7
S. oedipus-S. mystax	NA	1.1
S. oedipus-S. mystax-S. leucopus	10	(1.1)
Saguinus-Callithrix	26	(1.7)
(Saguinus-Callithrix)-Leontideus	32	NA
C. jacchus-Cebuella	7	(1.0)
Callithrix-Callimico	22	NA

The immunological distances were obtained using antisera to the albumin and transferrin of *C. jacchus*, the albumin of *S. oedipus*, and the transferrin of *S. leucopus*. NA = Not available.

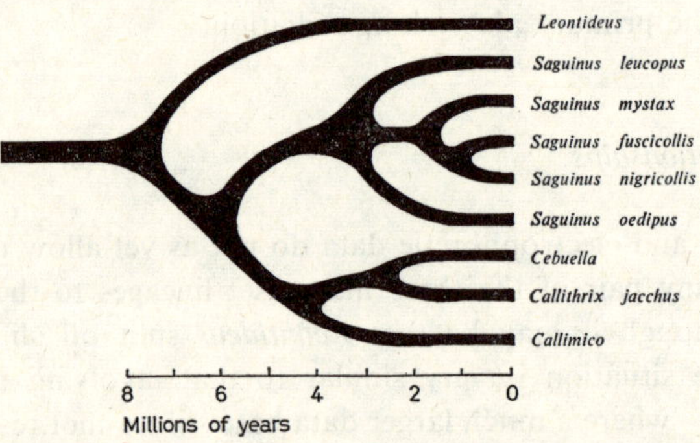

Fig. 2. Figure 1 is taken directly from CRONIN and SARICH [3]. This figure is an attempt to provide as precise a placement as currently possible from the various available molecular data sets of the marmoset lineages – with the changes from figure 1 being inspired, in the main, by the plasma protein electrophoretic work.

phoretic data. The uses of the former have been extensively discussed in print; the latter approach is relatively new [9].

The serum protein electrophoretic data (table I) lead to the following tentative conclusions (fig. 2). First, the immunological unity of *Saguinus* is reflected as a subjective pattern similarity among the various species of that genus, although the differences among *oedipus, mystax, leucopus,* and *fuscicollis-nigricollis* can be large enough to suggest divergence times of up to 3–4 million years. The closest relationship is between *fuscicollis* and *nigricollis* with a *D* value of 0.4. Next *mystax* joins the *fuscicollis-nigricollis* clade at a *D* value of ~0.7, and *oedipus* is appreciably more distant ($D \cong 1.1$). Our *leucopus* sample is older than these and has not maintained as much of its electrophoretic integrity, thus its placement is more tenuous. We have no data specifically associating it with either *oedipus* or *fuscicollis-nigricollis-mystax;* thus it is left as a third member of the *Saguinus* clade. The even older *Cebuella* sample (collected in 1966) is nonetheless in reasonably good condition, and its serum protein electrophoretic affinities with *Callithrix jacchus* are clear ($D \cong 1$), again supporting the immunology. Values of this magnitude, in our experience with primate sera, are typical of closely related species within a genus; for example, *Pan paniscus* and *troglodytes, Macaca maura* and *nigra,* or *Macaca arctoides* and *fuscata.* Thus one is led to wonder if, assuming the validity of at least some other *Callithrix* species, they might not be at least as far from *jacchus* as is *Cebuella* – in which case the generic status of the latter is put in serious jeopardy. Finally, after comparing the overall patterns subjectively, we suspect that *Callithrix* and *Saguinus* may be more closely related to one another than either is to *Leontideus* – a suspicion supported (though hardly conclusively) by the fact that the *Callithrix-Saguinus* albumin immunological distance of 21 units is less than the 25 units to *Leontideus* albumin.

Discussion

We see the marmosets as a very compact evolutionary unit of relatively recent origin, with all extant lineages still sharing a common ancestral lineage on the order of 7–10 million years ago. This implies that the marmoset grade of evolution cannot reasonably be seen as a retention of features primitive for the New World monkeys as a whole, but should be seen as a derived state developing along that common lineage subsequent to the basic New World monkey radiation and finally resulting in the adaptive radiation from which the modern lines all stem. The closest analogue among the primates would be the Cercopithecinae; among other mammals, the modern felines form a group of comparable diversity and recency of origin.

To continue to view the marmosets as primitive within the cladistic context provided by the molecular evidence would require that their features be those of the most recent common ancestor of all extant New World monkeys, thus negating the reality of a cebid clade and grade of organization. This appears to us to make unrealistic demands upon the relatively rare contributions of parallel and convergent evolution. It seems to us far easier to view that most recent common ancestor as a cebid, to see the cebid grade as the primitive one, to view many of the so-called 'primitive' marmoset features as simply results of their small size, and to accept the marmosets as one of the most recent and successful experiments in primate evolution.

Summary

We report here the results of comparative immunological and electrophoretic studies of the serum proteins of the New World monkeys. Specifically, we find that the New World monkeys share a long period of common ancestry with the Catarrhini and that the divergence between these two groups occurred some 35–40 million years ago. The extant New World monkey lineages are then seen as sharing a long period of common ancestry subsequent to that divergence, with their radiation beginning in the early Miocene. We see seven distinct lineages stemming from this radiation: (1) *Aotus,* (2) *Callicebus,* (3) *Cebus,* (4) *Saimiri,* (5) *Ateles-Lagothrix-Alouatta,* (6) *Pithecia-Cacajao* and (7) *Callimico-Callithrix-Cebuella-Saguinus-Leontideus.* Within those *Ateles* with *Lagothrix,* and *Callimico* with *Callithrix-Cebuella* form further subgroups. The marmoset radiation appears to have begun some 7–10 million years ago.

Acknowledgements

All of the experimental work was carried out in the laboratory of Dr. A.C. WILSON, Biochemistry Department, University of California, Berkeley. We thank him for making the facilities available. *Callithrix* and *Saguinus* samples used for electrophoretic comparisons were provided by Dr. N. GENGOZIAN of Oak Ridge Associated Universities. Other sources have been acknowledged in previous publications.

References

1 BABA, M.L.; GOODMAN, M.; DENE, H., and MOORE, G.W.: Origin of the Ceboidea from an immunological perspective. J. hum. Evol. *4:* 89–102 (1975).

2 CHAMPION, A.B.; PRAGER, E.M.; WACHTER, D., and WILSON, A.C.: Microcomplement fixation; in WRIGHT Biochemical and immunological taxonomy of animals, pp. 397–416 (Academic Press, London 1974).

3 CRONIN, J.E. and SARICH, V.M.: Molecular systematics of the New World monkeys. J. hum. Evol. *4:* 357–375 (1975).

4 CRONIN, J.E. and SARICH, V.M.: Molecular evidence for dual origin of mangabeys among Old World monkeys. Nature, Lond. *260:* 700–702 (1976).

5 HERSHKOVITZ, P.: Relationships, origins, and history of the ceboid monkeys and caviomorph rodents: a modern reinterpretation; in DOBZHANSKY, HECHT and STEERE Evolutionary biology, pp. 323–347 (Appleton Century Crofts, New York 1972).

6 MAXSON, L. and WILSON, A.C.: Convergent morphological evolution detected by studying proteins of tree frogs in the *Hyla exemia* subgroup. Science *185:* 66–68 (1974).

7 SARICH, V.M.: Pinniped origins and the rate of evolution of carnivore albumins. Systemat. Zool. *18:* 286–295 (1969).

8 SARICH, V.M.: Rates, sample sizes, and the neutrality hypothesis for electrophoresis in evolutionary studies. Nature, Lond. *265:* 24–28 (1977).

9 SARICH, V.M. and CRONIN, J.E.: Molecular systematics of the primates; in GOODMAN and TASHIAN Molecular anthropology, pp. 141–170 (Plenum Press, New York 1976).

10 SARICH, V.M. and WILSON, A.C.: Quantitative immunochemistry and the evolution of primate albumins. Microcomplement fixation. Science *154:* 1563–1566 (1966).

J.E. CRONIN, PhD, University of California, Berkeley Department of Anthropology, *Berkeley, CA 94720* (USA)

Prim. Med., vol. 10, pp. 20–29 (Karger, Basel 1978)

Callitrichids in Brazil and the Guianas: Current Conservation Status and Potential for Biomedical Research

RUSSELL A. MITTERMEIER, ADELMAR F. COIMBRA-FILHO and MARC G.M. VAN ROOSMALEN

Department of Anthropology and Museum of Comparative Zoology, Harvard University, Cambridge, Mass.; Department of Environmental Conservation, FEEMA, Rio de Janeiro, and LBB/STINASU, Paramaribo

Introduction

Although callitrichids have been widely used in biomedical research and were until recently imported in large numbers, we still know very little about the conservation status of wild populations of these animals. Recent surveys in Peru, Colombia, Bolivia, Brazil and the Guianas have given us some preliminary data on certain species, but the current situation of many others remains unknown. In this paper, we give a brief summary of the conservation status of all callitrichids occurring in Brazil, Surinam and French Guiana, where most of our recent work has taken place, and indicate the possibilities of each for biomedical research. We also discuss some relevant ecological data from long-term investigations conducted in Surinam, and mention some of the political problems involved in obtaining a supply of these animals.

Methods

Data on callitrichids in Brazilian Amazonia come from two surveys. The first was conducted by R.A.M. from July to November, 1973 and investigated 22 localities in middle and upper Brazilian Amazonia. The second was conducted by R.C. BAILEY (Harvard University) in November, 1973 and took place in the Rio Quixito, a small tributary of the Rio Itacuaí (Rio Javarí drainage) near the Brazilian-Peruvian border [12, 13]. Data on eastern Brazilian callitrichids come from field work conducted by A.F.C.-F. throughout the region, since the initiation of his studies of *Leontopithecus* in 1966 [2–4]. Information on *Saguinus midas midas,* the only callitrichid in the Guianas, comes from a brief survey in French Guiana in June, 1975 and from surveys and long-term studies carried out in Surinam by R.A.M. in April to June, 1975 and by R.A.M. and M.G.M.R. from January, 1976 to February, 1977. (Information on *S. m. midas* in Guyana, the third of the Guianas, may be found in MUCKENHIRN *et al.* [14] and is not discussed here.) The only locality investigated in French Guiana was the vicinity of Saül, a small village located in the interior, about 185 km southwest of Cayenne. In Surinam, survey work was conducted in nine different localities and long-term studies took place at the Raleighvallen-Voltzberg Nature Reserve along the Coppename River in central Surinam.

Results

Preliminary Findings in Brazil

Detailed ecological data are unfortunately not yet available for any Brazilian callitrichid and much survey work and long-term studies are still needed before we can have a clear picture of the status of all Brazilian species. However, the initial surveys conducted in both Amazonia and eastern Brazil make it possible to give at least a preliminary assessment of several species.

As far as we can tell, no callitrichid in Brazilian Amazonia is yet endangered and several species still appear to be widespread and abundant. However, a number of species with small geographic ranges must be considered vulnerable because the large-scale clear-cutting of forest accompanying Brazil's massive Amazonian development program threatens to destroy or is already destroying their habitat. In the past, Amazonian primates were for the most part protected from the encroachment of civilization by the sheer magnitude of the area. With the recent construction of an elaborate 14,000 km highway system, the remoteness of Amazonia is rapidly coming to an end.

In eastern Brazil, most callitrichids are already vulnerable or endangered. This part of the country has been settled for hundreds of years and habitat destruction, which has been taking place for a long time, is now more widespread than ever.

Hunting is fortunately not a problem for callitrichids in Brazil, since they are rarely shot for food. Callitrichids are sometimes captured to serve a local pet market, but numbers involved are small and of little consequence except for the most endangered species. For example, in the four month Brazilian Amazonia survey, we saw 36 monkeys being kept as pets and only one of these was a callitrichid *(Callithrix argentata argentata)*.

Table I. Conservation status of callitrichids in Brazil and the Guianas

Cebuella	
C. pygmaea	apparently common
Callithrix	
C. argentata argentata	common
C. argentata leucippe	vulnerable (possibly endangered?); small range, habitat being destroyed by clear-cutting to make way for pasture-land
C. argentata melanura	status unknown; probably common
C. humeralifer humeralifer	status unknown; possibly vulnerable because of small range
C. humeralifer chrysoleuca	status unknown; possibly vulnerable because of small range
C. jacchus	common; a widespread, adaptable species introduced in many areas outside its natural range
C. penicillata kuhlii	vulnerable; small range, habitat destruction widespread
C. penicillata penicillata	common; an adaptable, widespread species
C. geoffroyi	apparently common
C. flaviceps	endangered; very small range; habitat being destroyed by clear-cutting of forest
C. aurita aurita	endangered; small range, habitat destruction widespread
C. aurita caelestis	vulnerable; habitat destruction widespread
Leontopithecus	
L. rosalia rosalia	endangered; widespread habitat destruction in very small remaining range
L. rosalia chrysomelas	endangered; habitat destruction in very small range
L. rosalia chrysopygus	endangered; widespread habitat destruction has reduced this animal to two small, widely-separated populations
Saguinus	
S. bicolor bicolor	reportedly common, but possibly vulnerable; an adaptable species, but with a small range in close proximity to Manaus, second largest city in Amazonia; the status of this species should be investigated as soon as possible
S. bicolor martinsi	status unknown
S. bicolor ochraceous	status unknown
S. fuscicollis fuscicollis	common; one of the three most frequently exported Brazilian callitrichids, but apparently still abundant
S. fuscicollis acrensis	status unknown
S. fuscicollis avilapiresi	status unknown
S. fuscicollis crandalli	status unknown
S. fuscicollis cruzlimai	status unknown
S. fuscicollis fuscus	status unknown
S. fuscicollis melanoleucus	common?; local people report it common in the Rio Tarauaca region
S. fuscicollis weddelli	no data

Table I (Continuation)

S. imperator	status unknown
S. inustus	status unknown
S. labiatus labiatus	status unknown
S. labiatus thomasi	status unknown
S. midas midas	common; a widespread, adaptable species and the only callitrichid in the Guianas
S. midas niger	common; apparently surviving well in spite of small range in one of the most densely populated parts of Amazonia
S. mystax mystax	common; one of the three most frequently exported Brazilian callitrichids but apparently still abundant
S. mystax pileatus	status unknown
S. mystax pluto	status unknown
S. nigricollis	common; one of the three most frequently exported Brazilian callitrichids, but apparently still abundant

Taxonomy follows HERSHKOVITZ [9–11] except for genus *Callithrix*. We recognize five eastern Brazilian species where he recognizes only one.

Table I briefly reviews what is known of the conservation status of all Brazilian callitrichids. Each species is categorized as endangered, vulnerable, common or status unknown. Endangered species are animals on the verge of extinction and should not be considered for use in biomedical research. Vulnerable species still exist in numbers sufficient for long-term survival, but are diminishing rapidly because of man's activities. A continuing supply of vulnerable species for biomedical research is out of the question, but capture of small nuclei (perhaps ten pairs) for captive breeding programs would not be harmful to the survival of the species. Common species occur in sufficient numbers and are wide-ranging enough to withstand sustained yield cropping of a few hundred animals each year, but we definitely need more data on them before reasonably accurate, long-term sustained yield quotas can be determined. Although they are not presently in any danger, widespread habitat destruction such as is now taking place in many parts of Brazil could eventually place them in a precarious position as well. Species for which no data are available are categorized as status unknown. Field research is needed as soon as possible to determine their conservation status and possibilities for biomedical research.

Saguinus midas midas *in Surinam and French Guiana*

S. m. midas is the only callitrichid found in the Guianas and is one of the most abundant primate species there. It accounted for almost 25% of approximately 170 primate groups encountered during the course of field work, a higher percentage than for any other species. The animal appears to be quite flexible in choice of habitat, being found in high rain forest, coastal savanna forest, mountain savanna forest, liane forest and various secondary formations (but not riverside forest). On occasion, it is even seen on the ground, crossing roads. In the Voltzberg study area, it can be found in five different forest types and shows a preference for edge habitats of various kinds (fig. 1).

S. m. midas is also the least persecuted primate in the Guianas. In interviews conducted with 42 Surinamers (including Javanese farmers, Bushnegroes from four different tribes and Indians from five different tribes), only 11 (26%) said they would eat *Saguinus*. Most of these indicated that they would do so only if nothing else was available. This was by far the lowest percentage for a primate, the closest being *Saimiri* (50%). By contrast, more than 80% of those interviewed said they would eat some of the larger species like *Alouatta* and *Chiropotes*. Reasons for not eating *Saguinus* included its small size, its attractive appearance, and the fact that some people consider it crazy because it sometimes falls out of trees. Among the Bushnegroes, it is also taboo for anyone who has a twin sibling or twin children to eat *Saguinus*.

S. m. midas is sometimes kept as a pet in Surinam, but it is not as popular as several other species and the numbers involved are small and not likely to affect the species in any way. Of 84 monkeys being kept as pets in areas we investigated, only nine were *Saguinus*.

Discussion

Habitat destruction is without a doubt the major threat to the survival of callitrichids and other Neotropical primates, and if it continues at its present rate even the most common species will eventually be endangered. However, many callitrichid species appear to have a number of advantages over their cebid relatives. First of all, they are very adaptable animals as far as choice of habitat is concerned. Although widespread clear-cutting of forest is of course disastrous for all species, small-scale environmental alterations such as those accompanying slash and burn agriculture are not necessarily

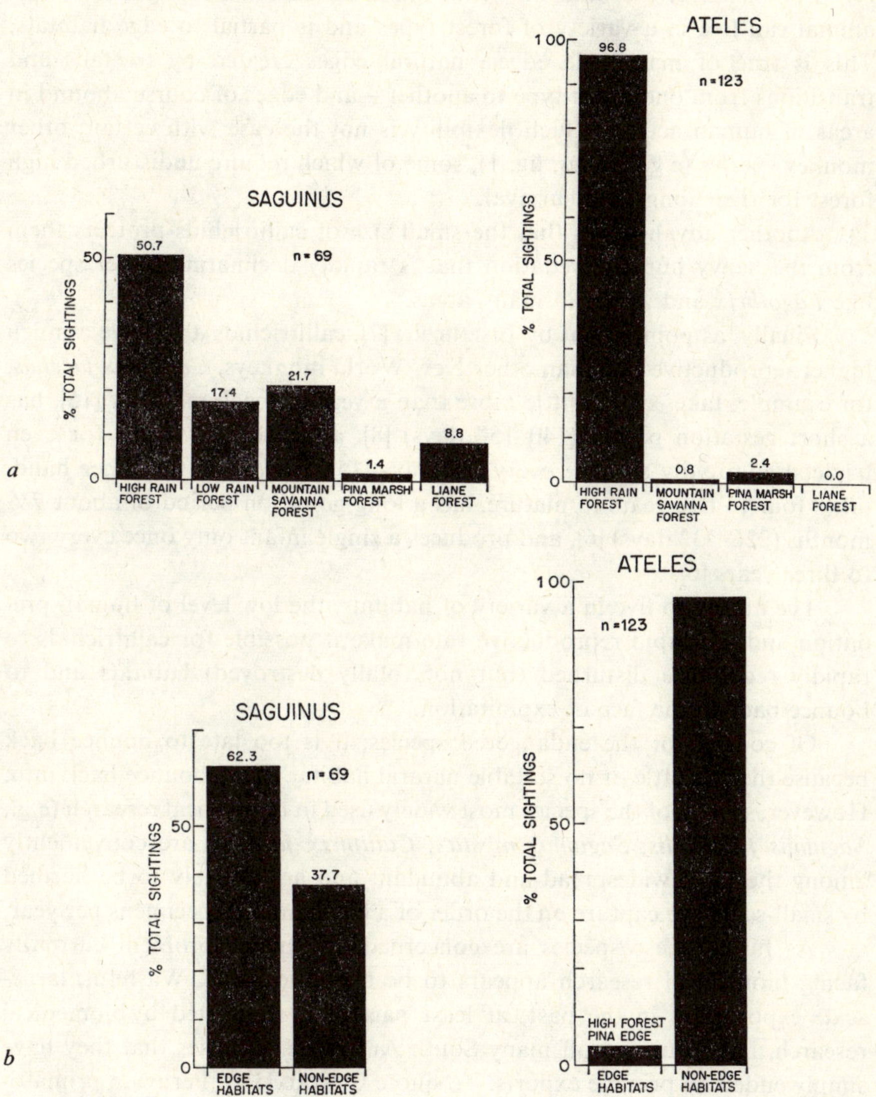

Fig. 1a. Habitat types utilized by *Saguinus midas midas* and *Ateles paniscus paniscus* in the Voltzberg study area, Raleighvallen-Voltzberg Nature Reserve, Surinam. *b* Utilization of edge habitats by *Saguinus midas midas* and *Ateles paniscus paniscus* in the Voltzberg study area, Raleighvallen-Voltzberg Nature Reserve, Surinam. All graphs based on first sightings.

detrimental to callitrichids and may even benefit them in the long run.

For instance, the data on *S. m. midas* in Surinam indicate that this animal can live in a variety of forest types and is partial to edge habitats. This is true of man-made edges, natural edges created by treefalls and transitions from one forest type to another – and edges of course abound in areas of human activity. Such flexibility is not the case with certain other monkey species (e.g., *Ateles,* fig. 1), some of which require undisturbed high forest for their long-term survival.

Another advantage is that the small size of callitrichids protects them from the heavy human predation that is rapidly decimating larger species like *Lagothrix* and *Ateles* in many areas.

Finally, as pointed out by Eisenberg [7], callitrichids also have a much higher reproductive rate than other New World monkeys. *Callithrix jacchus,* for example, takes only a little more than a year to reach maturity [15], has a short gestation period (140–150 days) [8], and produces twins (or even triplets) as rapidly as once every 160 days [15]. *Ateles,* on the other hand, takes four to five years to mature, has a long gestation period of about 7½ months (226–232 days) [6], and produces a single infant only once every two to three years [6].

The ability to live in a variety of habitats, the low level of human predation and the rapid reproductive rate make it possible for callitrichids to rapidly recolonize disturbed (but not totally destroyed) habitats and to bounce back in the face of exploitation.

Of course, for the endangered species, it is too late to bounce back because there is little or no suitable natural habitat left to bounce back into. However, several of the species most widely used in biomedical research (e.g., *Saguinus fuscicollis, Saguinus mystax, Callithrix jacchus*) are conveniently among the most widespread and abundant and are unlikely to be harmed by small-scale live capture on the order of a few hundred specimens per year.

As far as these species are concerned, the major problem currently facing biomedical research appears to be a political one. Wasteful, large-scale exportation in the past, at least part of it supported by biomedical research, has so turned off many South American countries that they have simply ended all primate exports. To quote Castro [1], a Peruvian primatologist, 'although the rational exploitation of such animals could obviously bring permanent benefits, to date man has only abused the resource... For the general national welfare, the trade in these monkey species could, no doubt, be better managed'.

The negative reaction of many former exporting countries is quite understandable. In most cases, primate export was conducted by profit-motivated businessmen, many of them foreigners, who made a lot of money for themselves, had little concern for the animals they shipped, and virtually no interest in conservation. Existing protective legislation was also frequently ignored.

This has now changed. South American countries have come to recognize primates as an important natural resource and it is unlikely that they will again permit uncontrolled exportation or the reopening of the privately-run primate trade.

It seems that there are now two equally reasonable legal alternatives open to biomedical research:

(1) Short-term, government-approved trapping expeditions like the Delta trapping program for *Saguinus mystax* [5]. In such operations, the research institutions requiring a particular species can determine in advance how many animals they need, contact a source country, make appropriate arrangements for the capture of the required stock, and then collect that number and no more. With a little advance planning, it should be possible to collect many of these animals from areas that are slated for destruction to make way for agriculture or pastureland, where the animals are doomed anyway. Hopefully, most of the animals obtained from the wild in this manner will be used to establish domestic captive breeding programs that will eventually supply most of the needs of research.

(2) Purchase of animals trapped from special management areas or raised in captive breeding programs in the source countries themselves. Some of the source countries like Peru [2] are starting experimental ranching and captive breeding projects of their own and these will also help supply the needs of biomedical research without harming natural populations.

Of course, it would probably also be constructive if biomedical institutions could take an active interest in conservation in the source countries by supporting conservation-oriented primate research. This would not only create a feeling of goodwill and convince source countries that biomedical research is sincerely concerned about primate conservation, but would also provide essential data on the behavior and ecology of the primates biomedical research is interested in.

Summary

Brazilian and Guianan callitrichids are categorized as endangered, vulnerable, common or status unknown. Endangered species should not be used in biomedical research and a continuing supply of wild-caught vulnerable species is also out of the question. Common species, on the other hand, could withstand limited sustained yield cropping without difficulty. Species categorized as status unknown should not be used until more data on them become available. Callitrichids have a higher reproductive potential than other New World monkeys, are for the most part quite adaptable, and are rarely hunted for food or captured as pets locally. Widespread habitat destruction is the major threat to their survival and species with naturally small ranges in areas of high human activity (e.g. *Leontopithecus, Callithrix flaviceps*) are in the greatest danger.

References

1 Castro, R.N.: Guidelines for the conservation of primates in Peru; in 1st Int.-Am. Conf. on Conservation and Utilization of American Nonhuman Primates in Biomedical Research, pp. 216–234 (Pan American Health Organization, Washington 1976).

2 Coimbra-Filho, A.F. and Mittermeier, R.A.: Distribution and ecology of the genus *Leontopithecus* Lesson, 1840 in Brazil. Primates *14:* 47–66 (1973).

3 Coimbra-Filho, A.F. and Mittermeier, R.A.: New data on the taxonomy of the Brazilian marmosets of the genus *Callithrix* Erxleben, 1777. Folia primatol. *20:* 241–264 (1973).

4 Coimbra-Filho, A.F. and Mittermeier, R.A.: Conservation of the Brazilian lion tamarins *(Leontopithecus rosalia)*; in Rainier and Bourne Primate conservation (Academic Press, New York, in press).

5 Dawson, G.A.: The Delta Regional Primate Research Center's *Saguinus mystax* trapping project: a field report (unpublished, 1975).

6 Eisenberg, J.F.: Communication mechanisms and social integration in the black spider monkey, *Ateles fusciceps robustus*, and related species. Smithson. Contr. Zool. *213:* 1–108 (1976).

7 Eisenberg, J.F.: Comparative ecology and reproduction of New World monkeys; in Kleiman Biology and conservation of the Callitrichidae (Smithsonian Press, Washington, in press).

8 Gengozian, N.; Smith, T.A., and Gosslee, D.G.: External uterine palpation to identify stages of pregnancy in the marmoset, *Saguinus fuscicollis* ssp. J. med. Primatol. *3:* 226–243 (1974).

9 Hershkovitz, P.: Metachromism or the principle of evolutionary change in mammalian tegumentary colors. Evolution *22:* 556–575 (1968).

10 Hershkovitz, P.: The evolution of mammals in Southern Continents. VI. The recent mammals of the Neotropical Region. A zoogeographic and ecological review. Q. Rev. Biol. *44:* 1–70 (1969).

11 Hershkovitz, P.: Notes on New World monkeys. Int. Zoo. Yb. *12:* 3–12 (1972).

12 Mittermeier, R.A. and Coimbra-Filho, A.F.: Primate conservation in Brazilian Amazonia; in Rainier and Bourne Primate conservation (Academic Press, New York, in press).

13 Mittermeier, R.A.; Bailey, R.C., and Coimbra-Filho, A.F.: Conservation status of the Callitrichidae in Brazilian Amazonia, Surinam and French Guiana; in Klei-

MAN Biology and conservation of the Callitrichidae (Smithsonian Press, Washington, in press).

14 MUCKENHIRN, N.; MORTENSEN, B.K.; VESSEY, S.; FRASER, C.E.O., and SINGH, B.: Report on a primate survey in Guyana, July-October, 1975. Report to Pan American Health Organization, Washington (1975).

15 PHILLIPS, I.R.: The reproductive potential of the common cotton-eared marmoset *(Callithrix jacchus)* in captivity. J. med. Primatol. *5:* 49–55 (1976).

R.A. MITTERMEIER, PhD, Museum of Comparative Zoology, Harvard University, *Cambridge, MA 02138* (USA)

Prim. Med., vol. 10, pp. 30–36 (Karger, Basel 1978)

Demography and Wildlife Management of Tamarins and Marmosets

P. G. HELTNE

Division of Comparative Medicine and Departments of Anatomy and Pathobiology, The Johns Hopkins University, Baltimore, Md.

Introduction

For some time in the future the cheapest source of any species of primate will be the animal born and reared in the wild. This will be equally true of galagos, tamarins, rhesus, orangs or any other primate. The very valid research uses of primates dictate that demands for primate imports will continue.

When, if ever, can a primate of any species be removed from a wild population? When that removal constitutes minimal damage to the present or future potential of the species or the local population. How then can we gauge when and where removal may be relatively harmless?

The first part of the answer to this question comes from a generalized consideration of demographic parameters: (a) average number of safe reproductive years for an adult female, (b) number of young per year per female, (c) sex ratio at birth, (d) relative effect of mortality on the sexes, (e) times at which mortality is assessed, and (f) number of years to maturity. HELTNE *et al.* [5] show how these parameters could be applied to data from field surveys of howler monkey populations. They developed microdemographic criteria which indicate whether or not a troop or a local population is replacing itself. The purpose of this paper is to meld the general demographic parameters with the special characteristics of callitrichid reproductive biology. This will generate models which suggest criteria for evaluating the health of a callitrichid population and whether or not animals may be harvested from a particular region.

The special characteristics of callitrichid reproductive biology are pair bonding, the high frequency of twinning, the capability of two gestations per year, and extensive paternal and sibling care of infants. Because marmosets are pair bonded and there is no evidence of excess production of males, the number of safe reproductive years for males as well as females must be considered. So must the number of years to maturity for both sexes.

Since callitrichids are capable of breeding twice a year, parameter (b) above is modified to the number of young per gestation per bond. Finally, indications are that many of the aspects of parental and particularly paternal care of infants must be learned through experience with younger sibs [2].

The Models

The development of microdemographic information is illustrated by the best-of-all-worlds model of a hypothetical callitrichid (model I of table I). The male and female become reproductively active at one year of age and have three safe reproductive seasons (1.5 years). The pair produces two young per gestation: one male and one female, both of which survive to reproductive age (no mortality is assessed). There is no differential effect of mortality on the sexes. Model I shows how a new pair replaces itself. From pair formation to replacement requires a minimum of 1.5 years, 0.5 years longer than the pre-adult period. The troop size (family and troop are used interchangeably in this paper) necessary for replacement is dictated by the number of young per gestation, the times at which mortality is assessed and the length of the pre-adult period. If the first two are held constant (as can be done during modelling) then replacement size depends on length of pre-adult period. So, if we expand the length of the pre-adult period to 1.5 years, eight is the number of animals required to indicate replacement in a family group. Also the minimum number of safe reproductive years expands to two (four seasons). A series of models can be developed increasing the length of the pre-adult period by half a year at a time. The trends continue steady and the results of this series of models gives the replacement troop sizes indicated in the second column of table II. We may also generalize that in this model series the minimum number of years required for a pair to replace itself is the length of the pre-adult period plus half a year.

Survival of both twins at every gestation is a very unlikely rate of reproductive success. Model II of table I indicates the somewhat more likely situation in which only one of the offspring of each gestation survives to adulthood with the mortality assessed at birth. Animals reaching reproductive

Table I. Models illustrating replacement conditions for an hypothetical callitrichid

Model I

Breeding season	Age of animals, years							
	0	0.5	1.0	1.5	2.0	2.5	3.0	3.5
0			MF					
1	MF			MF				
2	MF	MF			MF			
3	MF	MF	MF			MdFd		

Male and female become reproductively active at one year; both have three active reproductive seasons (1.5 years after reaching maturity); produce two infants per gestation with two gestations per year; 1:1 sex ratio among infants; no pre-adult mortality. 'd' indicates occurrence of a death; 'x' indicates expelled from troop.

Model II

Breeding season	Age of animals, years							
	0	0.5	1.0	1.5	2.0	2.5	3.0	3.5
0			MF					
1	M			MF				
2	F	M			MF			
3	M	F	Mx			MF		
4	F	M	Fx				MdFd	

Parameters as above except adults have four active reproductive seasons and only one infant survives past birth.

age are expelled. This leads to a lower level of family size required to indicate replacement but also necessitates that a breeding pair live one full year (two gestation periods) longer than the pre-adult period. Family-troop sizes which indicate a pair is apparently replacing itself are indicated in column 3 of table II. If death is not assessed at birth a larger troop size would be required to certify satisfactory replacement.

It might be argued that a male and female could produce one pair of twins, raise them to adulthood and thereby replace themselves. Numerically perhaps, but reproductively it is unlikely because it seems that much of parental and particularly paternal behavior must be learned in the process of caring for younger sibs. While the age pyramid of sibs may permit occasional vacant slots, it seems unlikely that it can be much reduced from the conditions listed here if the replacing animals are to have adequate socialization.

As the parental generation dies, it is possible that successively one of the pair of oldest sibs would take on the parental role, accept a mate from outside the family, and proceed to raise its younger sibs. Troop size required to indicate replacement would remain the same. This becomes an important factor when the models are viewed as populations.

The models of table I provide us with insight into the structure of a successful population as well as the structure of a successful family. If the stipulations of model I prevail exactly then every troop in the population would have either two, four, six members or transiently eight. If every adult who attains sexual maturity lives exactly 1.5 years more, the population in a forest or region would show some troops at replacement level and some not at replacement level. The two troops of lines 0 and 1 are obviously in this latter category. The troop of line 3 has achieved replacement size but this size is the same as that of line 2. However, such survival cannot be guaranteed in the wild; if the population of a forest or region is to be stable then the individuals who reach breeding age must have an *average* reproductive life of 1.5 years. Some adult pairs would survive beyond this age, expelling their maturing juveniles who would make up for the replacements not produced by troops who never get beyond the levels illustrated in lines 0, 1 and 2. Thus, under realistic assumptions, there must be an excess of apparently successful troops to ensure a stable population in a given forest or region. For the situation described in model I, three successful troops are required to balance those troops not at replacement level illustrated in lines 0, 1 and 2. These three replacement troops will all look like the troop of line 3, which has just replaced itself. However, these four apparently successful troops will be indistinguishable from that of line 2. Consequently a count of five apparently successful troops is needed to balance the two potentially unsuccessful troops of model I. This is perhaps a low figure because a portion of the three apparently successful troops will really be troops at the stage represented by line 2 of model I.

If we go through the same exercise for model II, six apparently successful troops are required to balance the troops represented in lines 0 to 2 of the model. One apparently successful troop is required to replace the male and one is required to replace the female of each of the potentially unsuccessful troops. In addition one successful troop would be needed to replace the partially successful troop of line 3. Line 4 shows a troop which has just replaced itself. However, all eight of these apparently successful troops would

Table II. Troop sizes and excess of apparently successful troops correlated with number of pre-adult years

Length of preadult period years	Size of apparently successful troops		Excess of apparently successful troops			
			2 young/gestation number of troops		1 young/gestation number of troops	
	2 young/ gestation	1 young/ gestation	obviously not at replacement	apparently at replacement	obviously not at replacement	apparently at replacement
1.0	6	4	2	5	2	10
1.5	8	5	3	6	3	12
2.0	10	6	4	7	4	14
2.5	12	7	5	8	5	16
3.0	14	8	6	9	6	18
3.5	16	9	7	10	7	20
4.0	18	10	8	11	8	22
4.5	20	11	9	12	9	24

Minimum safe adult reproductive years for apparently successful troop is one half year longer than the number of pre-adult years for two young per gestation and one year longer for one young per gestation.

be indistinguishable from the troops in lines 2 and 3. Thus a stable population under the conditions of model II would show a ratio of two troops obviously not at replacement level to ten troops apparently at replacement level. Once again this latter figure should be a little higher because of the impossibility of distinguishing truly successful troops from those which have not yet replaced themselves. The ratios from the two model series vary depending on the number of pre-adult years, and these figures appear in the third and fourth columns of table II. Importantly, for the model II situation, these figures cannot be attained unless the surviving adult or an older sib takes over the family-troop and raises its younger sibs with a mate from outside the troop.

Often in captivity the first born are not successfully reared by a marmoset family. In the modelling systems this would have the effect of lengthening the required number of safe reproductive years by one half year (one gestation period). The number of apparently successful troops in the ratio indicating a stable regional population would be increased by one. If the average number of adult reproductive years is greater than the minimum of 0.5 years more than the pre-adult period, then the number of excess troops apparently at replacement level is reduced by one for each additional six months added to the average reproductive span. It should be added that if mature animals do not leave the parenteral troop, their reproductive capacity seems inhibited. If these animals remain within the troop for six months, i.e. they shirk their responsibilities to form new troops, the numbers of troops apparently at replacement level should be increased in the ratio indicating a stable population.

Discussion

Table II represents the results from 16 different models like those in table I, varying the number of pre-adult years (and also therefore the average number of safe reproductive years for an adult). The likely range of pre-adult years is from two to three with lower reproductive success on the first gestation [6, 7]. Thus average size of a troop at replacement level could be between six and 14, with six to seven being the common average from field surveys [1, 3, 4, 8]. Slightly larger groups represent survival of both twins of one gestation. Much larger groups are probably temporary aggregations of two or more families. With one young per gestation, the ratio of troops required to indicate a stable population is somewhere in the range of four troops not at replacement level to 14 troops apparently successful to 6:18. This corresponds favorably with the field data DAWSON [1] gives for troop size distributions in *Saguinus oedipus geoffroyi*.

Captive studies point to the possibility that the average number of adult reproductive years may exceed the number of pre-adult years by more than the essential minimum [6]. This would reduce the required excess of apparently successful troops. However, we have no comparable information from the wild. In order to maintain a stable regional population, a conservative and conservationist position would be to say that apparently reproductively successful troops should outnumber potentially unsuccessful troops by 3–3.5 times. The excess of apparently successful troops could never be less than 1.5 times the number of troops obviously not at replacement level (unless triplets became the most common number at birth and they all lived to reproductive age). These figures give us a baseline for deciding whether a local population can sustain a harvest. Once a region has been harvested, subsequent surveys can check the results and obtain more refined estimates relating to numbers which could or ought not to be removed. Finally, the models point to those animals which are the best candidates for harvesting. These are entirely new pairs without young and the next to the oldest juvenile in a troop. For reproduction colonies removal of entirely new pairs without young will be least damaging to the population and still have a high likelihood of breeding. If terminal experimental subjects are required then the next to oldest juvenile should be the animal harvested from the troop. This will retain the breeders and the oldest juvenile animal already most prepared for future breeding, while providing a research animal who has already shown some hardiness. When callitrichids are trapped all other animals should be released in order to conserve the reproductive potential of the local population.

References

1 DAWSON, G.: Troop size, composition and instability in the Panamanian tamarin *(S. oedipus geoffroyi)*: ecological and behavioral implications. Paper presented at the Conference on the Biology and Conservation of the Callitrichidae, 1975.

2 EPPLE, G.: Parental behavior in *Saguinus fuscicollis* ssp. (Callithricidae). Folia primatol. *24:* 221–238 (1975).

3 FREESE, C.: Final report: a census of non-human primates in Peru; in Primate censusing in Peru and Colombia, pp. 17–41 (Pan American Health Organization, Washington 1975).

4 HELTNE, P.; FREESE, C., and WHITESIDES, G.: A field survey of non-human primate populations in Bolivia (Pan American Health Organization, Washington 1975).

5 HELTNE, P.G.; TURNER, D.C., and SCOTT, N.J., jr.: Comparison of census data on *Alouatta palliata* from Costa Rica and Panama; in THORINGTON and HELTNE Neotropical Primates: field studies and conservation, pp. 10–19 (National Academy of Sciences, Washington 1975).

6 HIDDLESTON, W.A.: The production of the common marmoset *Callithrix jacchus* as a laboratory animal. Paper presented at International Primatological Society Congress, 1976.

7 LUNN, S.F. and HEARN, J.P.: Breeding marmosets for medical research. Paper presented at International Primatological Society Congress, 1976.

8 WARNER, P.: The ecology and social behavior of the cotton-top tamarin in Colombia. Paper presented at the conference on the Biology and Conservation of the Callitrichidae, 1975.

P.G. HELTNE, Johns Hopkins University, Division of Comparative Medicine, 720 Rutland Avenue, *Baltimore, MD 21205* (USA)

Prim. Med., vol. 10, pp. 37–39 (Karger, Basel 1978)

Supply and Conservation Efforts for Nonhuman Primates

M. Moro

Regional Veterinary Advisor in Veterinary Medicine,
Pan American Health Organization, Washington, D.C.

Nonhuman primates are found throughout tropical America with the greatest concentration in the Amazon Region.

Different factors as indiscriminate hunting, changes in the environment as a result of agricultural and forest development and oil exploration have decimated primate population producing a depletion of these animal species that is becoming a threat both to biomedical research and to conservation of these animal species. The result of censusing done in Peru and Colombia has pointed to the urgency and importance of initiating activities for the establishment research and for testing purposes.

The Pan American Health Organization has been concerned with this problem and in order to find a rational approach to its solution has undertaken some joint activities with government authorities in Peru, Brazil and Colombia. During the years 1974–1975, in cooperation with the National Academy of Science, these activities dealt with the censusing of monkeys in Peru and Colombia. On June, 1975 the program was amplified to include not only population studies, but also the protection and management on an area basis and the breeding of indigenous species of importance for biomedical research purposes. The initial stages of the activities have been devoted to the establishment of a primate breeding station in Iquitos, Peru.

Primate Centers to be Established

So far, three centers in their natural habitat for primate conservation and breeding are being established in South America. The Primate Breeding and Conservation Center in Iquitos is the most advanced in its establishment and will be discussed later.

The Primate Breeding Center in Brazil will be established in Manaus, located on the Amazon River. It will be sponsored by the Ministry of Public Health and the Ministry of Agriculture through the National Institute of Amazonic Research (INPA) and it will be connected with other primate centers of Brazil. This breeding center of Manaus will be devoted to *Saimiri, Saguinus,* and *Lagothrix* species and is expected to have at least 500 animals of each kind.

The site for the Colombian Primate Center has not yet been chosen, but it appears that it will be near Cartagena, located on the Northern Atlantic Coast close to natural forests with endangered species of monkeys (*Aotus* and *Saguinus oedipus*). This primate center will be devoted to *Aotus* and possibly to *S. oedipus*.

The objectives of all these nonhuman primate breeding stations, besides breeding certain selected species, are to learn about the general behavior, reproduction, and diet of these species; determine the characteristics of the normal and pathological features of the species selected for reproduction; select geographical territories for the best maintenance and reproduction of these species; use the center as a pilot plan for studies on the conservation of the species of nonhuman primates; and also to use the center as a training exchange station for primatological research activities.

The Primate Conservation and Breeding Center of Iquitos, Peru

This Center is sponsored by the Ministry of Agriculture of Peru, through the Direction of Forest and Wildlife. This institution has signed an agreement with the Veterinary Institute of Tropical and High Altitude Research (IVITA) of San Marcos University to coordinate the local administration of the project. IVITA deals with the work in the Breeding Center and the Direction of Forest and Wildlife with the field work in general. So far, the quarantine facilities and two breeding units have been constructed. Next will come a veterinary clinic, an administration building, and eight more breeding units. The two monkey breeding units already constructed have 320 m^2 of roofed concrete floor area, totally enclosed with screen wire, and each possessing trapped entrance porches.

One of the monkey shelters now has 50 pairs of *Saguinus mystax* in suspended breeding cages measuring 1×1 m and 2 m in height. The cages are of wood and wire construction. Besides these cages, there is a large growing pen for juveniles.

The other monkey shelter will hold 255 squirrel monkeys in gang cages with approximately seven females per male, and there will be approximately 255 animals in each shelter. This station has two primatologists and a veterinarian along with supporting personnel.

The Direction of Forest and Wildlife of Peru with the technical assistance of the Primatologist of the Breeding Center is conducting studies in islands and protected areas of the forest. The preliminary studies done north of an island located in the Amazon River named 'Isla de Iquitos' (because it is close to the city), showed, according to the report of SOINI and MOYA [1] that in this island there exist a large number of *Saimiri sciureus* as well as some *Aotus trivirgatus*. The density of *S. sciureus* is of 2.6 ± 0.1 troops/km^2 (average 20 animals per troop) and there is an estimated population of 65–70 troops. It was also estimated that there is an increase of over 10% per year.

The density of *A. trivirgatus* in the north side of the Iquitos island is 10 families/km^2 (a pair with or without offspring per family).

Recent field studies carried out in the Amazon forest of Peru showed that the pygmy marmoset *(Cebuella pygmaea)*, heretofore considered an endangered species, exists in fair amounts. Apparently it was considered rare because of difficulties in detection.

References

1 SOINI, P. and MOYA, L.: Unpublished reports of the results of primate studies in the north of Iquitos Island, 1976.

M. MORO, MD, Regional Advisor in Veterinary Medicine, Department of Human and Animal Health, Pan American Health Organization, 525-23rd Street, NW, *Washington, DC 20037* (USA)

Session II: Breeding of Marmosets in Captivity

Chairman: J.K. HAMPTON, jr., San Luis Obispo, Calif.

Prim. Med., vol. 10, pp. 40–49 (Karger, Basel 1978)

Use of the Common Marmoset, *Callithrix jacchus,* in Reproductive Research

J.P. HEARN, D.H. ABBOTT, P.C. CHAMBERS, J.K. HODGES and S.F. LUNN

MRC Unit of Reproductive Biology, Edinburgh

Introduction

Early reports of the high fecundity of marmoset species in captivity and their small size, which makes them relatively easy to handle and inexpensive to house and feed [5, 8, 10], suggested that they might be suitable primates for laboratory studies related to human reproductive endocrinology and fertility control. Most of the laboratories using marmosets for biomedical research were then in the United States and were using species of tamarin monkeys. In Britain, however, KINGSTON [22] had evaluated the breeding success of several marmoset and tamarin species and concluded that the common marmoset, *Callithrix jacchus,* was perhaps the easiest to breed under controlled environmental conditions and that this species was amenable and robust once adapted to captivity.

The major drawback to using the common marmoset in reproductive research was that nothing was known of its reproductive endocrinology. Therefore our studies were initiated with two objectives. Firstly, to examine the reproductive biology and endocrinology of this primate as a potential model for human reproductive processes and secondly, to examine the effects of immunising marmoset monkeys against the β subunit of HCG and against LH-RH with a view to achieving long-term suppression of fertility.

In this communication we summarise current knowledge of the reproductive endocrinology of *C. jacchus* and the effects on fertility of immunising against HCG-β subunit and LH-RH.

Management

Full details of the management procedures developed in this colony were published elsewhere [15]. However there are a few points directly relevant to improving breeding performance of *C. jacchus* in captivity that are relevant here.

Marmosets can easily be caged in family groups and the contact that this allows between animals reduces the effects of boredom so often seen in colonies of larger monkeys that are caged singly. An occasional but regular change in environment also improves the general condition and increases the pregnancy rate in the colony. In the centre of each animal room we place a large exercise cage equipped with branches, ropes, etc. By using transparent flexible ducting the large cage can be connected to any individual cage in the room, allowing the animals freedom to run through the flexible ducting to the exercise cage. By connecting the ducting to a new cage each day all of the animals receive regular exercise and a change in environment, and the daily change of individuals using the exercise cage also provides a diversion for all of the other animals in the room.

Marmosets usually give birth to twins twice a year, but in about 25% of births triplets are produced. When triplets are born one invariably dies as the parents seem capable of rearing only two young at once. If from the second day of life each of the triplets is removed in rotation, placed in an incubator with a furry toy as a surrogate mother, and hand-fed for 24 h, this leaves the parents with only two young at any one time and yet does not remove any individual young from the parents for more than 24 h. This schedule can be maintained until the young are 40 days old, after which they can eat solid food and all three young can then be left permanently with their parents. In addition to ensuring survival of the triplets and so considerably boosting the productivity of the colony, this schedule does not deprive the young from the influence of their parents or interactions with their older siblings.

The practice in our colony is to leave young marmosets in their families until they are ten to 12 months old. By this time a further set of young have been born into the family and the older siblings gain experience of rearing these. At 10 to 12 months they are placed in peer groups of six unrelated animals (three females and three males) where they gain experience of social interactions for a further six to eight months. During this time a pair bond is formed between the dominant male and female in the group and the female becomes pregnant and gives birth. All of the members of the peer group assist in carrying and rearing the young, but only the dominant female breeds. Even though there are frequent copulations between other members of the group only the dominant female shows ovarian cyclicity.

At 15–18 months of age the peer groups are split up. The dominant pair and their young are recruited into the breeding nucleus of the colony and each of the other members of the peer group are paired with an experienced adult of the opposite sex. Most of the females become pregnant soon afterwards and if they do not their partners are changed until they do so.

The Ovarian Cycle

The female common marmoset has a clearly defined ovarian cycle of 16.4 ± 1.7 days when monitored by plasma levels of LH, progesterone and oestradiol 17β [14]. There is no clear cyclicity in their vaginal cytology and, in common with most South American primates, they do not menstruate. Although some cyclical variations in plasma diamine oxidase [9] and in prealbumin proteins from vaginal washings [16] have been reported, these are not precise enough to determine the stage of the cycle accurately. KLEIMAN [23] described increased mating at regular intervals in *Leontopithecus* sp. but *C. jacchus* copulates frequently throughout the cycle and the first half of pregnancy. Simple progesterone radioimmunoassay is at present the most accurate way of following the cycle although this does require frequent blood sampling.

The corpus luteum in the marmoset secretes oestradiol as well as progesterone. Levels of oestradiol in plasma during the cycle rise to 0.8–2.0 ng/ml at the mid-cycle peak and to 0.7–1.5 ng/ml during the luteal phase. However, the follicular phase of the cycle in the marmoset is short and the control of follicular development in this species has yet to be studied to determine its similarity to the human [2]. The luteal phase of the cycle is not prolonged by hysterectomy [12] implying that the corpus luteum is under a similar control to that in the human rather than that in the sheep, pig, guinea pig, etc. [11, 25].

Pregnancy

The width of the uterus across the fundus and the levels of progesterone, oestradiol 17β, oestrone, chorionic gonadotrophin, androstenedione and

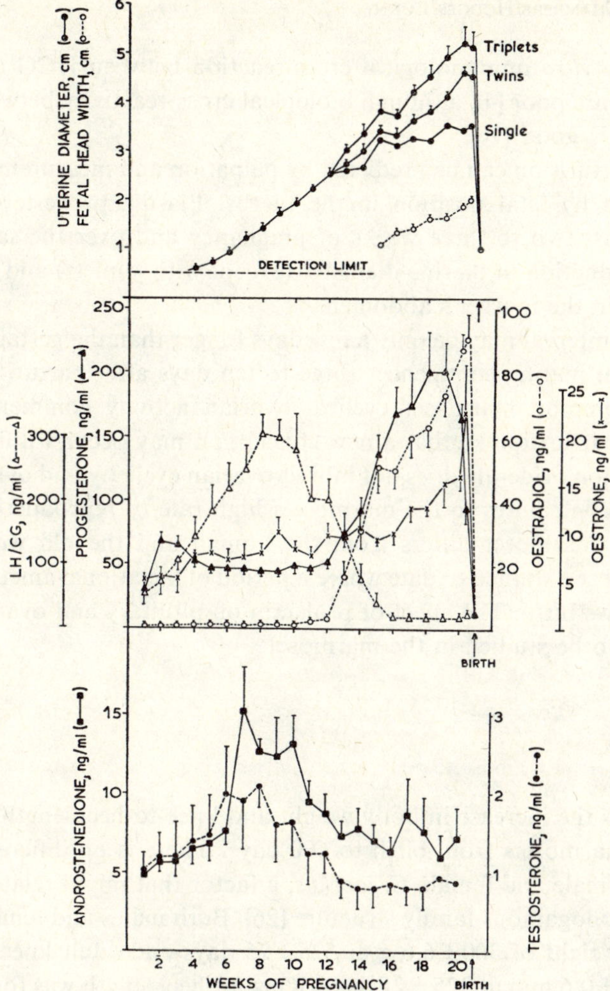

Fig. 1. The pregnant marmoset. Growth of the uterus during single, twin and triplet pregnancies, and growth of fetal head width (upper sector). Levels of LH/chorionic gonadotrophin, progesterone, oestrone, oestradiol (mid sector), and levels of androstenedione and testosterone (lower sector).

testosterone in peripheral plasma during pregnancy in *C. jacchus* are shown in figure 1. The gestation period is 144 days (range 141–146) calculated from the post-ovulatory rise in progesterone to the day of birth. Pregnancy can be diagnosed by the 4th week using transabdominal uterine palpation and by the second week using radioimmunoassays for LH/CG. Pregnancy cannot be accurately diagnosed using commercially available pregnancy kits, perhaps because the *in vitro* immunological cross-reaction between HCG and marmoset CG appears poor [4], although biological cross-reactivity between the two hormones is good [16].

The time of parturition can be predicted by palpation and measurement of fetal heads, and by fetal position in the uterus. Plasma progesterone declines over the last two to three weeks of pregnancy and over the same period there is a reduction in uterine diameter (except with triplets) and the fetal heads lie low in the mother's abdomen.

The interbirth interval may be only a few days longer than the gestation period. Post-partum mating commences three to ten days after parturition [14] and in the common marmoset cyclical ovarian activity commences immediately after parturition so that a new conception may occur within a few days [3]. Lactation evidently does not inhibit ovarian cyclicity and ovulation, a factor that contributes to the marmosets high rate of reproduction. In this respect the marmoset differs from the human and the old world monkeys that have been studied to date where a period of lactational amenorrhoea usually follows birth. The effect of prolactin on pituitary and ovarian sensitivity has yet to be studied in the marmoset.

Puberty

Figure 2 shows the increase in body weight and knee-to-heel length for male and female marmosets from birth to 600 days. There is no difference in growth between male and female *C. jacchus,* a factor that might relate to their apparently monogamous family structure [26]. Both males and females reach adult body weight of 300 ± 6.6 g at 525 ± 25 days and adult knee-to-heel length of 69.7 ± 0.6 mm at 325 ± 25 days. Knee-to-heel length was found to be a convenient and fairly accurate way of estimating the age of marmosets from birth to 300 days old since variation in this parameter is low. Measurements of the pudendal pads and plasma progesterone in females and the testis and plasma testosterone in males, showed that females reached sexual maturity (normal ovarian cyclicity) at about 400 days old, while males achieve adult testis size at 550 days and adult testosterone levels at about 350 days [1].

Neonatal male marmosets have elevated plasma testosterone levels between three and 80 days of age, reaching a peak at about 40 days. In this neonatal testosterone release they are similar to the human [7] although the function of this release has yet to be determined.

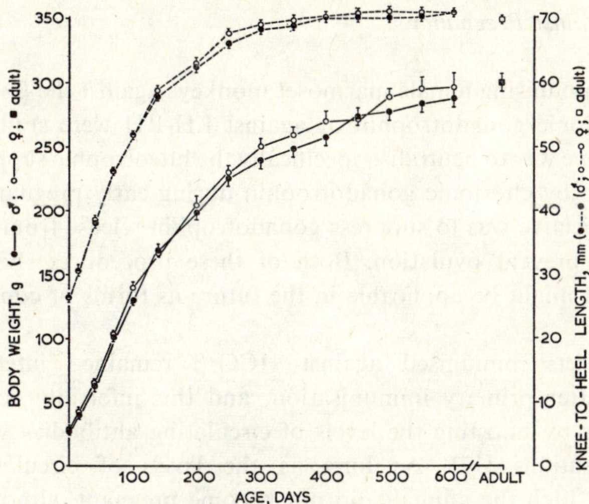

Fig. 2. Growth of marmosets in captivity.

Control of LH Secretion

The marmoset responds to an injection of oestradiol 17β with a release of LH from the pituitary gland, similar to the mid-cycle release of LH associated with ovulation. This release is termed 'positive feedback' of oestrogen on LH and in mammals other than primates it has been demonstrated only in the female, implying that there is a basic difference in the brain that distinguishes the female from the male. There is controversy about the capacity of the hypothalamus in the male primate to display positive feedback. Positive feedback has been demonstrated in castrate male rhesus monkeys on continuous oestrogen [21] and claimed in a proportion of normal [24] or homosexual men [6] but the results are inconclusive. Positive feedback has recently been clearly shown in both castrate and intact male marmoset monkeys [19] indicating that in this species there is no difference between the male and female. It remains to be seen whether these findings can be extended to man and other primates. Studies in progress in our laboratory are examining the roles of androgens, oestrogens and progesterone on hypothalamic and pituitary regulation of gonadotrophin release in male and female marmosets.

Immunisation against Pregnancy

The effects of immunising female marmoset monkeys against the β subunit of human chorionic gonadotrophin or against LH-RH were studied. In the first the objective was to neutralise specifically the luteotrophic support of the corpus luteum by chorionic gonadotrophin during early pregnancy. In the second the objective was to suppress gonadotrophin release from the pituitary gland and prevent ovulation. Both of these procedures should suppress fertility and might be applicable in the future as forms of contraception for humans.

Female marmosets immunised against HCG-β remained infertile for up to 2 years after primary immunisation, and this infertility could be prolonged further by boosting the levels of circulating antibodies with subsequent immunisations [17]. As long as the levels of circulating antibodies remained high the animals did not become pregnant, although they continued to cycle apparently normally, indicating that the antibodies to HCG-β were not cross-reacting with endogenous LH. If the antibody titres were allowed to fall, however, the animals experienced a series of early abortions that occurred progressively later in pregnancy as the antibody titres declined. Attempts to reverse the effects of immunisation with implants of progesterone during the first few weeks of pregnancy were not successful, implying that the antibodies, might be having a direct effect on the developing fetus as well as blocking progesterone production by the corpus luteum [13].

Female marmosets immunised against LH-RH also became infertile for one to two years. In these animals ovulation was suspended and there was no cyclical progesterone production indicative of a functional corpus luteum. When the levels of antibody were allowed to decline sufficiently to permit ovulation and formation of a corpus luteum the females returned to normal breeding and there was no evidence that low levels of circulating antibodies to LH interfered with pregnancy [20]. Thus the effects of immunisation against LH-RH were more simply reversed than those of immunisation against HCG-β, and carried less risk to future pregnancies.

From these studies it is clear that the marmoset, with its rapid rate of breeding, is an exceptionally suitable primate in which to investigate the effects of contraceptive treatments on fertility. Studies can be completed in a fraction of the time that they would take in the larger, more conventional laboratory primates. Since marmosets become pubertal at 18–20 months of age it is also possible, in a relatively short time, to study any possible effect that new hormonal or immunological contraceptive techniques might have on sexual development of fetuses born to treated mothers.

Discussion

Clearly, from the studies reported in this paper and the list of references, the history of the marmoset monkey in reproductive research is a very short one. Yet it is already becoming apparent that *C. jacchus* is an eminently suitable laboratory primate for research related to certain aspects of human reproduction.

A major disadvantage of the marmoset in reproductive research is the current inability to determine accurately the stage of the ovarian cycle without assistance of hormone radioimmunoassays. Studies in our laboratory to correlate hormone changes with possible changes in vaginal cytology and the appearance of the cervix are progressing but as yet there is no easy way. Consequently the marmoset is not at present a suitable species for precise studies into the control of cyclic ovarian function. Its small size precludes the withdrawal of large quantities of blood and if studies depend on daily sampling requiring more than 0.2 ml blood over any lengthy period the animals may become anaemic and stop cycling. Regular injections of Imferon (Fisons) or similar replacement of iron can prolong the time over which regular blood samples may be taken.

The high fecundity of *C. jacchus* in captivity makes it a suitable species for studies in pregnancy or the testing of antifertility agents. The early age at which puberty is reached makes this species potentially suitable for studies related to physical, endocrinological and behavioural maturation. It also enables early recruitment of colony-born young into the breeding nucleus, thereby ensuring that colonies of marmosets can be self-supporting and independent of further importations from the wild in a relatively short time for a relatively small cost. In a colony of 250–300 animals, over 70% of the animals can be used for research projects while the remaining 30% will, under the right conditions, produce sufficient young to allow a complete replacement of the colony every 4–5 years. Dealing with these numbers it is also possible to select high quality animals for the breeding nucleus.

In conclusion, although there is much basic work still to be done and many more areas to be explored, the future of the common marmoset as a model for studies in human reproductive research is certain.

Summary

A self-sustaining colony of common marmosets, *Callithrix jacchus*, was established for reproductive research. Methods of management were developed to ensure optimal breeding conditions for marmosets in captivity. Studies of the reproductive biology and endocrinology of this species suggest that the common marmoset may be a suitable model for the human in certain aspects of the ovarian cycle, pregnancy, puberty and the hypothalamic and pituitary control of luteinizing hormone secretion. The rapid rate of reproduction of *C. jacchus* in captivity allows a relatively large colony to be maintained at moderate cost and also makes this species a most suitable laboratory primate for longer term studies of the safety and efficacy of new contraceptive methods.

References

1 ABBOTT, D.H. and HEARN, J.P.: Puberty in the marmoset monkey, *Callithrix jacchus*. Aspects of physical and endocrinological development. J. Reprod. Fert. (in press, 1977).

2 BAIRD, D.T.; BAKER, T.G.; MCNATTY, K.P., and NEAL, P.: Relationship between the secretion of the corpus luteum and the length of the follicular phase of the ovarian cycle. J. Reprod. Fert. *45:* 611–619 (1975).

3 CHAMBERS, P.C. and HEARN, J.P.: Levels of progesterone, oestradiol 17β, oestrone, androstenedione and testosterone in peripheral plasma of the pregnant marmoset monkey, *Callithrix jacchus*. J. Endocr. (in press, 1977).

4 CHEN, H.C. and HODGEN, G.D.: Primate chorionic gonadotrophins. Antigenic similarities to the unique carboxyl terminal peptide of HCG-β subunit. J. clin. Endocr. Metab. *43:* 1414–1417 (1977).

5 DEINHARDT, J.B.; DEVINE, J.; PASSOVOY, M.; POHLMAN, R., and DEINHARDT, F.: Marmosets as laboratory animals. Lab. Anim. Care *17:* 11–29 (1967).

6 DORNER, G.; ROHDE, W.; STAHL, F.; KRELL, L., and MASIUS, W.G.: A neuroendocrine predisposition for homosexuality in men. Arch. sexual Behav. *4:* 1–8 (1975).

7 FORREST, M.G.; CATHIARD, A.M., and BERTRAND, J.A.: Evidence of testicular activity in early infancy. J. clin. Endocr. Metab. *37:* 148–151 (1973).

8 GENGOZIAN, N.: Marmosets: their potential in experimental medicine. Ann. N.Y. Acad. Sci. *162:* 336–362 (1969).

9 HAMPTON, J.K. and HAMPTON, S.H.: Some expected and unexpected characteristics of reproduction in Callithricidae. Lab. Anim. Handb. *6:* 235–240 (1975).

10 HAMPTON, J.K.; HAMPTON, S.H., and LANDWEHR, B.T.: Observations on a successful breeding colony of the marmoset, *Oedipomidas oedipus*. Folia primatol. *4:* 265–287 (1966).

11 HEAP, R.B. and PERRY, J.S.: The maternal recognition of pregnancy. Br. J. Hosp. Med., July 8–14 (1974).

12 HEARN, J.P.: The endocrinology of reproduction in the common marmoset, *Callithrix jacchus;* in KLEIMAN The biology and conservation of the Callithricidae (Smithsonian Press, Washington 1976).

13 HEARN, J.P.: Immunisation against pregnancy. Proc. R. Soc. Lond. B *195:* 149–160 (1976).

14 HEARN, J.P. snd LUNN, S.F.: The reproductive biology of the marmoset monkey, *Callithrix jacchus*. Lab. Anim. Handb. *6:* 191–202 (1975).

15 HEARN, J.P.; LUNN, S.F.; BURDEN, F.J., and PILCHER, M.M.: Management of marmosets for biomedical research. Lab. Anim. *9:* 125–134 (1975).

16 HEARN, J.P. and RENFREE, M.B.: Prealbumin proteins in the vaginal flushings of the marmoset, *Callithrix jacchus*. J. Reprod. Fert. *43:* 159–161 (1975).

17 HEARN, J.P.; SHORT, R.V., and LUNN, S.F.: The effects of immunising marmoset monkeys against the β subunit of HCG; in EDWARDS and JOHNSTON Physiological effects of immunity against hormones, pp. 229–247 (Cambridge University Press, Cambridge 1976).

18 HOBSON, B.M. and WIDE, L.: A comparison between chorionic gonadotrophins extracted from human, rhesus monkey and marmoset placentae. J. Endocr. *55:* 363–368 (1972).

19 HODGES, J.K. and HEARN, J.P.: A positive feedback effect of oestradiol on luteinising hormone release in the male marmoset monkey, *Callithrix jacchus*. J. Reprod. Fert. Lond. (in press, 1977).

20 HODGES, J.K. and HEARN, J.P.: Effects of immunisation against LHRH on reproduction of the marmoset monkey *Callithrix jacchus*. Nature, Lond. *265:* 746–747 (1977).

21 KARSCH, F.J.; DIERSCHKE, D.J., and KNOBIL, E.: Sexual differentiation of pituitary function: apparent difference between primates and rodents. Science *179:* 484–486 (1973).

22 KINGSTON, W.: Breeding of endangered species of marmosets and tamarins; in MARTIN Breeding endangered species in captivity, pp. 213–222 (Academic Press, London 1975).

23 KLEIMAN, D.: Social and sexual behaviour of *Leontopithecus rosalia* during the reproductive cycle, in KLEIMAN The biology and conservation of the Callithricidae (Smithsonian Press, Washington 1976).

24 KULIN, H.E. and REITER, E.O.: Gonadotrophin and testosterone measurements after estrogen administration to adult men, prepubertal and pubertal boys, and men with hypogonadotropism. Evidence for maturation of positive feedback in the male. Pediat. Res. *10:* 46–51 (1976).

25 SHORT, R.V.: Implantation and maternal recognition of pregnancy; in Ciba Foundation Symposium on Fetal Autonomy, pp. 2–26 (Churchill, London 1969).

26 SHORT, R.V.: Sexual selection and the descent of man; in CALABY Reproduction and evolution (Australian Academy of Science, Canberra 1977).

J.P. HEARN, PhD, Medical Research Council, MRC Reproductive Biology Unit, 2 Forrest Road, *Edinburgh, EHI 2QW* (Scotland)

Prim. Med., vol. 10, pp. 50–62 (Karger, Basel 1978)

Reproductive and Social Behavior of Marmosets with Special Reference to Captive Breeding[1]

GISELA EPPLE
Monell Chemical Senses Center, University of Pennsylvania, Philadelphia, Pa.

Introduction

Recent and severe limitations in the number of *Callithricidae* available for biomedical research have made it clear that captive breeding might soon be the only means by which sufficient numbers of animals can be supplied. Laboratories and zoos have been quite successful in breeding the more common and some of the rarer species imported from the wild [18, 21, 22, 24, 36]. This shows that feral-born individuals of several species reproduce well in captivity and raise their offspring without problems. Although several laboratories have meanwhile succeeded in producing second and third generations of callithricids [36], the reproductive success of these animals is not as good as one would hope [16, 21, 27, 36]. In some species this might be due to physiological causes, such as the possible sensitization of the mother to unidentified blood antigens [14, 36]. Behavioral deficiencies such as inadequate parental care or enhanced aggressiveness may also be responsible for part of these failures. It seems important, therefore, to not only study all aspects of reproductive physiology but also to obtain a detailed knowledge of the normal social and sexual behavior. In this contribution I shall briefly review some of our present knowledge of marmoset behavior in the wild and in captivity and try to point out the consequences some behavioral characteristics might have for captive breeding.

1 Personal studies were supported by Grant BNS 7606838 from the National Science Foundation and by a Research Career Development Award from the National Institute of Child Health and Human Development.

Social Structure and Hierarchies

To date we know little of the social structure and social behavior of any species of the *Callithricidae* in the wild. Recently, however, a number of field surveys and a few more detailed studies have been carried out. They show that the small group predominates among the *Callithricidae*. However, group sizes vary widely within and between species. Thus, single individuals, small groups of two to 12 and, occasionally, large groups of up to 40 animals were seen [3, 4, 19, 25, 35].

Only two field studies so far deal in some detail with social organization. DAWSON [6] found that groups of *Saguinus geoffroyi* in the Panama Canal Zone varied in size between one and 19 with an average of 6.93 animals. Groups of seven were most frequent. The groups contained an average of 2.4 adult males, 2.06 adult females, 1.15 immature males and 0.67 immature females, plus one set of infant twins. Although the number of individuals per group and their sex-age class proportion remained quite stable, animals emigrated from groups and immigrated into groups quite frequently. Immature males and females tended to emigrate from groups more frequently than mature individuals. It appears that juveniles may change group affiliation often and with a minimum of interference from the residents. When the stability of groups in moist lowlands and seasonally deciduous uplands was compared, it became apparent that the number of individuals associated with the group above and below the mean varied with the availability of moisture. Moreover, differences were observed in the size and usage of home ranges. The lowland group occupied and aggressively defended a territory of 26 ha while the group occupying the dry upland area used a much larger home range. This was not defended as a territory. There was evidence of a relatively high level of intragroup and intergroup aggression, separate male and female hierarchies within groups, and exclusive reproduction by only one female per group. The dominant pairs were apparently less likely to emigrate from the group. Thus the typical *S. geoffroyi* group consists of a dominant male, a dominant female, their dependent offspring and a variable number of more or less temporary group members, whose blood relationships are unknown.

NEYMAN [26] found that *Saguinus oedipus* groups in the coastal lowlands of Colombia consist of 3 to 13 animals. There was evidence for only one reproductively active female per group. Similar to the *S. geoffroyi* [6] the cottontop groups were rather unstable, with animals leaving groups and joining others. Apart from established groups which were clearly attached to a home range, transient individuals and groups were seen. In contrast to DAWSON's [6] findings, there was no evidence that juveniles were preferentially involved in these transits. However, females were more frequently found in transit than males. The three best known established groups occupied and defended home ranges of 7.8, 7.8 and 10 ha. Contacts between neighbors occurred irregularly and were accompanied by mild aggression. The behavior of encountering groups suggested the recognition of mutual territorial boundaries.

In the laboratory, family groups of marmosets and tamarins are usually much more stable and peaceful than groups consisting of several nonrelated animals of both sexes. This observation has led to the hypothesis that the basic social unit of the *Callithricidae* is an extended family group. Supposedly, it consists of a monogamous breeding pair which stays associated with several successive sets of their offspring for extended periods of time [2, 8, 15, 30]. The field studies reviewed above have made the family hypothesis somewhat unlikely. However, it remains to be seen how many of the transient animals are related to at least some of the residents they join and how other species behave in this respect.

In the laboratory, naturally growing families may increase to considerable size and live together peacefully for many years [2, 11, 30]. ROTHE [30], for instance, maintained a family of 20 *Callithrix jacchus*. Artificial groups, containing several nonrelated adults, on the other hand, are much more unstable. Although no overt aggressive behavior may be noticeable for several months, sudden fighting may erupt without prior warning. These fights can cause severe injuries and even death and usually make the removal of one of the combatants from the group necessary [2, 3, 8, 11, 15, 30].

Families as well as artificial groups tend to establish separate male and female hierarchies, with a 'quasi-monogamous' bond between the α male and the α female. In laboratory groups of *C. jacchus,* one adult male dominates all other adult males but usually does not compete with any of the females or with the juveniles of either sex. Likewise, one adult female dominates the other adult females, tolerating the males and all juveniles [8, 30]. EPPLE [8] found no hierarchy among subdominant group members but ROTHE [30] reports the existence of strictly linear male and female hierarchies in artificial groups and families. In artificial groups of *C. jacchus,* dominance interactions may be established by overt fighting and maintained by a variety of aggressive and submissive signals. Once established, however, groups are often stable for extended periods of time.

Saguinus fuscicollis groups also show parallel male and female hierarchies, which, however, may be more subtle among males than those of

C. jacchus. Epple [10] found that in artificial groups containing one adult female and two adult males there is close social and sexual association between the female and one of the males and suggested that dominance among nonrelated, adult males is often (but not always) expressed in the extent to which they associate with the dominant female rather than in overt aggression. In contrast to the relatively mild dominance interactions among males of artificial *S. fuscicollis* groups, dominance interactions are more frequent and serious among nonrelated females and usually result in vicious fighting and a break up of the group.

Families and artificial groups tend to behave very aggressively against conspecifics who are not members of their own social unit. Epple [9, 13] studied the behavior of established groups of *C. jacchus* and mated pairs of *S. fuscicollis* during a series of tests in which a strange conspecific (the 'intruder') was introduced to them. *C. jacchus* groups tolerated juvenile 'intruders' but aggressed quite severely against adults. The dominant group males mainly attacked adult male intruders while the dominant group females preferentially attacked adult female intruders. Although most members of the *Callithrix* groups participated in attacking and fighting strange conspecifics to a certain degree, the dominant pairs were much more active than subdominants of both sexes.

Pairs of *S. fuscicollis* cooperated in aggressing against adult strange males as well as strange females. Both partners showed more aggression against female intruders than against strange males. Females, however, were about four times as aggressive against female intruders than against male intruders, while males only doubled their aggression scores and showed more inter-individual variability.

It is our subjective impression that the extremely aggressive behavior of female *Saguinus* instigated a higher than spontaneous amount of aggression against female intruders in their male partners. During many of the social encounter tests reviewed above, both male as well as female subjects reacted with an increase in overt aggression as soon as their mate tried to establish any kind of contact with an intruder, particularly if the intruder was a female. Often, the subject would rush to the scene and squeeze itself between the visitor and the mate. The finding that permanent mates show a high correlation between their individual aggression scores [13] probably reflects the fact that they strongly influence each other's responses. The response of *S. fuscicollis* to juvenile intruders was not tested experimentally. Everyday laboratory experience in our colony, however, suggests that juveniles are usually accepted into a group.

Thus, in a laboratory situation, a mated pair of *S. fuscicollis* usually prevents positive social and sexual relations with adult strangers by aggressing against them and the female is the most active one in doing so, when the stranger is a female. These findings possibly indicate that aggression is one of the mechanisms in the enforcement of the pair bond.

Pair Bonding and Sexual Behavior

In many marmosets and tamarins a permanent pair bond exists between the breeding male and breeding female of families and between the α male and α female of artificial groups [2, 10, 11, 23, 30]. The existence of a pair bond does not necessarily demand total and exclusive sexual access to the partner. In artificial groups of *S. oedipus, S. fuscicollis* and *C. jacchus* sexual interactions may occur between all group members [8, 10, 15, 30]. However, quantitative as well as qualitative differences in the sexual and associated social behaviors of the various individuals within a group exist. ROTHE [30] found that in artificial groups of *C. jacchus* the highest ranking male and the highest ranking female were involved in the majority of all heterosexual activities in the group. Moreover, subdominant males never achieved intromission or ejaculation and dominant males never copulated with inferior females to ejaculation. In *S. fuscicollis* groups consisting of one female and two males, the majority of sexual interactions and of huddling and grooming occurred between the female and one of the males [10]. In family groups, members of the breeding pair possibly have exclusive sexual access to each other. ROTHE [30] reports that in *C. jacchus* families the highest ranking male and female copulate exclusively with each other, and we have never seen any of our *S. fuscicollis* mate with their offspring as long as the family contained both parents.

In most species studied so far, copulations are not strictly limited to any given time in the reproductive cycle of the female [2, 8, 15, 36]. In spite of this, some species show a more or less clearcut behavioral estrus, apparent in an increase in copulatory activities and a change in the frequencies of occurrence of some other behavioral patterns [2, 9, 23, 30]. In *C. jacchus* estrus usually occurs in the early postpartum period. The α male begins to show sexual interest in the female as early as two to three days following parturition. At this time, grooming, following, sniffing genitalia and scent marks, scent marking, arching and copulation attempts begin to increase. The female usually refuses mounting attempts by the male during the first few days postpartum but gradually becomes more receptive [9, 29, 30, 34]. ROTHE [29, 30] concluded that postpartum estrus lasts an average of 9.7 days. Within this period the sexual willingness of the female is restricted to five to six days, and only during this period copulations with intromission and ejaculation occur.

KLEIMAN [23] found that in captive groups of *Leontopithicus rosalia* sexual behavior occurred only sporadically during the nonbreeding season in summer and early autumn. At the onset of the breeding season in autumn, the females appeared to start cycling. Now sexual behavior was clumped into three to five day periods at 14–21 day intervals. At the time of estrus, as defined by maximum sexual activity of the breeding pair, there was a dramatic increase in the male's approaching and sniffing of the female. Males also exhibited increased grooming of the female shortly before the highest mounting activity was reached. Females, on the other hand, approached and sniffed males increasingly three to five days before estrus. They also showed an abrupt drop in scent marking activity during the time of estrus. Although copulations outside the estrus period are relatively infrequent, *C. jacchus* and *L. rosalia* copulate occasionally even during pregnancy [9, 23, 29, 34]. ROTHE [30] reports that *C. jacchus* relatively frequently shows a postconception estrus and KLEIMAN [23] found an increase in the copulatory activities of *L. rosalia* seven to eight weeks before parturition. Copulations during pregnancy also occur in several species of *Saguinus* [15, and personal observations]. ROTHE [30] suggests that in *C. jacchus* copulations outside the postpartum estrus period do not include intromission and ejaculation. In *S. fuscicollis* on the other hand, copulations are seen during pregnancy and spermatozoa are quite frequently found in vaginal lavages [EPPLE, unpublished].

In families as well as in artificial groups only the dominant female reproduces. Submissive females, however, often become pregnant almost immediately when they are removed from the group and paired with an adult male or when the dominant female is removed and they gain the α position [3, 7–9, 11, 30, 32]. ROTHE [30], however, reports that subdominant females may remain reproductively inhibited even after separation from the dominant female.

These observations strongly suggest that the dominant female inhibits reproduction in submissive females and that the presence of the mother inhibits it in her daughters. DAWSON's [6] and NEYMAN's [26] field studies indicate that this might be even true in wild marmosets and tamarins. Several mechanisms might be involved in this inhibition. ROTHE [30] suggests that subdominant *C. jacchus* females may not be copulated with to ejaculation or may not mate at the right time of the cycle. Hearn [17] found that only one female in a group of *C. jacchus* experiences ovulatory cycles. Our behavioral studies on *S. fuscicollis* indicate that the dominant female or mother prevents pair formation in all other females under the conditions of captivity [11]. It is possible that the dominant male affects the reproductive capacities of other males within the group, including his sons. Rothe [30] reports that subdominant *C. jacchus* males do not copulate to ejaculation. This possibly indicates similar inhibitory mechanisms among males as among females.

Parturition and Parental Behavior

To date Rothe's [28, 29, 31] and Stevenson's [32] studies on *C. jacchus* provide the only detailed data on parturition. Although their studies cannot be reviewed in detail, some results relevant to captive breeding shall be discussed.

Two findings appear of particular importance here. The first is the fact that many females seem to give birth in relative isolation. Rothe [28, 29] reports that the pregnant female tends to withdraw from the rest of her group approximately 1 h before the onset of labor. Under normal conditions she seems to give birth unattended by other members of the group and stays away from the common sleeping quarters for the rest of the night, sleeping and resting with the neonates. Rothe [28, 29, 31] regards the presence of other group members during some of the deliveries he observed as an artefact caused by illumination of the room for the purpose of observation. The parturient females actually attempted to threaten and push away curious group members who tried to contact them and the neonates. During the deliveries observed by Stevenson [32] the females did not withdraw from the rest of the group. This might reflect individual variability, differences in the living quarters between Rothe's and Stevenson's colonies or the fact that most of Stevenson's observations were done in an illuminated room.

The second major finding emerging from Rothe's [28, 29, 31] and Stevenson's [32] work is that neonates depend almost entirely on their own efforts to gain access to their mother's body and nipples. Thus, the chance to survive the first hours of extrauterine life is dependent upon the vigor of the neonate. As soon as they have been expelled far enough, they grasp their mother's fur with both hands performing rowing and rotating movements.

1 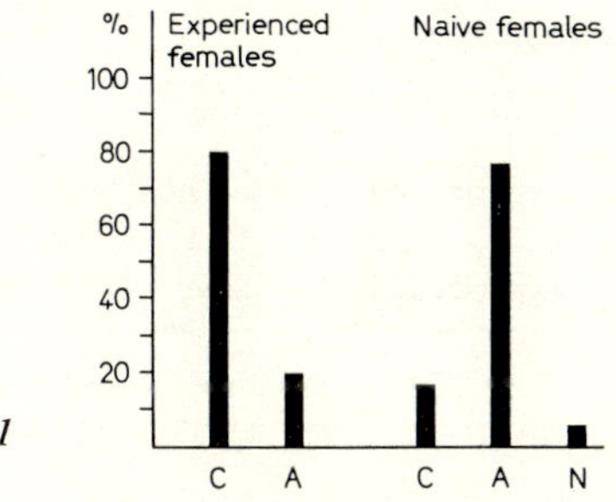

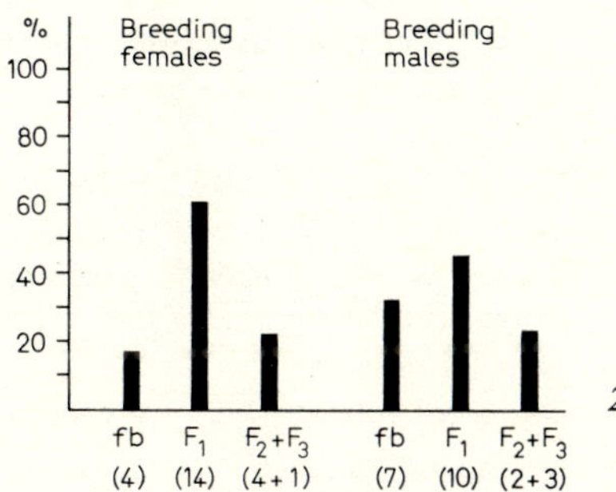 2

Fig. 1. Percentages of differentially experienced laboratory-born primiparas caring for their neonates (C), aggressing against them (A) or neglecting them (N).

Fig. 2. Percentages of feral-born (fb), F_1·F_2 and F_3 breeders in the colony.

Once contact is established they orient themselves toward the head of the mother. Immediately after being expelled fully, even if still attached to the unborn placenta, they start to climb along the mother's abdomen or back toward her nipples. In doing so, they receive no manual support from the mother. The mother licks the neonates and while doing so she may cup one or both hands under the infant's head. This possibly aids the baby in its initial orientation on the mother's body. Neonates who are too weak to move and vocalize are neglected by the mother after a short period of initial licking. Moreover, they are in great danger of being eaten by the group, particularly if they assume unnatural postures and do not vocalize [31]. During their first month, infant marmosets and tamarins spend most of their time being carried. In contrast to the majority of primate species the care and transport of the young is shared among all group members. Usually, mother, father, nonrelated adults and even juveniles participate in infant care. The degree to which they are involved, however, is subject to inter-individual and interspecies variability [1, 11, 20]. Several authors have recently pointed out that participation in infant care by juveniles not only benefits the parents but might be important in preparing the juveniles for their future role as parents [12, 18, 20]. Our observations on *S. fuscicollis* show that this experience is crucial for the development of adequate parental behavior, particularly in females of this species. As of February 1977, 22 laboratory-born females[2] have delivered full-term infants. Of these, 77.3%

2 We randomly crossbreed *S. f. fuscicollis* with *S. f. illigeri* and their hybrids.

Table I. Relationship between parental experience and infant care

	1-♀ naive 1-♂ naive	1-♀ naive f-♂ experienced	1-♀ experienced 1-♂ naive	1-♀ experienced f-♂ experienced	♀ pluriparous (6-f, 1-1) 1-♂ naive	f-♀ f-♂
Number of pairs caring for young	1 (2:2)	2 (4:3)	1 (1:1)	3 (6:6)	5 (11:11)	12 (21:21)
Number of pairs aggressing against young	9 (18:1)	4 (6:0)	1 (2:0)	0	2 (3:0)	0

Data include only the first live litter produced by females or sired by males born in the laboratory (1) and the first live captive born litter of feral parents (f). The ratio between the number of offspring born and the number of live young is given in parentheses.

never participated in the care of their siblings before becoming mothers, 22.7% did. Figure 1 illustrates the parental performance of these females with their first full-term infants. It shows that the neonates of 76.5% of all naive[3] primiparas were found dead or severely mutilated right after birth. Since parturition was not observed, stillbirths, perhaps caused by unknown blood factors [14, 36], cannot be excluded. There is evidence, however, that most of the infants were born alive and healthy and were killed by the mother or both parents [12].

Eighty percent of all experienced[3] primiparas, on the other hand, took adequate care of their offspring. These data very strongly suggest that participation in the care of infants prior to becoming a parent is essential for the development of adequate parental care. This experience is probably important for both parents, but the data presented in table I suggests that it is more crucial for the mother than for the father. Thus, all pairs consisting of feral-born parents and pairs consisting of experienced pluriparas and naive males gave adequate parental care to their first offspring sired by the male. In pairs with parentally naive fathers, the mothers seemed to receive less help in infant care from their mates during the first few days postpartum than in pairs where both parents were experienced. This indicates that previous parental experience is more crucial for the mother than for the father, perhaps because she is the one who has first access to the neonates.

3 Experienced and naive are defined as having taken care of siblings versus not having taken care of siblings during the first year of life.

Table II. Ratio between number of full-term young born by mothers with differing parental experiences between 1969 and March 1977 and young surviving for at least one month

l-mother naive (n = 17)	l-mother experienced (n = 5)	f-mother (n = 12)
83:36	33:25	117:95

l = Laboratory born, f = feral.

On the other hand, the neonates of naive primiparas seem to have a slightly better chance to survive if the father is experienced (table I).

The parental inadequacy of naive animals seems to correct itself somewhat as successive sets of infants are born. Fourteen of the primiparas listed in table I have reproduced repeatedly and all but one have raised some of their infants. We presently maintain 35 breeding pairs, in 11 of which either one or both partners are under two years of age and have not bred yet. Twenty-seven of the females have reproduced, most of them repeatedly and 22 males are proven breeders. Figure 2, showing the percentages of feral-born, F_1, F_2 and F_3 breeders presently in our colony, demonstrates that the majority of all breeding females and 45.5% of the breeding males are F_1. Although the productivity of naive, laboratory-born females was good, the ratio between the total number of offspring produced and the number of offspring surviving at least one month compares very unfavorably with that of experienced laboratory-born and of feral mothers (table II). Breeding pairs containing pluriparas who never participated in sibling care during their first year of life tend to kill one of their twins, occasionally both, chew the tails off the neonates or bite them more or less severely. Thus it appears that a permanent deficiency exists which can only be partly corrected as experience with their own offspring increases over the years. Similar observations are reported by Coimbra-Filho and Mittermeier [5] for *L. rosalia.* The tendency to aggress against neonates seems to be enhanced in a densely populated colony room. Removing 'killer' pairs from our colony room of 1,000 ft^2, housing 30 groups, and placing them into isolation shortly before parturition reduced the likelihood of the neonates being killed but did not eliminate it.

Participation in the care of siblings seems to be crucial for the development of adequate parental behavior of *S. fuscicollis* and also *L. rosalia* [5, 20].

It might not be equally important in *C. jacchus*. STEVENSON [33] reports that three naive *C. jacchus* primiparas reared their young without difficulties. PHILLIPS [27] on the other hand, also attributes the high mortality of infants of F_1 primiparas of *C. jacchus* to parental inexperience.

Synopsis

The field studies reviewed above raise some doubts about the laboratory concept of the extended family as the basic social unit of the *Callithricidae*. As DAWSON [6] suggests, wild groups might more closely approximate artificial laboratory groups. They probably consist of a dominant, monogamous breeding pair, its dependent offspring and separate hierarchies of subdominant males and females who stay associated with the group for various lengths of time. Some of these subdominants might be offspring or relatives of the breeding pair. As the field studies show, these groups are more or less open to immigrants coming from other groups. They possibly tolerate transient relatives more easily and for longer periods of time than nonrelated individuals. In spite of the relative tolerance of wild groups towards strange conspecifics, it appears most practical to maintain laboratory breeders as families and remove the offspring after they have participated in the care of their younger siblings. In this way their reproductive capacities can be utilized as soon as their parental behavior has developed adequately. Moreover, possible losses caused by keeping nonrelated adults of the same sex together are avoided.

As pointed out above, some species are very aggressive towards strange adult conspecifics and some seem to defend territories in the wild. It seems advisable therefore to house them in cages which provide a certain degree of isolation from neighboring groups. We have found this to be more important in *S. fuscicollis* than in *C. jacchus,* particularly in densely populated colony rooms. We therefore house our animals in cages which allow no visual contact with any other group, and by doing so have reduced the general level of excitement in the colony room. We believe that aggressive displays between groups are responsible for a large amount of redirected aggression between mates and for some of the abortions we have seen in our colony. Moreover, ROTHE's [28, 29] observation that the parturient female withdraws from her group and gives birth in relative isolation should be taken into consideration when designing breeding cages. Although not all individuals of all species might show this behavior [see 32] it seems to be widespread enough to be an important factor in breeding efficiency and might figure in some of the infanticides observed by us and other authors.

References

1 BOX, H.O.: A social developmental study of young monkeys *(Callithrix jacchus)* within a captive family group. Primates *16:* 419–435 (1975).

2 CHRISTEN, A.: Fortpflanzungsbiologie und Verhalten bei *Cebuella pygmaea* und *Tamarin tamarin.* Fortschritte der Verhaltensforschung. Z. Tierpsychol., suppl. 14 (1974).

3 COIMBRA-FILHO, A.F.: Mico-leão, *Leontideus rosalia* (Linnaeus, 1766), Situacão actual da especia no Brazil (Callithricidae-Primates). An. Acad. Brasil. Cîen. *41:* 29–52 (1969).

4 COIMBRA-FILHO, A.F. and MITTERMEIER, R.A.: New data on the taxonomy of the Brazilian marmoset of the genus *Callithrix*. Folia primatol. *20:* 241–264 (1973).

5 COIMBRA-FILHO, A.F. and MITTERMEIER, R.A.: Hybridization in the genus *Leontopithecus, L.r. rosalia* (Linnaeus, 1766), *L.r. chrysomelas* (Kuhl, 1820) (Callithricidae, Primates). Rev. Brasil. Biol. *36:* 129–137 (1976).

6 DAWSON, G.A.: Composition and stability of social groups of the tamarin, *Saguinus oedipus geoffroyi*, in Panama. Ecological and behavioral implications; in KLEIMAN Biology and conservation of the Callithrichidae (Smithsonian Press, Washington, in press).

7 DUMOND, D.F.: Comments on minimum requirements in the husbandry of the golden marmoset *(Leontopithecus rosalia)*. Lab. Primates Newsl. *10:* 30–37 (1971).

8 EPPLE, G.: Vergleichende Untersuchungen über Sexual- und Sozialverhalten der Krallenaffen (Hapalidae). Folia primatol. *7:* 37–65 (1967).

9 EPPLE, G.: Quantitative studies on scent marking in the marmoset *(Callithrix jacchus)*. Folia primatol. *13:* 48–62 (1970).

10 EPPLE, G.: Social behavior in laboratory groups of *Saguinus fuscicollis;* in BRIDGWATER Saving the lion marmoset. Proc. WAPT Golden Lion Marmoset Conference, pp. 50–58 (WAPT Oglebay Park, Wheeling 1972).

11 EPPLE, G.: The behavior of marmoset monkeys (Callithricidae); in ROSENBLUM Primate behavior, vol. 4, pp. 195–239 (Academic Press, New York 1975a).

12 EPPLE, G.: Parental behavior in *Saguinus fuscicollis* ssp. (Callithricidae). Folia primatol. *24:* 221–238 (1975b).

13 EPPLE, G.: Notes on the establishment and maintenance of the pair bond in *Saguinus fuscicollis;* in KLEIMAN Biology and conservation of the Callithrichidae (Smithsonian Press, Washington, in press).

14 GENGOZIAN, N.: A blood factor in the marmoset, *Saguinus fuscicollis*. Its detection, mode of inheritance and species specificity. J. med. Primatol. *1:* 272–286 (1972).

15 HAMPTON, J.K.; HAMPTON, S.H., and LANDWEHR, B.: Observations on a successful breeding colony of the marmoset, *Oedipomidas oedipus*. Folia primatol. *4:* 265–287 (1966).

16 HAMPTON, J.K.; HAMPTON, S.H., and LEVY, B.M.: Reproductive physiology and pregnancy in marmosets. Medical Primatology Proc. 2nd Conf. Exp. Med. Surg. Primates, pp. 527–535 (Karger, Basel 1971).

17 HEARN, J.P.: The endocrinology of reproduction in the common marmoset, *Callithrix jacchus;* in KLEIMAN Biology and conservation of the Callithrichidae (Smithsonian Press, Washington, in press).

18 HEARN, J.P. and LUNN, S.F.: The reproductive biology of the marmoset monkey, *Callithrix jacchus*. Lab. Anim. Handb. *6:* 191–202 (1975).

19 HERNÁNDEZ-CAMACHO, J. and COOPER, R.W.: Neotropical primates: aspects of habitat usage, population density and regional distribution in La Macarena, Columbia; in THORINGTON and HELTNE Neotropical primates. Field studies and conservation, pp. 35–69 (Nat. Acad. Sci., Washington 1976).

20 HOAGE, R.J.: Parental care in *Leontopithecus rosalia rosalia:* sex and age differences

in carrying behavior and the role of prior experience; in KLEIMAN Biology and conservation of the Callithrichidae (Smithsonian Press, Washington, in press).

21 KINGSTON, W.R.: The breeding of endangered species of marmosets and tamarins; in MARTIN Breeding of endangered species in captivity, pp. 213–222 (Academic Press, New York 1975).

22 KLEIMAN, D.G.: International studbook, golden lion tamarin, *Leontopithecus rosalia rosalia* (Nat. Zoo. Park, Washington 1976).

23 KLEIMAN, D.G.: Characteristics of reproduction and sociosexual interactions in pairs of lion tamarins *(Leontopithecus rosalia)* during the reproductive cycle; in KLEIMAN Biology and conservation of the Callithrichidae (Smithsonian Press, Washington, in press).

24 MALLINSON, J.J.C.: Breeding marmosets in captivity; in MARTIN Breeding endangered species in captivity, pp. 203–212 (Academic Press, New York 1975).

25 MOYNIHAN, M.: The New World Primates (Princeton University Press, Princeton 1976).

26 NEYMAN, P.F.: Aspects of the ecology and social organization of free-ranging cotton-top tamarins *(Saguinus oedipus)* and the conservation status of the species; in KLEIMAN Biology and conservation of the Callithrichidae (Smithsonian Press, Washington, in press).

27 PHILLIPS, I.R.: The reproductive potential of the common cotton-eared marmoset *(Callithrix jacchus)* in captivity. J. med. primatol. *5:* 49–55 (1976).

28 ROTHE, H.: Beobachtungen zur Geburt beim Weissbüscheläffchen *(Callithrix jacchus* Erxleben, 1777). Folia primatol. *19:* 257–285 (1973).

29 ROTHE, H.: Further observations on the delivery behavior of the common marmoset *(Callithrix jacchus)* Z. Säugetierk. *39:* 135–142 (1974).

30 ROTHE, H.: Some aspects of sexuality and reproduction in groups of captive marmosets *(Callithrix jacchus)*. Z. Tierpsychol. *37:* 255–273 (1975).

31 ROTHE, H.: Influence of newborn marmosets' *(Callithrix jacchus)* behavior on expression and efficiency of maternal and paternal care; KONDO, KAWAI and EHARA Contemporary primatology, pp. 315–320 (Karger, Basel 1975).

32 STEVENSON, M.F.: Birth and perinatal behaviour in family groups of the common marmoset *(Callithrix jacchus jacchus)*, compared to other primates. J. hum. Evol. *5:* 365–381 (1976).

33 STEVENSON, M.F.: Maintenance and breeding of the common marmoset with note on hand-rearing. Int. Zoo. Yb. *16:* 110–117 (1976).

34 STEVENSON, M.F. and POOLE, T.B.: An ethogram of the common marmoset *(Callithrix jacchus jacchus)*, general behavioural repertoire. Anim. Behav. *24:* 428–451 (1976).

35 THORINGTON, R.W.: Observations of the tamarin *Saguinus midas*. Folia primatol. *9:* 95–98 (1968).

36 WOLFE, L.G.; DEINHARDT, F.; OGDEN, J.D.; ADAMS, M.R., and FISHER, L.E.: Reproduction of wild-caught and laboratory-born marmoset species used in biomedical research *(Saguinus* sp., *Callithrix jacchus)*. Lab. Anim. Sci. *25:* 802–813 (1975).

G. EPPLE, Monell Chemical Senses Center, University of Pennsylvania, *Philadelphia, PA 19104* (USA)

Prim. Med., vol. 10, pp. 63–70 (Karger, Basel 1978)

Observations on a Colony of Cotton Eared Marmosets *Callithrix jacchus* with Some Plans for Future Expansion

Barnet M. Levy, Franklin J. Stein, Raymond F. Sis and Roscoe Lewis

The University of Texas Dental Branch, Dental Science Institute, Houston, Tex., and Texas A & M University, College of Veterinary Medicine, College Station, Tex.

It has been some 13 years since the method of marmoset colony management at the University of Texas Dental Branch was described [18]. About that time several other institutions were acquiring marmosets and publishing their methods of colony management [2, 5, 9, 13, 14, 18, 26, 30, 32]. Since only a few colonies were established, the papers on the subject of colony management dwindled, while many reports defining normal blood, urine and microbial values [1, 3, 6, 17, 31] and the reproductive physiology [15, 16, 25, 28] of marmosets appeared. The marmoset seemed to be an animal with special value in virological studies [4, 7, 8, 10–12, 19, 21–24, 29] and has proven to be the animal of choice for investigations in the field of periodontology [20] and teratology [26, 27].

The availability of this experimental animal for biomedical research has recently been curtailed. Some marmosets are being threatened with rapidly diminishing populations in their natural habitat, as a consequence of which some species have been placed on the 'endangered species' list and others on the 'threatened species' list. With an awakening new interest in the use of marmosets in biomedical research, it is obvious that husbandry and colony management will be an important aspect of laboratories wishing to utilize the animal for experimental purposes. As Phillips [26] pointed out, 'It is of prime importance that all users of simians should maintain breeding colonies, thereby obtaining their own replacement animals and avoiding further depletion of wild populations.' At this time there seems little evidence to support the contention that research use of the marmoset would be an important pressure in the depletion of the natural population; however, if

Fig. 1. Cages line entire wall of building or may be moved to establish any arrangement desired. Each cage is provided with a pump and bacterial filter in a standard holder.

we are to have an adequate supply of marmosets for biomedical research, it is imperative to share experiences, knowledge and future plans for the development of marmoset colonies.

The Colony at The University of Texas Dental Branch

The University of Texas Dental Branch, Dental Science Institute, has maintained marmosets since 1960. The colony maintenance routine as published [18] was followed until, with the assistance of JOHN and SUZANNE HAMPTON, a new cage design and housing system emerged. It consists of a barrier type cage with a clear plastic front door, a hinged access panel for removal of waste material, four perch bars and a nesting box. These units are 24 in wide, 22 in deep and 48 in high and are mounted on legs provided with casters to raise the height of the nesting box to approximately 78 in from the floor. The cage floor is constructed of 12 gauge welded stainless steel 1-in mesh welded in a frame of stainless steel rod. Each cage unit is

Fig. 2. Close-up of filter holder with its three plastic tubes.

supplied with air through a bacterial filter made for germ-free environments, placing the cages under slight positive pressure (fig. 1, 2). Temperature is maintained at 75–80 °C, humidity at 55–60%. Lights are controlled to provide a 12-hour light/dark cycle. This caging arrangement is most satisfactory in our colony of some 52 breeding pairs of *Callithrix jacchus*. The reproductive potential of *C. jacchus* is higher than that of virtually any other primate maintained in captivity [26]. In our colony the mortality rate of the F1 and F2 generations has been approximately 55% – with the neonatal mortality being highest in the first six weeks of life. The principal cause of death, while difficult to establish, appears to be parental rejection since there was little evidence of infectious disease during the neonatal period. These findings are consistent with the 50% pregnancy wastage reported for the Royal College of Surgeons *C. jacchus* colony by Phillips [26].

Since the use of marmosets for biomedical research will depend on the availability of laboratory-born animals, it seems imperative to establish new self-sustaining breeding colonies and to expand and improve existing colonies.

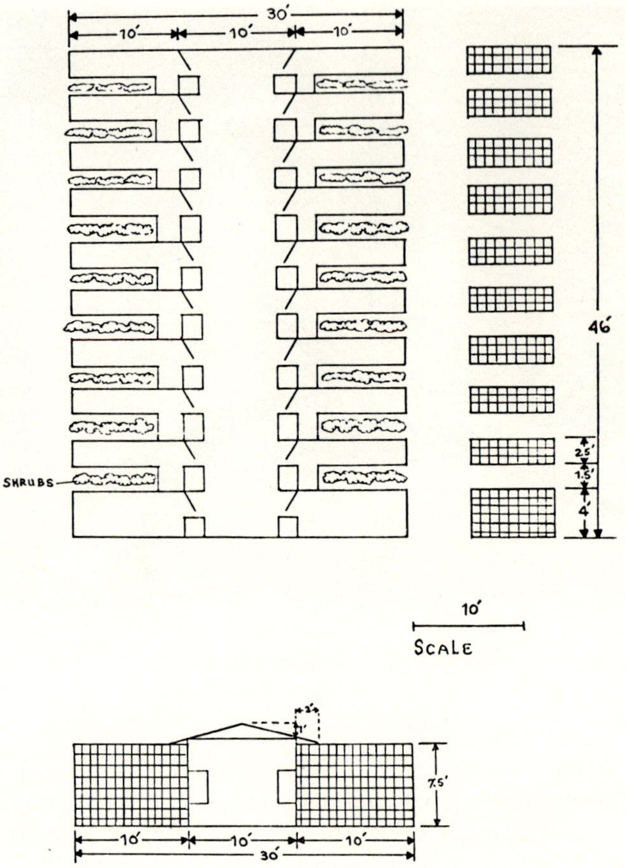

Fig. 3. Schematic of indoor-outdoor caging arrangement.

Accordingly, the University of Texas and the Texas A & M Systems have joined forces to develop a system of animal husbandry for the production of marmosets of the species *C. jacchus* and *Saguinus oedipus*. It is believed that an outdoor caging system with an available protected indoor environment will result in increased breeding and better laboratory animals. The purpose of the new program is to develop methods for the production of high quality *C. jacchus* and *S. oedipus* for appropriate biomedical research. Two caging systems will be tried and evaluated.

(1) An indoor-outdoor modular cage will be constructed of angle-iron,

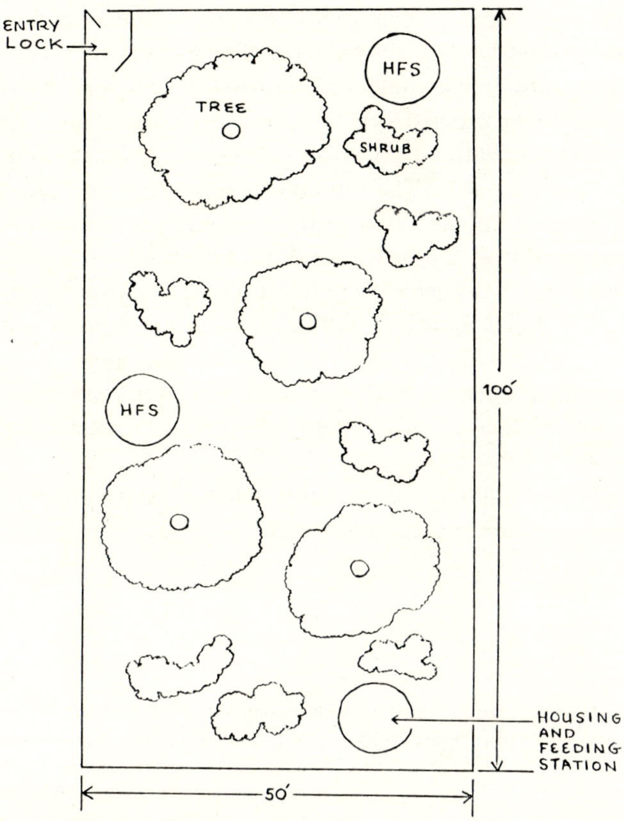

Fig. 4. Schematic of free-ranging enclosure.

aluminum screen doors, fiberglass panels and sheet metal. This unit will consist of a series of outdoor runs 10 ft long, 2.5 ft wide and 7.5 ft high, which will contain trees, shrubs and a nesting box. Each run will connect with an indoor cage 21 × 24 × 48 in located in a 10-ft working corridor running the length of the building. A nesting box will be placed in each indoor cage and in each outdoor run. There will be a 2-ft space containing shrubs between the runs. Pea gravel will be used as flooring for the outdoor run. The indoor portion of the cage will be built on a concrete slab (fig. 3). This type of indoor-outdoor caging allows some visual contact between animals, resembles a natural environment, is inexpensive, can be constructed on the campus locally and should be easy to maintain. These cages are a variation of a pilot module currently in use. Each unit consists of two cages 4 ft wide, 10 ft long and 8 ft high constructed side by side and separated by a double layer of 1 × 2 in welded wire with a 10-in space between the layers to minimize physical contact between animals. The top and three sides of one end of the unit are enclosed. Heat lamps are used in the enclosed portion to provide protection for animals during severe winter days.

(2) The second type of caging will be free ranging enclosures approximately 50 × 100 × 12 ft. Each enclosure will be constructed of posts and 1-in turkey wire and will be provided with individual housing units or nesting boxes that can be heated. Feeding stations will be placed in each enclosure. Trees and shrubs will be located inside the enclosure to provide a 'natural environment' (fig. 4).

Time will tell if our confidence in a natural outdoor environment for the production of laboratory marmosets is justified.

Summary

The current and proposed colony design and management are described. In an effort to increase the neonatal survival of marmosets, an outdoor-indoor housing arrangement is planned.

References

1 Anderson, E.T.; Lewis, J.P.; Passovoy, M., and Trobaugh, F.E., jr.: Marmosets as laboratory animals. II. The hematology of laboratory kept marmosets. Lab. Anim. Care *17:* 30–40 (1967).

2 Benirschke, K. and Richart, R.: The establishment of a marmoset breeding colony and its four pregnancies. Lab. Anim. Care *13:* 70–83 (1963).

3 Brown, L.R.; Handler, S.; Allen, S.; Shea, C.; Wheatcroft, M.G., and Frome, W.J.: Oral microbial profile of the marmoset. J. dent. Res. *52:* 815–822 (1973).

4 Cho, C.T.; Manohar, M.; Muangmanee, L.; Voth, D.W., and Liu, C.: Infectivity titration and immunofluorescent studies on experimental *Herpesvirus hominis* infection in marmosets. Fed. Proc. Fed. Am. Socs exp. Biol. *30:* 353 (1971).

5 Deinhardt, J.B.; Devine, J.; Passovoy, M.; Pohlman, R., and Deinhardt, F.: Marmosets as laboratory animals. I. Care of marmosets in the laboratory, pathology and outline of statistical evaluation of data. Lab. Anim. Care *17:* 11–29 (1967).

6 Deinhardt, F.; Holmes, A.W.; Devine, J., and Deinhardt, J.: Marmosets as

laboratory animals. IV. The microbiology of laboratory kept marmosets. Lab. Anim. Care *17:* 48–70 (1967).

7 DEINHARDT, F.; HOLMES, A.W.; CAPPS, R.B., and POPPER, H.: Studies on the transmission of human viral hepatitis to marmoset monkeys. I. Transmission of disease, serial passages, and description of liver lesions. J. exp. Med. *125:* 673–688 (1967).

8 DEINHARDT, F.; WOLFE, L.; NORTHROP, R.; MARCZYNSKA, B.; OGDEN, J., and DEINHARDT, J.: Induction of neoplasms by viruses in marmoset monkeys. J. med. Primatol. *1:* 29–50 (1972).

9 EPPLE, G.: Maintenance, breeding and development of marmoset monkeys in captivity. Folia primatol. *12:* 56–76 (1970).

10 FALK, L.A.: Oncogenic DNA viruses of nonhuman primates. A review. Lab. Anim. Sci. *24:* 182–192 (1974).

11 FALK, L.A.; WOLFE, L.G., and DEINHARDT, F.: Epstein-Barr virus. Experimental infection and lymphoma induction in marmoset monkeys. Fed. Proc. Fed. Am. Socs exp. Biol. *33:* 739 (1974).

12 FELSBURG, P.J.; HEBERLING, R.L.; BRACK, M., and KALTER, S.F.: Experimental genital herpes infection of the marmoset. J. med. Primatol. *2:* 50–60 (1973).

13 GENGOZIAN, N.: Marmosets. Their potential in experimental medicine. Ann. N.Y. Acad. Sci. *162:* 336–362 (1969).

14 HAMPTON, J.K., jr.; HAMPTON, S.H., and LANDWEHR, B.T.: Observations on a successful breeding colony of the marmoset, *Oedipomidas oedipus.* Folia primatol. *4:* 265–287 (1966).

15 HAMPTON, J.K., jr.; HAMPTON, S.H., and LEVY, B.M.: Reproductive physiology and pregnancy in marmosets. Medical Primatology 1970. Proc. 2nd Conf. Exp. Med. Surg. Primates, New York 1969, pp. 527–535 (Karger, Basel 1971).

16 HAMPTON, J.K., jr.; LEVY, B.M., and SWEET, P.M.: Chorionic gonadotrophin excretion during pregnancy in the marmoset, *Callithrix jacchus.* Endocrinology *85:* 171–174 (1969).

17 HOLMES, A.W.; PASSOVOY, M., and CAPPS, R.B.: Marmosets as laboratory animals. III. Blood chemistry of laboratory kept marmosets with particular attention to liver function and structure. Lab. Anim. Care *17:* 41–47 (1967).

18 LEVY, B.M. and ARTECONA, J.: The marmoset as an experimental animal in biological research. Care and maintenance. Lab. Anim. Care *14:* 20–27 (1964).

19 LEVY, B.M.; TAYLOR, A.C.; HAMPTON, S., and THOMA, G.W.: Tumors of the marmoset produced by Rous sarcoma virus. Cancer Res. *29:* 2237–2248 (1969).

20 LEVY, B.M.; DREIZEN, S., and BERNICK, S.: The marmoset periodontium in health and disease. Monogr. oral Sci., vol. 1, pp. 1–89 (Karger, Basel 1972).

21 MASCOLI, C.C.; ITTENSOHN, O.L.; VILLAREJOS, V.M.; ARGUEDAS, J.A.; PROVOST, P.J., and HILLEMAN, M.R.: Recovery of hepatitis agents in the marmoset from human cases occurring in Costa Rica. Proc. Soc. exp. Biol. Med. *142:* 276–282 (1973).

22 MELENDEZ, L.V.; HUNT, R.D.; DANIEL, M.D.; FRASER, C.E.O.; BARAHONA, H.H.; KING, N.W., and GARCIA, F.G.: *Herpesviruses saimiri* and *ateles* – their role in malignant lymphomas of monkeys. Fed. Proc. Fed. Am. Socs exp. Biol. *31:* 1643–1650 (1972).

23 PATTERSON, R.L.; KOREN, A., and NORTHROP, R.L.: Experimental rubella virus infection of marmosets (*Saguinus* species). Lab. Anim. Sci. *23:* 68–71 (1973).

24 PETERSON, D.A.; WOLFE, L.G.; DEINHARDT, F.; GAJDUSEK, D.C., and GIBBS, C.J., jr.: Transmission of kuru and Creutzfeldt-Jakob disease to marmoset monkeys. Intervirology *2:* 14–19 (1973/74).

25 PHILLIPS, I.R. and GRIST, S.G.: The use of transabdominal palpation to determine the course of pregnancy in the marmoset *(Callithrix jacchus)*. J. Reprod. Fert. *43:* 103–108 (1975).

26 PHILLIPS, I.R.: The reproductive potential of the common cotton-eared marmoset *(Callithrix jacchus)* in captivity. J. med. Primatol. *5:* 49–55 (1976).

27 POSWILLO, D.E.; HAMILTON, W.J., and SOPHER, D.: The marmoset as an animal model in teratological research. Nature, Lond. *239:* 460–462 (1972).

28 PRESLOCK, J.P.; HAMPTON, S.H., and HAMPTON, J.K., jr.: Cyclic variations of serum progestins and immunoreactive estrogens in marmosets. Endocrinology *92:* 1096–1101 (1973).

29 SHOPE, T.; DECHAIRO, D., and MILLER, G.: Malignant lymphoma in cotton-top marmosets after inoculation with Epstein-Barr virus. Proc. natn. Acad. Sci. USA *70:* 2487–2491 (1973).

30 STELLAR, E.: The marmoset as a laboratory animal. Maintenance, general observations of behavior and simple learning. J. comp. physiol. Psychol. *53:* 1–10 (1962).

31 WIENER, A.S.; MOOR-JANKOWSKI, J., and GORDON, E.B.: Marmosets as laboratory animals. V. Blood groups of marmosets. Lab. Anim. Care *17:* 71–76 (1967).

32 WOLFE, L.G.; DEINHARDT, F.; OGDEN, J.D.; ADAMS, M.R., and FISHER, L.E.: Reproduction of wild-caught and laboratory-born marmoset species used in biomedical research (Saguinus sp., *Callithrix jacchus*). Lab. Anim. Sci. *25:* 802–813 (1975).

B.M. LEVY, DDS, The University of Texas Dental Branch, PO Box 20068, *Houston, TX 77025* (USA)

Prim. Med., vol. 10, pp. 71–78 (Karger, Basel 1978)

Breeding of Marmosets in a Colony Environment[1]

N. Gengozian, J.S. Batson and T.A. Smith

Marmoset Research Center, Medical and Health Sciences Division,
Oak Ridge Associated Universities, Oak Ridge, Tenn.

Studies on the marmoset were initiated in our laboratory in 1961 with the specific aim of utilizing this primate in radiation biology. It soon became apparent that as an experimental subject this diminutive primate could offer advantages not observed with other primate species, an impression underscored by successful breeding that first year without any specific efforts made in reproduction [3]. During the past ten years our studies have been directed toward an understanding of the immunologic consequences of the natural blood chimerism that occurs in this species [2]. Although specific monies for a breeding program were not available during this time, greater attention to breeding was necessitated by our research goals, and indeed the concept of non-terminal experimentation was adapted early in the development of our program. The current restrictions on exportation of these animals from the countries of origin place their successful breeding at a high priority if we are to realize the contributions they are to make in biomedicine. The present report describes pertinent observations which may be of value to those planning to or presently engaged in breeding marmosets [4].

Materials and Methods

The predominant *Callitrichidae* species maintained in our colony has been *Saguinus fuscicollis*, with the subspecies *illigeri* comprising approximately 85% of the breeding

1 Under contract with the US Energy and Research Development Administration. Supported in part by United States Public Health Service Grant ROI AI 12007–10 from the Division of Allergy and Infectious Diseases and Grant HL 16757–01 from the Thrombosis and Hemorrhagic Diseases Branch, National Institutes of Health.

nuclei; the balance included *S.f. lagonotus*, *S.f. leucogenys*, and *S.f. nigrifrons*. (Data presented in figures 1 and 2 are, therefore, derived predominantly from *S.f. illigeri*.) More recently, we have gained experience with *Saguinus oedipus oedipus* and *Callithrix jacchus*, and information pertaining to these will be noted. Our husbandry methods were detailed in 1969 and have not changed significantly [1]. Caging, whether for experimental or breeding animals, was accomplished with pressed galvanized metal units measuring 18 × 20 × 42 in. Designed to hold six animals, particularly adolescents who take advantage of the large vertical space for exercise, these can be subdivided into cages measuring 18 × 18 × 20 in. Due to cage space limitations, pairs of animals are now housed in the latter size cages. The basic diet has consisted of a commercial primate chow supplemented with hard boiled eggs and fruit at regular periods during the week. In addition to vitamin D^3 in the chow, the animals are exposed daily to sunlamps for 30 min in the morning and afternoon. No special dietary formulas were used for the breeding females or their young following weaning. All young were separated from their parents at about three months of age and maintained as co-twin pairs or placed with other infants of the same age. These were then subsequently paired at about ten months of age with other unrelated animals or in a few cases with full sibs or co-twins.

Results and Discussion

All marmosets in our colony have been utilized in a variety of research projects, with primary efforts related to the immunology of blood chimerism. The breeding nuclei that have been established, therefore, were derived originally from animals that had been paired for mating but used experimentally during that time until the female became pregnant. The experiment was then interrupted and the animals used only after delivery if warranted by the study. Most often, since the majority of our experiments utilize *in vitro* procedures with blood or serum, this entailed additional bleeding; subsequently, however, the male and female were then set aside as a breeding pair and bled only if a specific need were presented. The reproductive data presented, therefore, were obtained under these conditions which should not be construed as being optimal for captive breeding.

Time to Pregnancy of Imported Animals

Of particular interest is the time after arrival to the colony that a female can be expected to become pregnant. Among *S. fuscicollis* (ssp. *illigeri*) 53 of 75 females (71%) paired with an adult male (same shipment) since their receipt into the colony became pregnant within the first year; with *S.o. oedipus*, 20 of 33 females (61%) became pregnant during the same time period. Greater than 90% of each group had experienced at least one pregnancy during the second year. We have had only eight pairs of *C. jacchus* and

Table I. Full-term interbirth intervals for a select group of *S. fuscicollis* marmosets[1]

Female number	Number of deliveries	Number of determinations[2]	Interbirth time, days	
			mean	range
1	14	12	188	162–261
2	10	5	191	163–234
3	16	7	209	169–244
4	16	10	211	168–294
5	15	12	232	171–304
6	16	8	235	181–306
7	11	6	287	162–367

[1] *S.f. illigeri.*

[2] Interbirth intervals determined only between live, full-term deliveries. Thus, if a live, full-term delivery was preceded or followed by an abortion, the time interval data were not used.

pregnancy occurred in seven during the first year and in the last shortly thereafter.

Interbirth Intervals

Early establishment of a closed colony would be quite dependent upon the rate at which these animals bred, particularly in view of the long gestation period (approximately 145 days) and age of sexual maturity (14–16 months). Among *S. fuscicollis,* the average interbirth interval in 60 deliveries (full-term live births) has been 222 days (range 162–367). In 27 *S.o. oedipus* deliveries, the average was 295 days (range 190–446), and with *C. jacchus,* in ten deliveries the mean interval time was only 188 days (range 126–335). These values, although limited for *S.o. oedipus* and *C. jacchus,* correlate with the experience of others for the three species [6–8]. The high rate of fecundity in *C. jacchus* is well documented by the data of HIDDLESTON [personal commun.], reporting an average interbirth interval of 178 days in 146 deliveries. In addition to species variation, we have noted individual females in the *S. fuscicollis* group to have their own characteristic breeding pattern (table I). Thus, one female in 12 deliveries with the same mate has shown an interbirth interval of 188 days, while another female (also with one mate), had an average interval time of 235 days for eight full-term deliveries. In a practical sense, therefore, one can, in select females of *S. fuscicollis,* anticipate

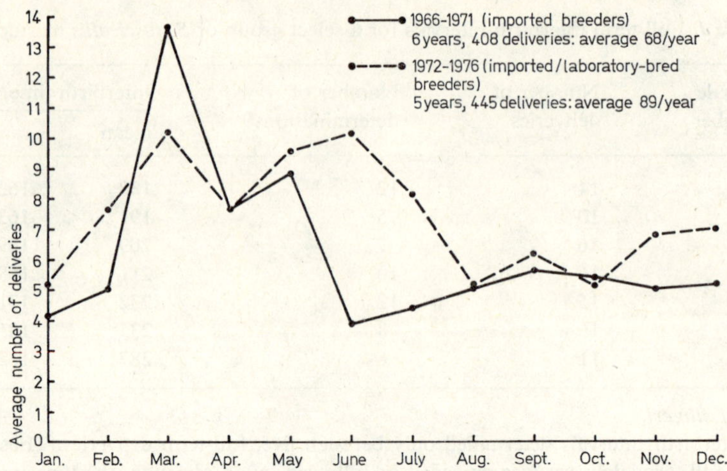

Fig. 1. 'Seasonal' breeding of *S. fuscicollis* marmosets in a colony environment. (As noted in Materials and Methods, data are derived primarily from *S.f. illigeri.*)

two pregnancies per year, while with *C. jacchus,* this appears to be the rule rather than the exception.

Estimation of Parturition Time

To facilitate colony management and anticipating utilization of fetal tissues of known ages for future studies, we explored external uterine palpation as a technique to identify pregnant marmosets at varying gestational ages. The data have been published elsewhere [5] and a brief account is given to indicate the simplicity and accuracy of the technique. Palpation of 92 females in varying stages of pregnancy permitted us to obtain uterine size data (diameter to nearest 0.5 cm) which could subsequently be related to days to delivery of live full-term twins. A graphic plot and an equation derived from this data allowed us to predict the parturition time of an additional 172 pregnant females. Of these, 12 or 7% delivered on the day predicted, 108 or 61% delivered within five days (±) of the anticipated parturition date and 142 or 83% within seven days (±). If one accepts a gestation time of 145 days, the earliest palpable uterus of 0.5 cm diameter represented detection of pregnancy 15 days after conception. The simplicity of the technique is underlined by the need to palpate a pregnant female only once to arrive at some estimation of the parturition date.

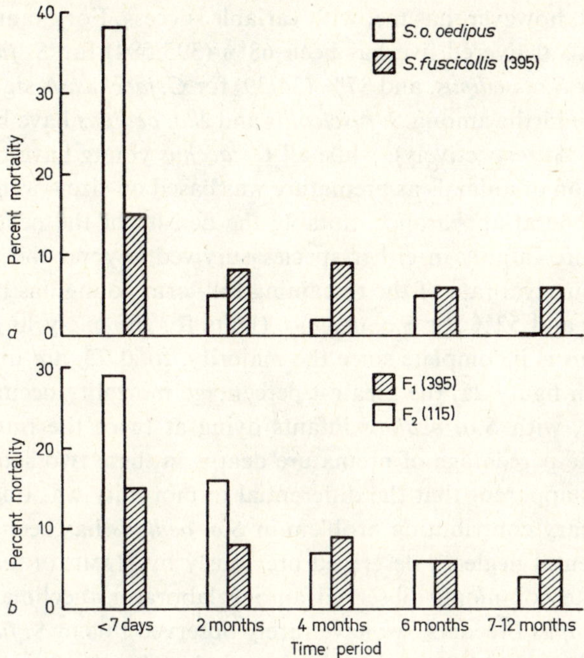

Fig. 2. a First year mortality pattern of colony-born (F_1) *S.o. oedipus* and *S. fuscicollis* marmosets. *b* First year mortality pattern of F_1 and F_2 colony-born *S. fuscicollis* marmosets. (As noted in Materials and Methods, data for *S. fuscicollis* are derived primarily from *S.f. illigeri.*)

Seasonal Breeding in the Laboratory

Upon establishment of breeding pairs, it soon became apparent that even under laboratory conditions, a seasonal breeding pattern similar to that occurring in the wild took place. Although deliveries occurred every month, peak numbers appeared in March, April and May (fig. 1). This effect was most noticeable for the first six years; in the succeeding five years, there appears to be a slight dampening to the peak periods. Although this latter group included laboratory-bred breeders (F_1), the decreased 'seasonality' effect was noted in both the imported and F_1 breeders.

Reproduction and Offspring Survival

Establishment of breeding pairs in the context of obtaining pregnancies is not difficult, as evidenced by the information above. Obtaining live young and extended survival, however, has met with variable success. For example, the percentage animals delivered live has been 68% (395/584) for *S. fuscicollis,* 70% (59/84) for *S.o. oedipus,* and 87% (34/39) for *C. jacchus.* A significant percentage of live births among *S. fuscicollis* and *S.o. oedipus* have been premature (11 and 15%, respectively), while all *C. jacchus* young have been full term. (Classification of animals as premature was based on size – weight/crown rump – and general appearance, notably the density of the pelage.) None of the premature infants in either species survived beyond the first week. The first year survival rate of the remaining full-term young has been 65% for *S. fuscicollis* and 52% for *S.o. oedipus.* (Data for the one-year survival rate of *C. jacchus* is incomplete since the majority, total 25, are under this age.) As shown in figure 2a, the greatest percentage mortality occurred during the first week, with *S.o. oedipus* infants dying at twice the rate of *S. fuscicollis.* Since the percentage of premature deaths in these two species was comparable, it is apparent that the differential in mortality was due to other factors. A primary contributing problem in *S.o. oedipus* has been the phenomenon of 'parental neglect', described previously by HAMPTON *et al.* [6]. While this is quite commonly observed among laboratory-acclimated, i.e. imported, *S.o. oedipus* breeders, we have rarely observed this in *S. fuscicollis* imported breeders. Indeed, in this species, neglect by the imported breeder invariably indicates a premature delivery or some other form of physiological distress which ultimately leads to death of the young. Numerous attempts to hand-rear such offspring have failed. In contrast, parental neglect among *S. fuscicollis* laboratory-bred or F_1 breeders has been observed with healthy, full-term young and hand rearing of these is usually met with success when attempted. The characteristic occurrence of parental neglect among these F_1 breeders may be a factor in the high mortality of the F_2 offspring in the first two months (see fig. 2b, F_1 vs. F_2 survival). Although a greater incidence (22 vs. 11%) of premature births among the F_2 generation can account for the higher mortality during the first week, the increased mortality of F_2 young during the first two months may be reflective of parental neglect. Thus, although we can recognize obvious refusal of parents to take care of their young, there may also be a more subtle form of inadequate parental care which has gone unnoticed, e.g. inability or failure to nurse properly. More recently, we have found foster nursing to be quite effective in maintaining survival of these neglected infants. This is accomplished by allowing the foster infant to play with the young of the newly intended parents (having only one young of comparable age) in a small container for approximately 30 min. Together they are then placed in the foster parents' cage, whereupon the male and female immediately begin their 'inspection' before acceptance. Failure to go through this infant play period has invariably resulted in marked aggression by the foster parents and death of the introduced infant if not removed immediately.

Age of F_1 Breeders at First Conception

Information pertaining to the age of F_1 breeders at first conception was obtained from 25 pairs of laboratory-bred males and females. In this group, the mean time to pregnancy occurred at 30.6 months of age, ranging from 18 months to 46 months (median time was 30 months).

Twinning Frequency and Sex of Offspring

In our laboratory the frequency of twin, triplet, and single births in 319 deliveries among *S. fuscicollis* has been 75, 5, and 20%, respectively. In 83 deliveries of *S.o. oedipus,* the litter size has been twins 69%, triplets 10%, and single births 19%. In each species, the sexes among the twins approximated that expected for fraternal twinning [4]. These data are comparable to that reported by other laboratories.

Conclusions

Breeding of *Callitrichidae* in a research-oriented laboratory environment has been successful even in the absence of any concerted effort or means of allowing for a strict geographic and experimental isolation of these animals. The relatively low frequency (70%) of full-term live deliveries among *S. fuscicollis* and *S.o. oedipus* marmosets may be more apparent than real since the data have included several animals that have never produced live young; in contrast, some breeders have consistently yielded greater than 90% live deliveries. In the absence of definitive answers concerning the reasons for the poor performance of some animals, a simple culling procedure to utilize these in experimental studies rather than retain them as breeders would appear to be expedient if we are to realize successive generations of laboratory-bred offspring. The high mortality attendant during the first two months among the F_2 offspring of *S. fuscicollis* would appear to be correctable once we have established that variables other than parental neglect can be excluded. Recognition of these factors and greater information concerning the behavioral aspects of breeding in a laboratory environment should permit one to fully exploit the potential already demonstrated.

References

1 GENGOZIAN, N.: Marmosets. Their potential in experimental medicine. Ann. N.Y. Acad. Sci. *162:* 336–367 (1969).

2 GENGOZIAN, N.: Immunology and blood chimerism of the marmoset; in DEINHARDT and GENGOZIAN Marmosets in experimental medicine. Prim. Med., vol. 10, pp. 173–183 (Karger, Basel 1978).

3 GENGOZIAN, N.; BATSON, J.S., and SMITH, T.A.: *Tamarinus nigricollis* as a laboratory primate; in Bone marrow therapy and protection in irradiated primates. Proc. Symp. 1962 (Radiobiological Institute TNO, 1962).

4 GENGOZIAN, N.; BATSON, J.S., and SMITH, T.A.: Breeding of the marmoset in the laboratory; in KLEIMAN Smithsonian Institute Journal. Proc. Conf. The Biology and Conservation of the Callitrichidae, Front Royal 1975 (Smithsonian Institute Press, Washington, in press).

5 GENGOZIAN, N.; SMITH, T.A., and GOSSLEE, D.G.: External uterine palpation to identify stages of pregnancy in the marmoset, *Saguinus fuscicollis* ssp. J. med. Primatol. *3:* 236–243 (1974).

6 HAMPTON, J.K., jr.; HAMPTON, S.H., and LANDWEHR, B.T.: Observations on a successful breeding colony of the marmoset, *Oedipus oedipus*. Folia primatol. *4:* 265–287 (1966).

7 HAMPTON, J.K., jr.; HAMPTON, S.H., and LEVY, B.M.: Reproductive physiology and pregnancy in marmosets. Medical Primatology 1970. Proc.2nd Conf. Exp. Med. Surg. Primates, New York 1969, pp. 527–535 (Karger, Basel 1971).

8 WOLFE, L.G.; OGDEN, J.D.; DEINHARDT, J.B.; FISHER, L., and DEINHARDT, F.: Breeding and hand-rearing marmosets for viral oncogenesis studies. Breeding primates, pp. 145–157 (Karger, Basel 1972).

N. GENGOZIAN, PhD, Marmoset Research Center, Medical and Health Sciences Division, Oak Ridge Associated Universities, *Oak Ridge, TN 37830* (USA)

Prim. Med., vol. 10, pp. 79–83 (Karger, Basel 1978)

Breeding *Saguinus* and *Callithrix* Species of Marmosets under Laboratory Conditions

J. D. Ogden, L. G. Wolfe and F. W. Deinhardt

Departments of Microbiology, Rush-Presbyterian-St. Luke's and University of Illinois Medical Centers, Chicago, Ill.

Marmosets belong to the infraorder *Platyrrhini* in the family *Callitrichidae* [12]. They are small monkeys native to Central and South America. Their small size, ease of handling, low maintenance cost, high reproductive rates with frequent twinning, blood chimerism, and susceptibility to various human and animal viruses including human hepatitis A and several tumor-inducing agents rapidly increased their use in biomedical research [1–11, 15–25]. In the early 1970s, the supply of marmosets seemed plentiful and efforts to establish self-sustaining breeding colonies were deemed unnecessary by many investigators. However, the embargoes on exportation imposed by the countries of origin during the last few years and the real need for conservation made it necessary to establish breeding colonies to provide sufficient numbers of animals for the various experimental studies.

Our original colony of marmosets was established in 1961 to determine the feasibility of their use in biomedical research [4] and breeding colonies were established in 1963, 1966 and 1974, respectively. The breeding colonies were established for three major reasons: (1) to supply newborn and juvenile animals for oncogenesis studies; (2) to provide experimental animals and future breeding stock free of parasites and diseases indigenous to the animals in the wild; and (3) to protect against the possibility of eventual exhaustion of the feral sources. Such colonies in addition provide animals of known parentage and date of birth.

This paper briefly summarizes the breeding performance of our colony for the years 1974 through 1976; previous publications from our laboratory have described earlier results and marmoset husbandry [24, 25]. Our colony currently numbers some 960 marmosets, with about 400 animals housed as permanent breeding pairs in three geographically separate areas.

Table I. Monthly deliveries of breeding colonies and experimental animals 1974–1976

	Total	Live	Stillborn	Live, %
January	89	56	33	63
February	119	72	47	60
March	125	87	38	69
April	126	69	57	59
May	104	65	39	63
June	79	43	36	54
July	88	51	37	59
August	90	56	34	63
September	65	33	32	51
October	73	50	23	71
November	64	41	23	64
December	99	67	32	67

A total of 215 live births were observed in the breeding and experimental colonies in 1974. This increased to 229 in 1975 and to 246 in 1976. Our current breeding population consists primarily of two species of white-lipped (WL) marmosets *(Saguinus fuscicollis* subspecies and *S. nigricollis),* cotton-topped (CT) marmosets *(S. oedipus oedipus),* and common or cotton-eared (CE) marmosets *(Callithrix jacchus jacchus).* We recently acquired a small number of red-bellied marmosets *(S. labiatus labiatus)* but more time is needed to evaluate them both as breeders and as biomedical research animals. The species are identified according to the classification of HERSHKOVITZ [13, 14].

Seasonal breeding is not as pronounced in the laboratory as in the wild due to standardized conditions (12 h of light and 12 h of dark) but an increase in the number of deliveries is noted in the early months of the year (table I). Approximately 90% of our pairs have bred successfully, with the first delivery occurring about one year after pairing (range 5.2–36 months) [24]. Previously we had attempted to pair only wild-caught male/female pairs with wild-caught animals (WC/WC) and similarly colony-born marmosets were paired among themselves (CB/CB). Realizing the dwindling supply of wild-caught animals, we are currently pairing some wild-caught adults (whenever possible experienced breeders) with juvenile to mature colony-born animals in an effort to improve the breeding efficiency of the colony-born animals. It is hoped that this will produce experienced, colony-born breeders which might later be paired with younger, inexperienced animals, thus promoting efficient breeding of colony-born animals. As noted on

Table II. Marmoset reproduction breeding colonies 1974–1976

Marmoset pairs (M/F)	Fertile pairs / Total pairs	Live births	Live births/pair/year[1]
WC/WC	130/147 (88.4%)	450	2.04
WC/CB	19/25 (76.0%)	45	1.75
CB/WC	11/13 (84.6%)	35	1.40
CB/CB	29/37 (78.4%)	58	1.16

[1] Animal year commences following first delivery.

Table III. Reproduction of *Callithrix jacchus jacchus* 1974–1976

18/19 pairs fertile (94%)	
41 deliveries	
63 live births	12% singles
	66% twins
	22% triplets
73% live offspring	
4 live births/pair/year	

table II, the WC/WC pairs are the most efficient breeders and the CB/CB the least, with the mixed pairs intermediate. A total of 588 live births occurred in our breeding colonies during 1974 through 1976 (table II). Of all the species (and subspecies) that we have studied, the most prolific breeder is the CE marmoset *(C.j. jacchus)* (table III).

Successful maintenance of a breeding colony depends upon the efficient breeding of colony-born animals. During this reporting period (1974–1976), 74 live second generation animals were born, of which 50 are still alive.

Survival of colony-born animals, from birth to two years is presented in table IV. The main loss of colony-born animals occurs within the first 30 days and mainly involves premature or parentally-rejected offspring.

To sustain productive marmoset breeding colonies, additional studies of their nutrition, housing and reproductive physiology are urgently needed else this group of animals will slowly diminish to a point where their utilization in biomedical research will either cease completely or will become impractical because of lack of adequate supply.

Table IV. Survival of marmosets born in breeding colonies during 1974

	White-lipped	Cotton-topped	Cotton-eared
Live births	116 (23)	28 (15)	7 (1)
1 week	90	19	6
1 month	87	16	5
6 months	81	15	4
1 year	64	11	4
2 years	57	10	4
2-year survival total, %	49.1	35.7	57.1
2-year survival less rejects, %	59.1	76.9	66.7

( ) = parentally-rejected offspring (are included in total numbers).

References

1 BENIRSCHKE, K. and RICHART, R.: The establishment of a marmoset breeding colony and its four pregnancies. Lab. Anim. Care *13:* 70–83 (1963).

2 CHO, C.T.; MANAHAR, M.; MUANGMANEE, L.; VOTH, D.W., and LIU, C.: Infectivity titration and immunofluorescent studies on experimental herpesvirus infection in marmosets. Fed. Proc. Fed. Am. Socs exp. Biol. *30:* 352 (1971).

3 DEINHARDT, F.: Herpesvirus saimiri; in The herpesviruses, pp. 595–625 (Academic Press, New York 1973).

4 DEINHARDT, F. and DEINHARDT, J.: The use of platyrrhine monkeys in medical research; in Some recent developments in comparative medicine, No. 17. Symposia of the Zoological Society of London, pp. 127–152 (Academic Press, New York 1966).

5 DEINHARDT, J.B.; DEVINE, J.; PASSOVOY, M.; POHLMAN, R., and DEINHARDT, F.: Marmosets as laboratory animals. I. Care of marmosets in the laboratory, pathology and outline of statistical evaluation of data. Lab. Anim. Care *17:* 11–29 (1967).

6 DEINHARDT, F.; HOLMES, A.W.; CAPPS, R.B., and POPPER, H.: Studies on the transmission of human viral hepatitis to marmoset monkeys. I. Transmission of disease, serial passages, and description of liver lesions. J. exp. Med. *125:* 673–688 (1967).

7 DREIZEN, S.; LEVY, B.M., and BERNICK, S.: Diet-induced atherosclerosis in the marmoset. Proc. Soc. exp. Biol. Med. *143:* 1218–1223 (1973).

8 FALK, L.A.: Oncogenic DNA viruses of nonhuman primates. A review. Lab. Anim. Sci. *24:* 182–192 (1974).

9 GENGOZIAN, N.: Marmosets: their potential in experimental medicine. Ann. N.Y. Acad. Sci. *162:* 336–362 (1969).

10 HAMPTON, J.K., jr.; HAMPTON, S.H., and LANDWEHR, B.T.: Observations on a successful breeding colony of the marmoset, *Oedipomidas oedipus*. Folia primatol. *4:* 265–287 (1966).

11 HAMPTON, J.K., jr.; HAMPTON, S.H., and LEVY, B.M.: Reproductive physiology and pregnancy in marmosets; in Medical primatology 1970, pp. 527–535 (Karger, Basel 1971).

12 HERSHKOVITZ, P.: A new genus of late Oligocene monkey *(Cebidae, Platyrrhini)* with notes on postorbital closure and platyrrhine evolution. Folia primatol. *21:* 1–35 (1974).

13 HERSHKOVITZ, P.: Metachromism or the principle of evolutionary change in mammalian tegumentary colors. Evolution *22:* 556–575 (1968).

14 HERSHKOVITZ, P.: Taxonomic notes on tamarins, genus *Saguinus* (*Callithricidae*, primates), with descriptions of four new forms. Folia primatol. *4:* 381–395 (1966).

15 HODGEN, G.D.; WOLFE, L.G.; OGDEN, J.D.; ADAMS, M.R.; DESCALZI, C.C., and HILDEBRAND, D.F.: Diagnosis of pregnancy in marmosets: hemagglutination inhibition test and radioimmunoassay for urinary chorionic gonadotropin. Lab. Anim. Sci. *26:* 224–229 (1976).

16 KRAUS, B.S. and GARRETT, W.S.: Cleft palate in a marmoset: report of a case. Cleft Palate J. *5:* 340–345 (1968).

17 LEVY, B.M.; DREIZEN, S., and BERNICK, S.: The marmoset periodontium in health and disease. Monogr. oral Sci., vol. 1, pp. 1–89 (Karger, Basel 1972).

18 MAHOUY, G.B.: The marmoset – a tool for the study of transplantation mechanisms; in Medical primatology 1972, pp. 117–124 (Karger, Basel 1972).

19 PATTERSON, R.L.; KOREN, A., and NORTHROP, R.L.: Experimental rubella virus infection of marmosets (*Saguinus* species). Lab. Anim. Sci. *23:* 68–71 (1973).

20 PETERSON, D.A.; WOLFE, L.G.; DEINHARDT, F.; GADJUSEK, C., and GIBBS, C.J.: Transmission of kuru and Creutzfeldt-Jakob disease to marmoset monkeys. Intervirology *2:* 14–19 (1973/1974).

21 PROVOST, P.J.; ITTENSOHN, O.L.; VILLAREJOS, V.M.; ARGUEDAS, G., and HILLEMAN, M.R.: Etiologic relationship of marmoset-propagated CR326 hepatitis A virus to hepatitis in man. Proc. Soc. exp. Biol. Med. *142:* 1257–1267 (1973).

22 RABIN, H.: Assay and pathogenesis of oncogenic viruses in nonhuman primates. Lab. Anim. Sci. *21:* 1032–1049 (1971).

23 STANBRIDGE, T.N. and PRESTON, N.W.: Experimental pertussis infection in the marmoset: type specificity of active immunity. J. Hyg. *72:* 213–228 (1974).

24 WOLFE, L.G.; DEINHARDT, F.; OGDEN, J.D.; ADAMS, M.R., and FISHER, L.E.: Reproduction of wild-caught and laboratory-born marmoset species used in biomedical research *(Saguinus* sp., *Callithrix jacchus)*. Lab. Anim. Sci. *25:* 802–813 (1975).

25 WOLFE, L.G.; OGDEN, J.D.; DEINHARDT, J.B.; FISHER, L., and DEINHARDT, F.: Breeding and hand-rearing marmosets for viral oncogenesis studies; in Breeding primates, pp. 145–157 (Karger, Basel 1972).

J.D. OGDEN, DVM, Department of Microbiology, Rush-Presbyterian-St. Luke's Medical Center, *Chicago, Ill. 60612* (USA)

Prim. Med., vol. 10, pp. 84–87 (Karger, Basel 1978)

Notes on Comparative Reproduction of Selected *Callitrichidae* in the Laboratory

J.L. Cicmanec and A.K. Campbell

Litton Bionetics, Inc., Department of Laboratory Animal Medicine and Science, Kensington, Md.

Introduction

Callitrichids have been used in experimentation for more than 50 years, and in fact, laboratory reproduction was recorded as early as 1927 [7]. Intensive breeding efforts have been made since 1961 in at least four separate laboratories [1–3, 8]. Recent investigations by Hearn and Lunn [5] and Hampton *et al.* [4] have defined reproductive physiology for some members of *Callitrichidae*. Surprisingly, certain endocrinologic events of marmosets more closely parallel events of human parturition than do those of great apes or Old World primate species [6]. We would like to add some practical comparative features among selected marmoset species to this repository of information on callitrichid reproduction.

Materials and Methods

Most of the callitrichids of our breeding colony were acquired from commercial suppliers (South American Primates, Inc., Miami, Fla., Tarpon Zoo, Tarpon Springs, Fla.) between March 1972 and November 1973. Black and red tamarins *(Saguinus nigricollis)*, saddle-backed tamarins *(Saguinus fuscicollis)*, and cotton-topped marmosets *(Saguinus oedipus oedipus)* were the initially acquired species of our colony. Common marmosets *(Callithrix jacchus jacchus)* and *Saguinus weddelli* were acquired from another commercial supplier (Primate Imports Corporation, Port Washington, N.Y.) in 1975.

The newly received callitrichids are placed in quarantine for 90 days. During this time, they are skin-tested for tuberculosis, fecal samples are analyzed for the presence of internal parasites, and appropriate anthelmintics are administered. Most callitrichids are housed as monogamous breeding pairs during the quarantine period. *C. j. jacchus* are housed as small groups of four to five animals per group for the initial month before breeding pairs are established.

Breeding pairs are housed in stainless steel cages 55 by 60 cm and 70 cm high that are provided with metal perches. The animals are fed a mush composed of primate biscuits, applesauce, and fruit. Physical examinations, including abdominal palpation of females for detection of pregnancy, are performed monthly.

Results and Discussion

The reproductive performance for the various species of callitrichids which we maintain are summarized in table I. Conception rates vary widely among the species and subspecies, and although our numbers of some subspecies are low, we feel that reproductive performance is representative of their groups. GENGOZIAN [2] and WOLFE *et al.* [9] have observed that *S. fuscicollis illigeri* show better reproductive performance in the laboratory than other subspecies of *S. fuscicollis,* and our results support this view. Laboratory reproduction of *C. j. jacchus* has been observed to be better than *Saguinus* spp, and our data support this finding.

Table I. Callithrichid reproductive performance

	Number of pairs	Conception rate	Delivery rate (total live births)	Live births/ pair/year	Average months to first conception in laboratory
Saguinus f. lagonotus	5 (3)	1.01	0.66	1.23	17
Saguinus f. illigeri	11	1.56	0.56	1.64	7
Saguinus f. nigrifrons	11	1.48	0.55	1.41	7
Saguinus oedipus	8	1.35	0.75	2.09	17
Saguinus weddelli	11	3.14	0.90	5.07	11
Callithrix jacchus	12 (9)	2.07	0.68	2.08	9

Note: The number in parentheses is the number of breeding pairs that have actually produced live infants.

Table II. Approximate uterine size during gestation in *S. o. oedipus*

Uterine size and fetal definition	Days prior to birth
1 cm, spherical	98 (84–114)
2 cm, spherical	67 (61–73)
2.5 cm, spherical	57 (49–69)
3.5 × 7 cm	33 (15–47)
4 × 8 cm, fetal head 1 cm	28 (5–50)
Fetal head 1–5 cm	18 (6–34)

Conception rates and delivery rates for *S. weddelli* have been very encouraging; however, production for a more prolonged period of time will not be as high as our initial findings indicate.

Abdominal palpation of females for the detection of pregnancy is a useful practice for predicting delivery date and as a means of monitoring the advancement of pregnancy. The results of our data for *S. o oedipus* are presented in table II. Additional changes, such as vulvar relaxation and behavioral changes, are also useful for indicating the date of birth.

Previous reports have stressed that there is not a seasonal pattern of births for callitrichids maintained in captivity [4, 9], and our experience would support this. As an interesting note, 33 of 67 deliveries within our colony have occurred in March, May, and August.

Analyses of abortion and stillbirth data indicate that most *Saguinus* females are as likely to abort at any time during gestation. The one exception to this is *S. o oedipus* in which all four abortions occurred during the last third of gestation. Among *C. j. jacchus,* all six abortions occurred during the first third of gestation. The exact causes remain unknown.

Summary

Analysis of reproductive performance for four species of callitrichids maintained in our laboratory shows conception rates to vary from 3.14 for *S. weddelli* to 1.01 for *S. fuscicollis lagonotus.* Delivery rates vary from 0.55 for *S. fuscicollis nigrifrons* to 0.90 for *S. weddelli.* There is no seasonal pattern of births for any of the species maintained in the laboratory. Abdominal palpation of females has proven useful for predicting the date of parturition and assessing development of the fetus. Abortions occur among all species and tend to occur during the first third of gestation in *C. jacchus* and the last third of gestation in *S. oedipus.*

References

1 Epple, G.: Maintenance, breeding, and development of marmoset monkeys (Callitrichidae) in captivity. Folia primatol. *12:* 56–76 (1970).

2 Gengozian, N.: Marmosets: their potential in experimental medicine. Ann. N.Y. Acad. Sci. *162:* 336–367 (1969).

3 Hampton, J.K., jr.; Hampton, S.H., and Landwehr, B.T.: Observations on a successful breeding colony of the marmoset, *Oedipomidas oedipus.* Folia primatol. *4:* 265–287 (1966).

4 Hampton, J.K., jr.; Hampton, S.H., and Levy, B.M.: Reproductive physiology and pregnancy in marmosets. Medical Primatology 1970. Proc. 2nd Conf. Exp. Med. Surg. Primates, New York 1969, pp. 527–535 (Karger, Basel 1971).

5 Hearn, J.P., and Lunn, S.F.: The reproductive biology of the marmoset monkey *Callithrix jacchus.* Lab. Anim. Handb. *6:* 191 (1975).

6 Lanman, J.T.: Parturition in nonhuman primates. Biol. Reprod. *16:* 28–38 (1977).

7 Lucas, N.S.; Hume, E.M., and Smith, H.H.: On the breeding of the common marmoset *(Hapale jacchus)* in captivity when irradiated with ultraviolet rays. Proc. zool. Soc. Lond. *30:* 447–451 (1927).

8 Wolfe, L.G.; Ogden, J.D.; Deinhardt, J.R.; Fisher, L., and Deinhardt, F.: Breeding and handrearing marmosets for viral oncogenesis studies; in Beveridge Breeding primates, pp. 145–157 (Karger, Basel 1972).

9 Wolfe, L.G.; Deinhardt, F.; Ogden, J.D., *et al.:* Reproduction of wild-caught and laboratory-born marmoset species used in biomedical research *(Saguinus* sp., *Callithrix jacchus).* Lab. Anim. Sci. *25:* 802–813 (1975).

J.L. Cicmanec, MD, Litton Bionetics, Inc., Department of Laboratory Animal Medicine and Science, *Kensington, MD 20795* (USA)

Prim. Med., vol. 10, pp. 88–95 (Karger, Basel 1978)

A Comparison of Captive Breeding Performance and Offspring Survival in the Family *Callitricidae*[1]

S. H. Hampton, M. J. Gross and J. K. Hampton, jr.

Kineto Matic, Route I, Box 63c, Templeton, Calif.

Introduction

This report covers a history of marmoset breeding at several locations over an extended time period. The first animals were acquired in August, 1960, at Tulane University Medical School, New Orleans, La. In July, 1966, the Tulane animals were incorporated into a marmoset colony which had been established in October, 1961, at the University of Texas Dental Science Institute, Houston, Tex. From 1966 to January, 1974, all animals were housed at the Dental Science Institute in Houston. From 1974 to December, 1976, a small colony was maintained at York College of the City University of New York, Jamaica, N. Y. The husbandry practices and other aspects of this marmoset group have been reported previously [1–3, 6–12].

Prior to 1966 and after 1974, the number of colony breeding pairs was small and variable. During 1966–1974, the number of breeding pairs at any one time averaged about 75 pairs, with a maximum of 100 and a minimum of 50 pairs. During the total period, 1,039 offspring were born of 632 pregnancies. The total marmoset inventory during the period was 3,814 animals, but many of this number were used for experimental purposes and did not enter the breeding program.

The nomenclature of Hershkovitz [8–11] was used for identification of the animals in this colony.

1 This work was supported in part by PHS Research Grants No. HD-08695, DE-02232, and RR08153.

Breeding Performance by Genus

All genera of marmosets *(Callimico, Callithrix, Cebuella, Saguinus,* and *Leontideus)* were included in the colony, but *Callithrix* and *Saguinus* were represented in far larger numbers. Approximately 90% of all animals acquired by means other than breeding (i.e., purchase and donation) were of these two genera. Within the genus *Callithrix, C. jacchus jacchus* was best represented and produced 87% of *Callithrix* offspring; the remaining 13% were produced by *C. argentata argentata* or *C. jacchus penicillata.* The *Saguinus* species in greatest number in the colony was *S. oedipus oedipus* which produced 75% of the *Saguinus* offspring; the remaining 25% were of the species *S. fuscicollis, S. nigricollus, S. leucopus, S. mystax* or *S. oedipus geoffroyi.* No breeding occurred in the genus *Leontideus.* Breeding performance of those genera represented in small numbers has been reported previously [7].

Table I presents data on the number of pregnancies and the number of offspring per pregnancy in each genus. Litter size was known for 579 of the 632 pregnancies. In the 579 pregnancies, 51% were twins, 40% were single

Table I. Relationship of litter size and genus

Genus	Pregnancy litter size				Total, %
	single birth	twin birth	triplet birth	quadruplet birth	
Callithrix	72	141	37	0	250 (43)
Row, %	29	56	15		
Col., %	35	47	68		
Cebuella	3	5	1	0	9 (2)
Row, %	33	56	11		
Col., %	2	2	2		
Saguinus	128	150	16	1	295 (51)
Row, %	44	51	5		
Col., %	63	51	30		
Subtotal, %	203 (37)	296 (53)	54 (10)	1	554
Callimico	25	0	0	0	25 (4)
Row, %	100				
Col., %					
Total, %	228 (40)	296 (51)	54 (9)	1	579

Table II. Relationship of condition of young at birth and genus

Genus	Condition of young at birth			Total, %
	live born	still-born	abortion	
Callimico	19	4	4	25 (3)
Row, %	76	16	8	
Col., %	3	3	1	
Callithrix	314	75	62	451 (46)
Row, %	70	16	14	
Col., %	53	52	26	
Cebuella	8	7	1	16 (2)
Row, %	50	44	6	
Col., %	1	5	0	
Saguinus	253	57	172	482 (49)
Row, %	52	12	36	
Col., %	43	40	73	
Total, %	594 (61)	143 (15)	237 (24)	974

births, 9% were triplet births and there was one quadruplet pregnancy. Single, twin, and triplet pregnancies occurred in all genera except *Callimico,* where all pregnancies were single births. If *Callimico* pregnancies are omitted, 53% of pregnancies resulted in twin births. Again omitting *Callimico,* over 60% of all single births occurred in the genus *Saguinus* and approximately two-thirds of all triplet births were of the genus *Callithrix.*

The condition of young at birth is shown in table II. Only 61% of offspring were born alive; the remaining 39% of offspring were full-term but stillborn or were aborted prematurely. The genus *Callimico* showed the highest percent live born young (76%). *Callithrix* delivered 70% of its young as live born, while *Saguinus* produced only 52% of its young live born. In order to compare pregnancy outcomes in *Callithrix* and *Saguinus,* the two groups in which large numbers of pregnancies occurred, the number of each type of pregnancy expected was compared to that observed, using a χ^2 analysis (table III). *Callithrix* produced a significantly larger number of live born young ($p<0.01$) and a significantly smaller number of aborted young ($p<0.01$) than did *Saguinus.*

Mortality of live born offspring during their first two years of life is shown in table IV. At the end of the first week of life, 43% of *Saguinus* offspring were dead, whereas, in *Callithrix* only 20% were dead. However, at the end of the first two years of life, a greater percent of *Saguinus* offspring

Table III. Comparison of pregnancy outcome in *Callithrix* and *Saguinus*

Genus	Condition of young at birth			Total, %
	live born	stillborn	abortion	
Callithrix	314 (274)[1]	75 (64)	62 (113)	451
Saguinus	253 (293)	57 (68)	172 (121)	482
Total	567	132	234	933
	$\chi^2 = 11.3$ (2)	$\chi^2 = 3.7$ (2)	$\chi^2 = 44.5$ (2)	
	$p<0.01$	$p<0.20$	$p<0.01$	

[1] ( ) = Expected: row total × column total/total.

Table IV. Length of life of liveborn offspring

Length of life	*Callimico*	*Callithrix*	*Cebuella*	*Saguinus*
Day 0	1 (5%)[1]	0	0	8 (3%)
Week 1	6 (32%)	63 (20%)	3 (37%)	108 (43%)
Week 4	6 (32%)	121 (39%)	4 (50%)	135 (53%)
Month 6	13 (68%)	223 (71%)	6 (75%)	183 (72%)
Month 12	15 (79%)	264 (84%)	7 (87%)	195 (77%)
Month 24	15 (79%)	293 (93%)	7 (87%)	204 (81%)
Cohort total	19	314	8	253

[1] ( ) = Cumulative percent dead.

were still living (19%) than in other genera; only 7% of *Callithrix* offspring survived the first two years of life.

Of the 1,039 offspring born, 427 were male, 375 were female ($\chi^2_{(1)}$ = 3.4, $p<0.10$), and 237 were of unknown sex. The offspring of unknown sex were frequently the result of parental mutilation. Premature birth of an infant increased the chance that the infant would be wholly or partially consumed by parents and thus be unsexable. Of the 237 individuals which could not be sexed, 92.3% were stillborn or prematurely aborted.

Although young were born during every month of the year, the greatest number of births occurred in March and the smallest number in November (table V). The peak birth months for *Saguinus* were February and April,

Table V. Month of birth of full-term offspring

Month	*Callithrix*	*Saguinus*	All Genera[1]
January	27	28	57
February	29	37	74
March	42	36	81
April	34	37	76
May	34	32	69
June	35	26	63
July	51	25	78
August	36	11	52
September	37	17	57
October	29	26	57
November	16	17	35
December	19	18	38
Total	389	310	737
	$\chi^2 = 31.8$ (11)	$\chi^2 = 32.3$ (11)	$\chi^2 = 40.46$ (11)
	$p<0.01$	$p<0.01$	$p<0.01$

[1] Includes *Callimico, Cebuella, Callithrix* and *Saguinus.*

while in *Callithrix* the peak birth month was July. Animals in a captive environment, under conditions of constant day length and humidity might be expected to produce term young at a constant rate throughout the year. However, deviation from that expectation produced a χ^2 value which was significant ($p<0.01$) in *Callithrix, Saguinus,* and in all genera combined. Thus, full-term births are not evenly distributed but are more likely to occur during certain months of the year.

The mean interbirth intervals were as follows: *Callimico,* 187 days (three females); *Callithrix,* 265 days (44 females); *Cebuella,* 185 days (two females); *Saguinus,* 301 days (50 females). Although the gestation period for Callitricidae has been reported at approximately 145 days [4, 5], most interbirth intervals are considerably longer; thus, even if a post-partum estrus occurs, pregnancy often does not result.

Offspring Survival

Because of the large number of non-living offspring produced and the high mortality rates of colony born young during the first two years of life,

Table VI. Relationship of condition at birth to month of birth

Period of birth	Condition at birth		Total
	liveborn	non-living[1]	
January–February–March	183 (31%)	91 (24%)	274
April–May–June	169 (28%)	102 (27%)	271
July–August–September	149 (25%)	93 (24%)	242
October–November–December	93 (16%)	94 (25%)	187
Total	594	380	974
	$\chi^2 = 31.6$ (3) $p < 0.01$	$\chi^2 = 1.12$ (N.S.) (3)	

[1] Includes stillborn and aborted young.

the reproductive performance of captive marmosets was far below its potential. In order to investigate possible causes of high pre-natal and post-natal mortality, factors that might influence these mortality rates were studied.

The relationship of the condition of the young at birth to the month of birth is shown in table VI. The greatest percent of live births, as compared to other pregnancy outcomes, occurred in the first quarter of the year (31%), the smallest percent in the last quarter (16%). Although non-living young are equally likely to result during any quarter of the year, live born young are significantly more likely during the early part of the year ($p < 0.01$).

Litter size, as related to condition at birth, is shown in table VII. Of all live born infants, 60% were of twin pregnancies, 22% were single born and 18% were triplet born. The litter size with the highest percent live born was triplet (66%). The litter size with the highest percent stillborn was also triplet (20%) but the most prevalent litter size in abortions was single births, making up 29%.

In the colony, a rather small number of females produced approximately two-thirds of all offspring. Eighty-four mothers produced 651 offspring. No significant differences were seen in life expectancy of offspring associated with maternal parity through litters one to five.

The effect of litter size on life expectancy is shown in table VIII. It appears that twins and triplets have better survival chances than single born young. Only 23% of the single births were alive the end of the first month,

Table VII. Relationship of condition at birth to litter size

Litter size	Condition at birth			Total
	live born	stillborn	abortion	
Single birth	131	28	64	223
Row, %	59	12	29	
Col., %	22	20	27	
Twin birth	354	83	150	587
Row, %	60	14	26	
Col., %	60	58	63	
Triplet birth	109	32	23	164
Row, %	66	20	14	
Col., %	18	22	10	
Total	594	143	237	974

Table VIII. Relationship of liter size[1] and length of life

Length of life	Litter size		
	single birth	twin birth	triplet birth
Day 0	132 (59%)[2]	264 (45%)	67 (41%)
Week 1	156 (70%)	337 (57%)	98 (60%)
Week 4	169 (77%)	392 (67%)	113 (69%)
Month 6	200 (90%)	493 (84%)	138 (84%)
Month 12	208 (93%)	527 (90%)	149 (91%)
Month 24	215 (97%)	549 (94%)	153 (93%)
Cohort total	131	354	109

[1] Includes abortions, stillborn and live born.
[2] ( ) = Cumulative percent dead.

while 33% of twins and 31% of triplets survived the first month. At the end of two years, 97% of the single births, 94% of twin births, and 93% of triplets, were dead.

Data regarding the relationship between sex of offspring and (a) mortality during the first two years of life and (b) litter size, show no significant differences in any categories.

Acknowledgments

The authors wish to thank the following persons for their technical assistance: STEVEN BUSH, Social Science Computer Laboratory, Austin, Tex.; Mr. KENNETH DARLING, University of Texas, Houston, Tex.; Ms. CLAIRE CAMPBELL, York College of the City University of New York.

References

1 HAMPTON, J.K., jr.: Laboratory requirements and observations of *Oedipomidas oedipus*. J. phys. Anthrop. *22:* 239–244 (1964).

2 HAMPTON, J.K. and HAMPTON, S.H.: Marmosets (Hapalidae): breeding seasons, twinning and sex of offspring. Science *150:* 915–917 (1965).

3 HAMPTON, J.K.; HAMPTON, S.H., and LANDWEHR, B.: Observations on a successful breeding colony of the marmoset, *Oedipomidas oedipus*. Folia primatol. *4:* 265–287 (1966).

4 HAMPTON, J.K.; HAMPTON, S.H., and LEVY, B.M.: Reproductive physiology and pregnancy in marmosets; in Medical Primatology 1970. Proc. 2nd Conf. Exp. Med. Surg. Primates, pp. 527–535 (Karger, Basel 1971).

5 HAMPTON, J.K.; LEVY, B.M., and SWEET, P.M.: Chorionic gonadotropin excretion during pregnancy of the marmoset, *Callithrix jacchus*. Endocrinology *85:* 171–174 (1968).

6 HAMPTON, S.H. and HAMPTON, J.K.: Rearing marmosets from birth by artificial laboratory techniques. Lab. Anim. Care *17:* 1–12 (1967).

7 HAMPTON, S.H.; HAMPTON, J.K., and LEVY, B.M.: Husbandry of rare marmoset species; in BRIDGWATER Saving the lion marmoset, pp. 70–87 (Viking Services, Minnesota 1972).

8 HERSHKOVITZ, P.: On the identification of some marmosets, family Callithricidae (Primates). Mammalia *30:* 327–332 (1966).

9 HERSHKOVITZ, P.: Taxonomic notes on tamarins, genus *Saguinus* (Callithricidae, Primates) with descriptions of four new forms. Folia primatol. *4:* 381–395 (1966).

10 HERSHKOVITZ, P.: Metachromism or the principle of evolutionary change in mammalian tegumentary colors. Evolution *22:* 556–575 (1968).

11 HERSHKOVITZ, P.: The evolution of mammals on southern continents. VI. The recent mammals of the neotropical region: a zoogeographic and ecological review. Q. Rev. Biol. *44:* 1–70 (1969).

12 LEVY, B.M. and ARTECONA, J.: The marmoset as an experimental animal in biological research: care and maintenance. Lab. Anim. Care *14:* 20–27 (1964).

Dr. SUZANNE H. HAMPTON, Kineto Matic, Route I, Box 63c, *Templeton, CA 93465* (USA)

Session III: Oncology

Chairman: H. Rabin, Frederick, Md.

Prim. Med., vol. 10, pp. 96–118 (Karger, Basel 1978)

Overview of Viral Oncology Studies in *Saguinus* and *Callithrix* Species[1]

L. G. Wolfe and F. Deinhardt

Departments of Microbiology, Rush-Presbyterian-St. Luke's and University of Illinois Medical Centers, Chicago, Ill.

Introduction

During the early 1960s, when the emphasis on primate research was gaining momentum, marmoset monkeys along with several other nonhuman primate species were selected for evaluation as potential models for viral carcinogenesis. Marmosets were chosen primarily because of their unique biologic characteristics, small size, relatively low procurement and maintenance costs, ease of handling and reproductive capabilities in captivity, namely high production rate and short generation time. Subsequently, the susceptibility of certain species of marmosets to many different oncogenic animal viruses has been evaluated and although the percentage of agents that have induced tumors has been relatively low, marmosets have proven more susceptible to a broader range of viruses than any other monkey species evaluated. So far they have proven susceptible to two groups of viruses, i.e., type C RNA sarcoma viruses (family: retraviruses; subfamily: oncornaviruses) and lymphotropic herpesviruses. Because of their documented susceptibility, they should be a species of choice for evaluating the oncogenicity of all viruses considered potentially oncogenic.

This communication reviews the experimental tumor models studied in marmosets and relates relative susceptibility of different marmoset species whenever applicable. With few exceptions, the species studied, dictated primarily by availability, have been limited to white-lipped (WL) *(Saguinus fuscicollis* subspecies, *S. nigricollis)*, cotton-topped (CT) *(S. oedipus oedipus)* and common (CM) *(Callithrix jacchus jacchus)* marmosets.

1 Supported in part by Public Health Service Contract No 1 CP 33219 from the Division of Cancer Cause and Prevention, National Cancer Institute, National Institutes of Health.

Table I. Viral oncogenesis in marmosets

Tumor induction	No tumor induction
Type C sarcoma viruses	oncornaviruses
Rous, feline, simian, HL-23V (simian-like)	avian leukosis, murine leukemia/sarcoma, mammary tumor, feline lymphoma, RD-114, gibbon ape lymphoma, Mason-Pfizer monkey virus, baboon endogenous virus
Herpesviruses	
H. saimiri, H. ateles, Epstein-Barr	DNA viruses
Baboon herpesvirus	
	Marek's disease, adenoviruses, Yaba virus, polyoma, SV-40, JC, BK

Viruses Evaluated for Oncogenicity in Marmosets

Susceptibility of marmosets to tumor induction by RNA and DNA tumor viruses is summarized in table I. Three sarcoma models (avian, feline and simian) and three models of lymphoproliferative disease *(Herpesvirus saimiri, Herpesvirus ateles* and Epstein-Barr virus) have been well established and will be reviewed in this report. Demonstration of oncogenicity for other viruses in table I represent recent findings and are presented elsewhere in this volume [5, 14, 37]. Oncogenicity studies of viruses which failed to induce tumors (and which are not discussed in this report) were performed, almost without exception, in newborn and/or adult WL marmosets, without immunosuppressive conditioning.

Experimental Tumor Models

Rous Sarcoma Virus (RSV)

Oncogenicity of RSV in marmosets, as in other mammals, is limited to subgroup D strains. WL and CT marmosets are highly susceptible to the Schmidt-Ruppin strain of RSV (SR-RSV) [11, 15, 52, 74, 75, 83, 97, 100] and less so to other strains [74]. CM also are susceptible to SR-RSV but the number of animals studied in parallel is too small for comparison of relative susceptibility [50, 100]. Neonatal or juvenile WL and CT animals inoculated intramuscularly (i.m.) or subcutaneously (s.c.) with SR-RSV (>10^5 focus forming units FFU on chicken cells or $>10^5$ tumor-inducing units in chickens), without immunosuppressive conditioning, consistently develop tumors at inoculation sites within three to six weeks. Over 80% of adult WL [52, 83, 97] and CT [74] marmosets also develop tumors when inoculated under similar conditions although the latent periods are sometimes longer than observed in neonates. In young animals the tumors are always progressors and the disease fatal, within three to six weeks after detection of the tumor, whereas in adults the clinical course is usually longer and 20–30% of the tumors eventually regress. Serial transmission of progressor tumors (tumors composed of recipient cells) can be accomplished by passage in chimeric twins or unrelated animals, without immunosuppressive conditioning. Comparing the general susceptibility of marmosets to other primate species, based on incidence, latent period, and behavior of the tumors, marmoset monkeys are the most susceptible to SR-RSV [75, 97] followed in descending order of susceptibility by squirrel monkeys, Old World monkeys and baboons, and finally, by galagos [97]. In preliminary studies using adult WL marmosets, BCG immunization and/or administration of viable, non-tumorigenic, allogeneic marmoset cells transformed *in vitro* by SR-RSV had no effect on tumor incidence, latent period or clinical course [83].

Based on limited data WL and CT marmosets are seemingly much more susceptible to SR-RSV than to two other strains of RSV, namely the Bryan high titer strain (Br-RSV), containing avian myeloblastosis virus as a helper, and the Carr-Zilber strain (CZ-RSV). CZ-RSV induced tumors in only two of eight adult CT marmosets (both animals treated with betamethasone)[74] and Br-RSV induced a sarcoma in only one of four neonatal WL marmosets (latent period of several months).

Progressor tumors induced by i.m. or s.c. inoculation of RSV are highly vascular rhabdomyosarcomas [50], fibrosarcomas or undifferentiated sarcomas; large multinucleated giant cells in regional lymph nodes and tumor metastasis to regional lymph nodes and/or lung are frequent findings. In one study intraosseus inoculation of CM induced osteosarcomas [50].

Type C virus particles [15, 50] or viral structural antigens have not been detected in marmoset tumors induced by SR-RSV, in tumor cell lines [52] or in marmoset cells transformed *in vitro* by SR-RSV [15, 54, 83]. The RSV genome can be recovered from early passages of about 50% of the tumor cell lines, by co-cultivation with leukosis-free C/O chick embryo fibroblasts [52], but not from cells transformed *in vitro* [15, 54].

Despite the apparent absence of virus expression in the tumor cells, approximately 50% of the animals develop serum neutralizing antibodies and an occasional animal with a long clinical course develops complement-fixing antibodies against core antigens [15, 52]. As measured by ^{51}Cr-release microcytotoxicity assays [42] approximately 25% of animals inoculated with cell-free virus develop transient, low levels of serum cytotoxic antibodies but no lymphocytotoxicity against RSV transformed marmoset cells [83]. In contrast marmosets either inoculated with tumorigenic doses of autologous or allogeneic transformed cells (2.0×10^7 cells i.m.) or immunized with autologous transformed cells (1.0×10^6 cells i.m.) and subsequently challenged with tumorigenic doses develop, independent of tumor formation, both serum cytotoxic antibodies and cytotoxic lymphocytes that react in microcytotoxicity assays against autologous cells transformed not only by SR-RSV but also by feline sarcoma virus (FeSV) [16, 42]. These reactivities often disappear once a rapidly growing tumor develops.

Feline Sarcoma Virus

WL and CT marmosets are susceptible to two strains of FeSV, Snyder-Theilen (ST-FeSV) [90] and Gardner-Arnstein (GA-FeSV) [35], whereas the susceptibility of *Callithrix* species has not been evaluated. WL and CT marmosets appear equally susceptible although the number of CT marmosets studied is relatively small [15, 97, 102]. Oncogenicity of a third strain of FeSV (McDonough) [56, 57] was not demonstrated in two WL marmosets studied. Primate species other than marmosets have only been inoculated with ST-FeSV and based on limited data, marmosets and squirrel monkeys appeared more susceptible than *Macaca* sp. or *Cercopithecus talapoin* [77, 92, 97].

Marmosets <1 month old inoculated intraperitoneally (i.p.) or i.m. with $\geq 4.0 \times 10^3$ FFU [55] of virus regularly develop sarcomas [102]. Latent periods range from two to four weeks and the animals die from progressive neoplastic disease one to four months after inoculation (PI). Animals inoculated when two months old with ST-FeSV were less susceptible than neonates and of marmosets inoculated when four months old or older, only an aged marmoset (>9 years old) developed a tumor (regressor). Although only a single observation, the susceptibility of aged marmosets might be comparable to animals approximately two months old. Serial passages of FeSV-induced sarcomas can be readily established in neonatal or adult siblings or unrelated animals, without immunosuppressive conditioning, and cytogenetic studies of recipient tumors at different stages of development indicate that donor cells contribute to the early proliferative response but are later completely overgrown by transformed recipient cells [102].

The morphology of the tumors is virus-dependent, i.e., ST-FeSV induces undifferentiated sarcomas or poorly-differentiated fibrosarcomas whereas GA-FeSV consistently induces well-differentiated fibrosarcomas. The morphology of marmoset tumor cells grown *in vitro* closely resembles the respective tumor morphology and the two strains also induce morphologically-distinct foci in cat and marmoset cell lines [55, 56]. Metastasis to lung or regional lymph nodes occurs in 10–20% of the animals and regional lymph nodes, without metastatic fibrosarcoma, frequently contain large multi-nucleated giant cells, identical to those in RSV-inoculated animals.

Type C virus particles are not produced by tumors induced by cell-free virus, but inoculation of the tumor cells into other marmosets or propagation *in vitro* results in complete virus expression. The virus thus produced is oncogenic in marmosets and cats and transforms marmoset and cat cells *in vitro*. FeLV p30 antigen is present in marmoset tumors induced either by cell-free preparations or minced neoplastic tissue, irrespective of complete virion expression, indicating at least partial expression of the virus genome in all tumors [97, 102].

About 70% of tumor-bearing marmosets with PI courses of six weeks or longer develop antibodies to FeLV p30 [80, 102] and those with virus-producing, transplanted tumors also develop neutralizing antibodies [56] and antibodies to feline oncornavirus-associated cell membrane antigen (FOCMA) [21, 80]. A correlation between progressive or regressive tumor behavior and titers of neutralizing or FOCMA-specific antibodies has not been evident. Adult marmosets inoculated with tumorigenic doses of autologous or allogeneic, ST-FeSV transformed cells develop serum cytotoxic antibodies and cytotoxic lymphocytes that react against both RSV- and FeSV-transformed cells in ^{51}Cr-release microcytotoxicity assays [16, 42].

Simian Sarcoma Virus, Type 1 (Lagothrix) *(SSV-1)*

Type C virus particles were demonstrated electron microscopically in a spontaneous fibrosarcoma of a woolly monkey (*Lagothrix* sp.) [91] and the virus isolated from this tumor was designated SSV-1 [98]. SSV-1 is defective for virus replication [1, 4, 85, 86], requiring the presence of nontransforming associated virus [6, 79, 101]. Oncogenicity of SSV-1 has been demonstrated in neonatal WL, CT [15, 97, 98] and white-moustached [93] *(Saguinus mystax)* marmosets although only one CT and one white-moustached marmoset were studied. In a comparative study of marmosets, squirrel monkeys *(Saimiri sciureus)* and a galago *(Galago crassicaudatus)*, marmosets were the most susceptible although the number of animals was too small to yield definitive conclusions [93].

Neonatal WL marmosets are much less susceptible to sarcoma induction by SSV-1 than by SR-RSV or FeSV. About 50% of marmosets inoculated with 5.0×10^4 to 1.0×10^6 FFU [101] of cell-free virus develop tumors at the inoculation site after latent periods of four to six weeks. The tumors are either very slowly progressive, well-differentiated fibrosarcomas, with clinical courses up to one year, or fibromas that eventually regress within three to six months. Tumor metastasis (lung, subcutis) has only been observed in one animal which had a PI course of one year. Intramuscular inoculation of unrelated marmosets with minced neoplastic tissue or with marmoset fibroblasts transformed *in vitro* induces tumors which are composed of recipient cells, as determined by sex chromosomal analyses. Administration of anti-marmoset thymocyte serum usually results in enhanced growth of these tumors and sometimes to development of widespread metastases [40, 97]. The tumors and tumor cell lines are virus producers and the virus transforms susceptible cells *in vitro* and is oncogenic *in vivo*. Tumor-bearing marmosets develop antibodies to SSV-1 p30 protein and serum neutralizing antibody titers of 1:32 to 1:256.

The intracerebral (i.c.) route of inoculation is much more sensitive than the i.m. route for demonstrating oncogenicity of SSV-1. In a preliminary study [41] gliomas were induced in six of ten neonatal WL marmosets inoculated i.c. (right frontal lobe of cerebrum) with 0.1 ml of SSV-1 (2.0×10^3 FFU), a dose which did not induce tumors in five neonatal WL marmosets inoculated i.m. Of the six animals with tumors, four died at 2.5, 6.5, 16 and 20 months PI respectively and two were sacrificed at 24 months PI. All ten animals developed serum neutralizing antibodies and antibodies to other virus structural proteins. The antibody titers in tumor-bearing marmosets persisted throughout the course of the disease and the neutralizing antibody titers (up to 1:8,192) were severalfold higher than in animals with sarcomas or in the four animals that did not develop gliomas. Type-C virus particles were observed in two of two brain tumors studied by electron microscopy and virus produced by tumor cell lines was indistinguishable biologically from SSV-1 stock virus. Virus also was isolated from brain tissue distant from tumor in one animal and from cerebrospinal fluid (CSF) of one animal but not from CSF of three animals and from urine of two animals. CSF collected at necropsy from three animals with brain tumors contained SSV-1 antibodies, but the titers were six-to eightfold lower than in sera.

Tumors ranged in size from 2.0×1.0 mm to 2.0×1.0 cm and were located in cerebral hemispheres, diencephalon and/or midbrain, with the midbrain being the most common site. Astrocytomas, grades II–III, were

diagnosed in three animals and the tumor in one of these animals has been previously described as morphologically more similar to human glioblastoma multiforme than any other viral-induced tumor studied [7]. The more malignant tumors were characterized by densely and sparsely cellular areas of pleomorphic glia, greatly increased vascularity, prominent endothelial and adventitial cell proliferation, thrombosis, edema, hemorrhage and necrosis. Some tumors were composed entirely of astroglia whereas others were admixtures of astroglia and oligodendroglia. In two animals the tumors, grades II–III, were characterized by prominent perivascular cuffs of lymphoid cells and the serum antibody titers had declined in these animals, suggesting they may have been undergoing regression. A glial scar was present in the sixth animal at the approximate site of inoculation, and the lesion was interpreted as the site of a regressed tumor because of the presence of hydrocephalus, a common finding in tumor-bearing animals, and because of antibody titers which correlated with those of tumor-bearing animals. Internal hydrocephalus was present in four other animals and was caused by occlusion of the aqueduct of Sylvius or the 4th ventricle by neoplastic tissue. Evidence of tumor development or glial scars was absent in four animals, two of which died of nonspecific causes approximately six months PI and the other two were sacrificed 24 months PI.

Simian Herpesviruses: Herpesvirus saimiri *(HVS)*, Herpesvirus ateles *(HVA)*

HVS was first isolated in 1968 [59] and HVA in 1970 [62, 66] from primary kidney cell cultures of healthy squirrel *(Saimiri sciureus)* and spider *(Ateles geoffroyi)* monkeys, respectively. Subsequently, only lymphoid cells have been identified as target cells for these two viruses under natural or experimental conditions *in vivo,* but both HVS and HVA can be grown in a number of fibroblastic or epithelial cell cultures *in vitro* with development of typical herpesvirus cytopathic effect [10, 12, 24, 27, 99]. Therefore, the original isolates probably came from lymphocytes present in the kidney cell cultures. Serologic and virus isolation (co-cultivation of lymphoid cells with permissive cell monolayers) studies of adult animals maintained in captivity for less than one month revealed that >85% of squirrel monkeys and 50–60% of spider monkeys *(Ateles paniscus hybridus, A.p. robustus)* were infected latently (no overt disease detectable) with the respective viruses [24, 26, 27]. The physicochemical and biological properties *in vitro* of HVS and HVA are similar to other herpesviruses and have been reported in detail [10, 12, 24, 60, 62, 64, 66]. All HVS isolates appear identical antigenically [10] whereas antigenic differences have been found between the original HVA isolate and subsequent isolates [12, 24]; HVS and HVA isolates, with the possible exception of the original isolate [62, 66], share cross-reacting antigens [12, 22, 24]. HVA transforms marmoset (CT, WL, CM) lymphoid cells *in vitro* and the lymphoblastoid cell lines possess T-cell characteristics [12, 28, 31]; so far, transforming activity of HVS *in vitro* has not been demonstrated unequivocally although persistent infections are readily established [84].

HVS and HVA are highly oncogenic in experimentally-infected marmosets of all ages, without immunosuppressive conditioning, having consistently induced lymphomas or lymphocytic leukemias in all species evaluated. Oncogenicity of HVS has been demonstrated in CT [10, 27, 33, 39, 61, 99], WL [10, 27, 99], CM [45, 103] and white-moustached (one animal) [75] marmosets, with CT and WL marmosets the most thoroughly studied. CT and WL marmosets inoculated parenterally develop peripheral lymphadenopathy within ten days to three weeks PI and the neoplastic disease follows a progressive course, with appearance of absolute lymphocytosis (maximal values usually within a range of 12,000–24,000/mm^3) and abnormal circulating cells (5–50% lymphoblasts or atypical cells) by four to six weeks PI. The duration of the PI course for CT marmosets is highly predictable, i.e., the animals are dead within four to six weeks PI whereas the clinical course in WL marmosets is less predictable and sometimes extends for three to four months (no indication of species or subspecies differences among WL marmosets). In a study to determine the minimal lethal dose, CT marmosets inoculated i.m. were susceptible to $\geq$3 and WL marmosets to $\geq$23 plaque forming units (PFU) of virus [22]; tumor development and clinical course were not dose dependent [22, 65]. Transplantation of viable tumor cells into unrelated animals of either species induces a neoplastic disease indistinguishable from that induced by cell-free virus and the tumors are composed of recipient cells, determined by chromosomal analysis [53]. Tumors induced by either cell-free virus or viable cells have been considered multiclonal in origin [8, 53], based on circumstantial evidence wherein cell lines established from tumors of hematopoietic-chimeric marmosets contain both xx and xy cells. Horizontal transmission of HVS infection and/or neoplastic disease from experimentally-infected marmosets to uninoculated cagemates has not been demonstrated [99], and attempts to transmit infection mechanically by blood-sucking insects were unsuccessful [32]. The susceptibility of adult common marmosets, previously reported to develop latent HVS infection but not lymphoma [45, 49], was re-evaluated in a recent study [103] and in contrast to previous reports, all animals inoculated i.m. with $\geq$100 PFU of virus developed lymphomas and died about three weeks PI. Of the species other than marmosets which are susceptible to HVS [62] owl monkeys *(Aotus trivirgatus)* are seemingly the best choice. In contrast to marmosets they are not consistently susceptible to lymphoma or leukemia induction, however, and they have extremely variable clinical courses [2, 9, 39, 63].

Oncogenicity of cell-free HVA has been demonstrated in adult CT [12, 22, 24, 38, 46, 62, 64, 66], WL [12, 22, 24] and CM [46, 49] marmosets. CT and WL marmosets died of malignant lymphoproliferative disease two to six weeks PI (usually within four weeks) and CM marmosets four to 15 weeks PI. In contrast CM marmosets inoculated with viable tumor cells from a CT marmoset developed persistent HVA infections but no overt disease during a seven month observation period [46]. In one study [38] HVA-infected CT marmosets consistently developed evidence of leukemia, in addition to solid lymphomas, with numerous lymphoblasts in the peripheral circulation, whereas in another study [24] no peripheral leukemia developed. As the viral inocula used in the two studies were of different origin, perhaps the characteristics of HVA-induced disease vary somewhat depending on the isolate. Based on limited comparative data, the susceptibility of CT and WL marmosets appears identical [22] and slightly greater than CM marmosets (longer survival periods). HUNT *et al.* [38] reported horizontal transmission of HVA-induced disease to one of four contact-cagemate controls (CT marmoset). In addition to marmosets, HVS/HVA, antibody-negative squirrel and owl monkeys also are susceptible to HVA infection but squirrel monkeys develop no overt illness and owl monkeys, in limited studies, apparently develop a mild, nonfatal lymphoproliferative disease [46, 66].

Pathologic findings characteristic of HVS- and HVA-induced lymphoproliferative disease have been described [10, 12, 24, 39, 43, 46, 99, 103] and are very similar or indistinguishable for the two viruses, with only minor species-dependent differences. Characteristic macroscopic findings include generalized lymphadenopathy (two to five times normal size), splenomegaly, hepatomegaly, enlarged hemorrhagic thymuses, tonsils, Peyer's patches, adrenal glands, and salivary glands, focal grayish mottling of kidney cortices, and reddish-gray stippling at intramuscular sites of inoculation. Hepatomegaly, a consistent finding in CT marmosets, is sometimes inapparent in WL marmosets, splenomegaly which is severe in CT marmosets (spleen sizes up to ten times normal size) is mild in WL marmosets, and the lesions in adrenal glands, involving both cortex and medulla in CT marmosets are primarily restricted to medulla in WL marmosets. In the limited study of common marmosets, macroscopic lesions of liver and spleen are very similar to those of CT marmosets and lesions of adrenal glands similar or identical to those of WL marmosets. Microscopically, the experimentally-induced diseases are most often classified as diffuse lymphomas, poorly differentiated lymphocytic type, although stem cells sometimes predominate in HVS-induced lesions of CT marmosets and small, well-differentiated lymphocytes in HVS-induced lesions of WL marmosets. The macroscopic lesions as described above reflect *in situ* proliferation of lymphoid cells and/or infiltration of neoplastic cells, causing replacement of normal parenchyma and tissues, with hemorrhage present in most affected organs. Focal areas of necrosis within proliferative lesions are present to variable degrees in HVA- and HVS-induced diseases and are probably caused by compression and/or thrombosis of small vessels. Perivascular infiltrates of neoplastic cells are widely disseminated in the body, including organs and tissues that appear macroscopically normal. In our experience, infiltrates of polymorphonuclear leukocytes (neutrophils and eosinophils) are frequently found in HVA-induced lesions but not in HVS-induced lesions. Lung involvement occurs in both HVS- and HVA inoculated animals but the lesions are usually much more severe in HVA-inoculated animals, often resulting in grossly apparent pneumonia. In two CT marmosets inoculated i.c. with HVS, brain lesions (lymphoid infiltrates in meninges and choroid plexus) and parenchymal lesions were indistinguishable from those in i.m. inoculated animals.

Evidence for virus expression *in vivo,* i.e., viral inclusion bodies, virus particles, and/or viral antigens, has not been detected, either in neoplastic cells or in parenchymal cells. HVS or HVA genomes can be demonstrated in lymphoid cells of infected animals, however, by molecular hybridization methods or by examination of cells cultured *in vitro* for virus expression. By co-cultivation of circulating lymphocytes with permissive cell monolayers virus can be recovered from the circulating lymphocytes at about two weeks PI until death [22, 24, 99, 103]. In quantitative studies of HVS-inoculated *Saguinus* sp. [22], it was found that one in about 10^6 circulating mononuclear cells carried HVS at two to three weeks PI whereas in terminal stages, as many as one in three cells expressed virus. Co-cultivation of lymphoid tissue, or any tissue containing neoplastic lymphoid cells, with permissive cells results in very rapid expression of virus, i.e., expression of viral antigens and production of infectious virions within 24–72 h. Short-term cultures of lymphoid cells from the peripheral circulation [29] or solid tissues, established alone without co-cultivation, express viral antigens (and occasionally complete virus) very rapidly; several virus-producing, continuous lymphoblastoid cell lines have been established from such cultures and have been well characterized [12, 24, 45, 73, 76, 78]. Most lymphoblastoid cell lines produce small amounts (10^1–10^3 PFU/ml) of infectious virus and 1–10% of the cells express early antigen (EA) and late antigen (LA) for the first year or so in culture; after one to two years, virus expression often stops even though the viral genomes can still be demonstrated in the cells by molecular hybridization methods [12, 33]. All continuous lymphoblastoid cultures established from HVS and HVA inoculated aninals so far have some properties characteristic of T-lymphocytes [10, 22, 24]. Based on preliminary data, T-lymphocytes are most likely the target cells for HVS in the natural host, squirrel monkeys [104].

HVS- and HVA-infected marmosets develop serum neutralizing antibodies and antibodies against other viral antigens [12, 22, 24, 44–46, 99, 103]. Antiviral antibodies which usually appear at about the same time, or one to two weeks after virus-carrying lymphocytes become detectable in peripheral blood, develop two to three weeks PI in HVA-infected CT and WL marmosets, two to three weeks PI in HVS-infected CT and CM marmosets, and three to five weeks PI in HVS-infected WL marmosets. Antibodies to HVS EA and LA appear either at the same time or to EA slightly later, and the titers against both antigens usually increase until death.

Within the past three years studies of the HVS and HVA experimental marmoset models have been directed primarily toward methods of prevention. One approach has involved evaluation of infectivity and oncogenicity of an attenuated variant of HVS (A-HVS), isolated after serial passage of wild-type HVS in cell culture at 39 °C and subsequent cloning at 34 °C [81]. Intramuscular inoculation of A-HVS ($\geq 6 \times 10^2$ PFU) resulted in development of persistent infections (virus-carrying lymphocytes, seroconversion) in all CT and CM marmosets and in one of six WL marmosets, without development of overt disease [25, 81, 103]. CT marmosets persistently infected with A-HVS and subsequently challenged with $\geq$770 PFU of oncogenic HVS were not protected against lymphoma induction although induction of disease was delayed and the survival times were about eight weeks longer than in control animals [25]. In vaccine studies preparations of inactivated oncogenic HVS or HVA (heat- and formaldehyde-treated), inoculated into CT marmosets, induced high titers of neutralizing antibodies but not disease, and the animals were protected against lymphoma induction when subsequently challenged i.m. with $\geq$215 LD_{50} of HVS or 316 LD_{50} of HVA, respectively [47, 48]. Virus could not be recovered from peripheral leukocytes of vaccinated and challenged monkeys by co-cultivation methods.

Vaccinated animals, however, were not resistant against higher doses of virus or tumor induction by tumor cell transplantation although survival times were slightly longer than in control animals. BCG immunization of WL marmosets at various intervals before and after i.m. inoculation of 5×10^3 infectious units of HVS had little or no effect on lymphoma incidence, disease course, and anti-HVS antibody response [82].

Epstein-Barr Virus (EBV)

EBV, a human lymphotropic herpesvirus, is a ubiquitous, horizontally-transmitted virus in humans that has the capacity to cause lymphoproliferation both *in vivo* and *in vitro*. Although primary infection with EBV early in life usually causes only subclinical or mild clinical disease, it has been established as the cause of heterophile-positive infectious mononucleosis and has been implicated as an etiologic factor in African patients with Burkitt's lymphoma and in African and Chinese patients with nasopharyngeal carcinoma (review [67]). Continuous lymphoblastoid cell lines can be established readily *in vitro* by exposure of lymphoid cells from EBV-negative donors to selected strains of EBV and by cultivation of lymphoid cells from EBV-positive donors, including healthy people. Different strains or biotypes of EBV have been identified by properties *in vitro* [71] but the relationship, if any, of *in vitro* properties to biological properties *in vivo* in man has not been defined.

Marmoset lymphoid cells, like human cells from EBV-negative donors, can be readily transformed or immortalized by exposure to EBV *in vitro*. Strains of EBV which have transformed marmoset lymphocytes include: (1) infectious mononucleosis strains – Hawley [34, 70, 72], Lederer [70], Kaplan [18, 30], B95–8 (CT marmoset cells transformed by Hawley strain) [34, 70, 72]; (2) Burkitt's lymphoma strains – Olare [69], Nyevu [69]; (3) nasopharyngeal carcinoma strain – M81 (common marmoset cells transformed by Ly-28 cell line) [17]. CT [30, 34, 58, 69, 70, 72], WL [30, 58] and CM [105] marmoset cells have been transformed by one or more of the EBV strains and continuous lymphoblastoid cell lines established; all cell lines possess some B-lymphocyte characteristics [30]. Continuous marmoset lymphoblastoid cell lines have been well characterized [30, 34, 69, 72], and compared to human or other primate cells transformed by EBV. Marmoset lymphoblastoid cell lines transformed by B95–8 virus are much higher yielders of viral structural antigen-positive cells and of extracellular, biologically active virus [34, 69, 70] than any other human or primate cells studied.

Induction of lymphoproliferative disease in marmosets, first reported by SHOPE *et al.* [89], has been demonstrated in adult CT marmosets by the B95–8 [13, 34, 67, 89] and Kaplan [96] strains and in adult CM marmosets by the B95–8 strain [23]. Adult CT and WL marmosets inoculated with the M-81 strain seroconverted but did not develop clinical disease [58]. Several other strains and preparations have been inoculated into CT and WL marmosets without success [13, 30, 34, 96, 100] although the quality of the preparations were often in doubt [100]. Based on small samples, CT marmosets appear slightly more susceptible to B95–8 than common marmosets whereas the susceptibility of WL marmosets to B95–8 has not been evaluated. Age-dependent susceptibility or horizontal transmission have not been evaluated. An alternative species to marmosets for EBV studies *in vivo* has not been identified although based on limited studies, owl monkeys [19, 20] and gibbon apes [94, 95] are worthy of future investigation.

CT marmosets inoculated with $\geq 10^3$ transforming units of B95–8 virus or $1.0–3.0 \times 10^8$ autologous transformed cells, in divided doses (i.v., i.m., i.p.), develop a spectrum of responses including inapparent infections, regional lymphadenopathy or generalized lymphoproliferative disease [13, 34, 89]. With an occasional exception death caused by the experimentally-induced disease occurs four to eight weeks PI. In some animals enlarged lymph nodes apparent at four to eight weeks PI subsequently return to normal size and the animals recover completely [34, 88]. In one study daily immunosuppressive therapy (azathioprine, prednisolone) during the first three weeks PI increased the incidence of host response [34, 89]. The incidence of lymphoproliferative disease induced in CM marmosets by 10^4 transforming units of B95–8 virus, inoculated by multiple routes, was comparable to the incidence in CT marmosets and the animals usually died seven to eight weeks PI [23]. Administration of horse anti-human thymocyte serum to CM marmosets for 14 days PI had no apparent effect on incidence or clinical course. One of three CT marmosets inoculated with Kaplan virus developed clinical signs of lymphoma seven weeks PI and was sacrificed two weeks later [96]. With one exception [88], hematologic abnormalities (absolute lymphocytosis, immature lymphocytes in peripheral blood) have not been reported in any marmosets inoculated with EBV.

Macroscopic pathologic findings usually include peripheral and/or visceral lymphadenopathy (lymph nodes draining sites of inoculation usually most severely affected), mild splenomegaly and enlarged thymuses. Microscopically, the normal tissue in grossly affected lymphoid organs is replaced to various degrees by the lymphoproliferative response (lymphoreticular cell hyperplasia or lymphoma) and lymphoreticular cell infiltrates, especially perivascularly, are usually widely distributed in other tissues of the body [13, 23, 34, 89, 96]. Lesions in CT marmosets have been diagnosed either as lymphoid hyperplasia or lymphomas, stem cell type [34, 89] or poorly differentiated lymphocytic type [13], and lesions in common marmosets as lymphoid hyperplasia [23]. Compared to CT marmosets, the lesions in common marmosets were usually composed of a more heterogeneous population of cells, i.e., small lymphocytes, lymphoblasts, immunoblasts and cells with a plasmacytoid appearance. Viral inclusion bodies, viral capsid antigens or herpesvirus particles have not been detected in any tissues examined. By indirect immunofluorescence tests, Epstein-Barr viral nuclear antigen (EBNA) was demonstrated in neoplastic cells from one CT marmoset inoculated with B95–8 virus [68] but not in tissues of common marmosets [23]. EBV DNA was demonstrated by molecular hybridization techniques in tumor material from the CT marmoset with Kaplan virus-induced lymphoma [96]. Continuous lymphoblastoid cell lines have been established from tissues of CT marmosets with B95–8 or Kaplan virus-induced disease and the cultured cells produce virus-specified antigens and infectious virus [13, 34, 68, 96] and possess characteristics compatible with B-lymphocytes [13].

The majority of CT marmosets inoculated with B95–8 virus develop antibodies to viral capsid antigen (VCA) four to six weeks PI, including some which do not develop detectable signs of disease [13, 34, 68, 89]. Most CT marmosets also develop antibodies to EA, membrane antigen and EBNA. Common marmosets usually develop low titers of anti-VCA antibodies four to six weeks PI but antibodies to EA or EBNA have not been detected [23]. Except for one report of complement-fixing antibodies in wild-caught and laboratory-born marmosets [36], naturally-occurring EBV antibodies have not been detected in marmosets.

Discussion and Conclusions

Marmosets are highly susceptible to tumor induction by type C sarcoma viruses and oncogenic herpesviruses, reflected by the six experimental tumor model systems discussed in this report and by the recent findings reported by others at this symposium [5, 14, 37]. Studies of species susceptibility to the three sarcoma viruses and three herpesviruses have been limited primarily to CT, WL and/or CM marmosets and although the susceptibility of each species has not been tested in all six systems, each species has proven susceptible when evaluated, with only minor differences in relative susceptibility. In all three species RSV induces sarcomas and HVS and HVA induce lymphomas. FeSV and SSV–1 induce sarcomas and/or fibromas (SSV–1) in CT and WL marmosets whereas the susceptibility of CM marmosets has not been evaluated. The B95–8 strain of EBV causes inapparent infections, lymphoid hyperplasia, or lymphomas in CT marmosets and severe lymphoid hyperplasia in CM marmosets whereas WL marmosets have not been evaluated. WL marmosets have been studied most extensively in the oncornavirus systems, CT and WL marmosets with the simian herpesviruses and CT marmosets with EBV. No differences in susceptibility of the different species and subspecies of WL marmosets have been detected in any of the systems evaluated. Adult marmosets, without immunosuppressive conditioning, are susceptible to disease induced by RSV, HVS, HVA and EBV whereas animals older than two to three months, except for aged animals, are not susceptible to FeSV. Young animals are slightly more susceptible to RSV than adults whereas newborns and adults are equally susceptible to HVS; only neonates have been studied with SSV–1 and only adults with HVA and EBV. Males and females are equally susceptible in all six models. Immunosuppressive conditioning apparently slightly increases the incidence of EBV-induced disease in CT marmosets [34, 89] and it enhances the malignant behavior (growth rate, metastases) of SSV–1-induced sarcomas. The pathogenesis and pathology of the tumors induced by oncornaviruses are basically identical in the different species whereas the diseases induced by the herpesviruses are more variable from species to species. Of the oncornaviruses, RSV is the most oncogenic and SSV–1 the least oncogenic in marmosets. Characteristics of virus expression in the oncornavirus and herpesvirus models, *in vivo* and *in vitro,* are virus-dependent and not host-dependent. Humoral immune responses to viral antigens are identical in the different species whereas cell-mediated immune responses have not been evaluated sufficiently to permit comparisons.

Data on relative susceptibility indicate that marmosets are more susceptible to oncornaviruses and oncogenic herpesviruses than other species of nonhuman primates and are the primate species of choice for evaluating 'new' animal or human virus isolates for oncogenicity. Based on our experience with SSV–1 wherein the i.c. route of inoculation was much more sensitive than the i.m. route, isolates of undefined or marginal oncogenicity, at least SSV–1-like viruses [51], should be inoculated i.c. and evaluated for induction of brain tumors. The best alternative primate species to marmosets for experimental viral oncology research, based on current data, is the squirrel monkey for oncornaviruses and the owl monkey for herpesviruses. For obvious reasons greater efforts should be made in the future toward identification and establishment of comparable models in lower mammalian species.

Indigenous oncornaviruses [3] or oncogenic herpesviruses have not been isolated from marmosets. However, oncornavirus particles were observed in placentas of CT marmosets [87] and recent work in our laboratory suggests that perhaps endogenous oncornavirus information was rescued during passage of a sarcoma virus *in vivo* [5].

Marmosets and marmoset cell lines are a valuable resource for viral carcinogenesis and co-carcinogenesis studies. Marmoset fibroblastic cells *in vitro* are highly susceptible to transformation by type C sarcoma viruses and provide excellent indicator cell lines for focus assays of transformation, for detection of nontransforming viruses (use of transformed, nonproducer cells carrying sarcoma genomes [4]), for evaluation of cell-mediated immune responses of inoculated animals, and for basic studies of cell biology and molecular biology associated with virus-host cell interactions. Marmoset lymphoid cells *in vitro* are susceptible to transformation by primate herpesviruses, and depending on the virus, the continuous lymphoblastoid cells have characteristics of B-lymphocytes (EBV), T-lymphocytes (HVA) or null cells. The availability of autologous marmoset cells transformed by these two herpesviruses provides a tool for basic virologic and immunologic studies of transformed cells *in vitro* and provides target cells for studying cell-mediated immune responses of inoculated animals. *In vivo,* the RSV, FeSV, HVS and HVA models are well characterized and provide potential models for studies of basic tumor cell biology, therapy and prevention. Vaccine studies using preparations of inactivated HVS or HVA have been reported [47] and should be extended to the evaluation of preparations free of nucleic acids. The SSV–1 and EBV models currently require further study and characterization but potentially will be experimental models of choice in the future, including models for selected studies of therapy and prevention. Interactions of RNA and DNA tumor viruses in tumor induction and pathogenesis and viral-chemical co-carcinogenesis are other areas of investigation for the future. Many basic questions pertaining to marmoset susceptibility remain unanswered, not the least of which include the reason(s) for their relatively high susceptibility to oncogenic viruses compared to other species, for restriction of their susceptibility so far to type C sarcoma viruses and primate herpesviruses, and for their susceptibility to lymphoma induction by herpesviruses but not by type C lymphoma viruses.

Summary

Marmoset monkeys are highly susceptible to tumor induction by type C sarcoma viruses and primate lymphotropic herpesviruses. Six experimental models were reviewed, three sarcoma models induced by Rous, feline or simian sarcoma viruses and three models of lymphoproliferative disease induced by *Herpesvirus saimiri, H. ateles* or Epstein-Barr virus. Relative susceptibility of cotton-topped *(Saguinus oedipus oedipus)*, white-lipped *(S. nigricollis, S. fuscicollis* subspecies) and common *(Callithrix jacchus jacchus)* marmosets to the different viruses was compared.

References

1 AARONSON, S.A.: Biologic characterization of mammalian cells transformed by a primate sarcoma virus. Virology *52:* 562–567 (1973).

2 ABLASHI, D.V.; LOEB, W.F.; VALERIO, M.G.; ADAMSON, R.H.; ARMSTRONG, G.R.; BENNETT, D.G., and HEINE, V.: Malignant lymphoma with lymphocytic leukemia in owl monkeys induced by *Herpesvirus saimiri*. J. natn. Cancer Inst. *47:* 837–856 (1971).

3 BENVENISTE, R.E.; HEINEMANN, R.; WILSON, G.L.; CALLAHAN, R., and TODARO, G.J.: Detection of baboon type C viral sequences in various primate tissues by molecular hybridization. J. Virol. *14:* 56–67 (1974).

4 BERGHOLZ, C.M.; WOLFE, L.G., and DEINHARDT, F.: Establishment of simian sarcoma virus, type 1 (SSV-1) transformed nonproducer marmoset cell lines. Int. J. Cancer *20:* 104–111 (1977).

5 BERGHOLZ, C.M.; WOLFE, L.G., and DEINHARDT, F.: Oncogenicity of a C-type virus, HL-23V in marmosets and characterization of virus isolated from an HL-23V-induced marmoset tumor. A comparison with SSV-1/SSAV-1; in DEINHARDT and GENGOZIAN Marmosets in experimental medicine. Prim. Med., vol. 10, pp. 135–141 (Karger, Basel 1978).

6 BERGHOLZ, C.M.; WOLFE, L.G.; DEINHARDT, F.; THAKKAR, B., and MARCZYNSKA, B.: Oncogenicity in marmosets of HL-23V, a type C oncornavirus isolated from human leukemic cells, and comparison with simian sarcoma virus, type 1 (SSV-1/SSAV-1). J. natn. Cancer Inst. *58:* 1041–1046 (1977).

7 BIGNER, D.D. and PEGRAM, C.N.: A review of virus-induced experimental brain tumors and of the putative associations of viruses with human brain tumors. Adv. Neurol. *15:* 57–83 (1976).

8 CHU, E.W. and RABSON, A.S.: Chimerism in lymphoid cell culture line derived from lymph node of marmoset infected with *Herpesvirus saimiri*. J. natn. Cancer Inst. *48:* 771–775 (1972).

9 CICMANEC, J.L.; LOEB, W.F., and VALERIO, M.G.: Lymphoma in owl monkeys *(Aotus trivirgatus)* inoculated with *Herpesvirus saimiri:* clinical, hematologic and pathologic findings. J. med. Primatol. *3:* 8–17 (1974).

10 DEINHARDT, F.: *Herpesvirus saimiri;* in The herpesviruses, chapter 18, pp. 595–625 (Academic Press, New York 1973).

11 DEINHARDT, F.: Neoplasms induced by Rous sarcoma virus in New World monkeys. Nature, Lond. *210:* 443 (1966).

12 DEINHARDT, F.W.; FALK, L.A., and WOLFE, L.G.: Simian herpesviruses and neo-

plasia; in Advances in cancer research, vol. 19, pp. 167–205 (Academic Press, New York 1974).

13 DEINHARDT, F.; FALK, L.; WOLFE, L.G.; PACIGA, J., and JOHNSON, D.: Response of marmosets to experimental infection with Epstein-Barr virus; in Oncogenesis and herpesviruses II. IARC Sci. Publ. No. 11, pp. 161–168 (Lyon 1975).

14 DEINHARDT, F.; FALK, L.; WOLFE, L.G.; NONOYAMA, M.; SCHUDEL, A.; LAPIN, B., and YAKOVLEVA, L.: Susceptibility of marmosets to the EBV-like baboon herpesvirus; in DEINHARDT and GENGOZIAN Marmosets in experimental medicine. Prim. Med., vol. 10, pp. 163–170 (Karger, Basel 1978).

15 DEINHARDT, F.; WOLFE, L.; NORTHROP, R.; MARCZYNSKA, B.; OGDEN, J.; MCDONALD, R.; FALK. L.; SHRAMEK, G.; SMITH, R., and DEINHARDT, J.: Induction of neoplasms by viruses in marmoset monkeys. J. med. Primatol. *1:* 29–50 (1972).

16 DEINHARDT, F.; WOLFE, L.; FALK, L.; JOHNSON, T.; JOHNSON, D., and MASSEY, R.: Immunological control of virus-induced tumors in primates; in Comparative leukemia research 1973. Leukemogenesis. Biblthca haemat. No. 40, pp. 639–648 (University of Tokyo Press, Tokyo/Karger, Basel 1975).

17 DE THÉ, G.: Personal commun.

18 DIEHL, V.; HENLE, G.; HENLE, W., and KOHN, G.: Demonstration of a herpes group virus in cultures of peripheral leukocytes from patients with infectious mononucleosis. J. Virol. *2:* 663–669 (1968).

19 EPSTEIN, M.A.: EB virus in the owl monkey *(Aotus trivirgatus)*. Lab. anim. Sci. *26:* 1127–1130 (1976).

20 EPSTEIN, M.A.; HUNT, R.D., and RABIN, H.: Pilot experiments with EB virus in owl monkeys *(Aotus trivirgatus)*. I. Reticuloproliferative disease in an inoculated animal. Int. J. Cancer *12:* 309–318 (1973).

21 ESSEX, M.; KLEIN, G.; DEINHARDT, F.; WOLFE, L.G.; HARDY, W.D., jr.; THEILEN, G.H., and PEARSON, L.D.: Induction of feline oncornavirus associated cell membrane antigen in human cells. Nature new. Biol. *238:* 187–189 (1972).

22 FALK, L.A.: Oncogenic DNA viruses of nonhuman primates. A review. Lab. Anim. Sci. *24:* 182–192 (1974).

23 FALK, L.; DEINHARDT, F.; WOLFE, L.; JOHNSON, D.; HILGERS, J., and DE THÉ, G.: Epstein-Barr virus: experimental infection of *Callithrix jacchus* marmosets. Int. J. Cancer *17:* 785–788 (1976).

24 FALK, L.A.; NIGIDA, S.M.; DEINHARDT, F.; WOLFE, L.G.; COOPER, R.W., and HERNANDEZ-CAMACHO, J.I.: *Herpesvirus ateles:* properties of an oncogenic herpesvirus isolated from circulating lymphocytes of spider monkeys (*Ateles* sp.). Int. J. Cancer *14:* 473–482 (1974).

25 FALK, L.; SCHAFFER, P.; WRIGHT, J.; DEINHARDT, F.; WOLFE, L., and BENYESH-MELNICK, M.: Experimental infection of marmoset and squirrel monkeys with attenuated HVS; in Comparative leukemia research 1975. Bibltheca haemat., No. 43, pp. 336–338 (Karger, Basel 1976).

26 FALK, L.; WOLFE, L., and DEINHARDT, F.: Epidemiology of *Herpesvirus saimiri* infection in squirrel monkeys; in Medical primatology 1972, Part III, pp. 151–158 (Karger, Basel 1972).

27 FALK, L.A.; WOLFE, L.G., and DEINHARDT, F.: Isolation of *Herpesvirus saimiri* from blood of squirrel monkeys *(Saimiri sciureus)*. J. natn. Cancer Inst. *48:* 1499–1505 (1972).

28 FALK, L.; WOLFE, L., and DEINHARDT, F.: Transformation *in vitro* with *Herpesvirus ateles;* in Oncogenesis and herpesvirus II. IARC Sci. Publ. No. 11, pp. 379–384 (Lyon 1975).

29 FALK, L.A.; WOLFE, L.G.; HOEKSTRA, J., and DEINHARDT, F.: Demonstration of *Herpesvirus saimiri*-associated antigens in peripheral lymphocytes from infected marmosets during *in vitro* cultivation. J. natn. Cancer Inst. *48:* 523–530 (1972).

30 FALK, L.A.; WOLFE, L.; DEINHARDT, F.; PACIGA, J.; DOMBOS, L.; KLEIN, G.; HENLE, W., and HENLE, G.: Epstein-Barr virus: transformation of non-human primate lymphocytes *in vitro*. Int. J. Cancer *13:* 363–376 (1974).

31 FALK, L.; WRIGHT, J.; WOLFE, L., and DEINHARDT, F.: *Herpesvirus ateles:* transformation *in vitro* of marmoset splenic lymphocytes. Int. J. Cancer *14:* 244–251 (1974).

32 FISCHER, R.G.; FALK, L.A.; RYTTER, R.; BURTON, G.J.; LUECKE, D.H., and DEINHARDT, F.: *Herpesvirus saimiri:* viability in four species of hematophagous insects and attempted insect transmission to marmosets. J. natn. Cancer Inst. *52:* 1477–1481 (1974).

33 FLECKENSTEIN, B.; BORNKAMM, G.W., and WERNER, F.J.: The role of *Herpesvirus saimiri* genomes in oncogenic transformation of primate cells; in Comparative leukemia research 1975. Biblthca haemat., No. 43, pp. 308–312 (Karger, Basel 1976).

34 FRANK, A.; ANDIMAN, W.A., and MILLER, G.: Epstein-Barr virus and nonhuman primates: natural and experimental infection. Adv. Cancer Res. *23:* 171–201 (1976).

35 GARDNER, M.B.; ARNSTEIN, P.; RONGEY, R.W.; ESTES, J.S.; SARMA, P.S.; RICKARD, C.G., and HUEBNER, R.J.: Experimental transmission of feline fibrosarcoma to cats and dogs. Nature, Lond. *226:* 807–809 (1970).

36 GERBER, P. and LORENZ, D.: Complement-fixing antibodies reactive with Epstein-Barr virus in sera of marmosets and prosimians. Proc. Soc. exp. Biol. Med. *145:* 654–657 (1974).

37 HEBERLING, R.L.; KALTER, S.S., and EICHBERG, J.W.: Studies with cell-associated baboon type C virus pseudotype of murine sarcoma virus in marmosets; in DEINHARDT and GENGOZIAN Marmosets in experimental medicine. Prim. Med., vol. 10, pp. 142–148 (Karger, Basel 1978).

38 HUNT, R.D.; MELENDEZ, L.V.; GARCIA, F.G., and TRUM, B.F.: Pathologic features of *Herpesvirus ateles* lymphoma in cotton-topped marmosets *(Saguinus oedipus)*. J. natn. Cancer Inst. *49:* 1631–1639 (1972).

39 HUNT, R.D.; MELENDEZ, L.V.; KING, N.W.; GILMORE, C.E.; DANIEL, M.D.; WILLIAMSON, M.E., and JONES, T.C.: Morphology of a disease with features of malignant lymphoma in marmosets and owl monkeys inoculated with *Herpesvirus saimiri*. J. natn. Cancer Inst. *44:* 447–465 (1970).

40 JOHNSON, D.; WOLFE, L.G., and DEINHARDT, F.: Unpublished data.

41 JOHNSON, L., jr.; WOLFE, L.G.; WHISLER, W.W.; NORTON, T.; THAKKAR, B., and DEINHARDT, F.: Induction of gliomas in marmosets by simian sarcoma virus, type 1 (SSV-1). Abstract. Proc. Am. Ass. Cancer Res. *16:* 119 (1975).

42 JOHNSON, T.R.; MASSEY, R.J., and DEINHARDT, F.: Lymphocyte and antibody cytotoxicity to tumor cells measured by a micro-^{51}chromium release assay. Immunol. Commun. *1:* 247–261 (1972).

43 KING, N.W. and MELENDEZ, L.V.: The ultrastructure of *Herpesvirus saimiri*-induced lymphoma in cotton-topped marmosets *(Saguinus oedipus)*. Lab. Invest. *26:* 682–693 (1972).

44 KLEIN, G.; PEARSON, G.; RABSON, A.; ABLASHI, D.V.; FALK, L.; WOLFE, L.; DEINHARDT, F., and RABIN, H.: Antibody reactions to *Herpesvirus saimiri* (HVS)-induced early and late antigens (EA and LA) in HVS-infected squirrel, marmoset and owl monkeys. Int. J. Cancer *12:* 270–289 (1973).

45 LAUFS, R. and FLECKENSTEIN, B.: Susceptibility to *Herpesvirus saimiri* and antibody development in Old and New World monkeys. Med. microbiol. Immunol. *158:* 227–236 (1973).

46 LAUFS, R. and MELENDEZ, L.V.: Oncogenicity of *Herpesvirus ateles* in monkeys. J. natn. Cancer Inst. *51:* 599–608 (1973).

47 LAUFS, R. and STEINKE, H.: Vaccination of nonhuman primates against malignant lymphoma. Nature, Lond. *253:* 71–72 (1975).

48 LAUFS, R. and STEINKE, H.: Vaccination of nonhuman primates with killed oncogenic herpesviruses. Cancer Res. *36:* 704–706 (1976).

49 LAUFS, R.; STEINKE, H.; STEINKE, G., and PETZOLD, D.: Latent infection and malignant lymphoma in marmosets *(Callithrix jacchus)* after infection with two oncogenic herpesviruses from primates. J. natn. Cancer Inst. *53:* 195–199 (1974).

50 LEVY, B.M.; TAYLOR, A.C.; HAMPTON, S., and THOMA, G.W.: Tumors of the marmoset produced by Rous sarcoma virus. Cancer Res. *29:* 2237–2248 (1969).

51 LIEBER, M.M.; SHERR, C.J.; TODARO, G.J.; BENVENISTE, R.E.; CALLAHAN, R., and COON, H.G.: Isolation from the Asian mouse *Mus caroli* of an endogenous type C virus related to infectious primate type C virus. Proc. natn. Acad. Sci. USA *72:* 2315–2319 (1975).

52 MARCZYNSKA, B.; DEINHARDT, F.; SCHULIEN, J.; TISCHENDORF, P., and SMITH, R.D.: Rous sarcoma virus-induced tumors in marmosets: characteristics of tumor cells and host-cell-virus interrelationships. J. natn. Cancer Inst. *51:* 1255–1274 (1933).

53 MARCZYNSKA, B.; FALK, L.; WOLFE, L., and DEINHARDT, F.: Transplantation and cytogenetic studies of *Herpesvirus saimiri*-induced disease in marmoset monkeys. J. natn. Cancer Inst. *50:* 331–337 (1973).

54 MARCZYNSKA, B.; TREU-SARNAT, G., and DEINHARDT, F.: Characteristics of long-term marmoset cell cultures spontaneously altered or transformed by Rous sarcoma virus. J. natn. Cancer Inst. *44:* 545–572 (1970).

55 McDONALD, R.; WOLFE, L.G., and DEINHARDT, F.: Feline fibrosarcoma virus: quantitative focus assay, focus morphology and evidence for a 'helper virus'. Int. J. Cancer *9:* 57–65 (1972).

56 McDONALD, R.; THAKKAR, B.; WOLFE, L.G., and DEINHARDT, F.: Characteristics of three strains of feline fibrosarcoma virus grown in cat and marmoset monkey cells. Int. J. Cancer *17:* 396–406 (1976).

57 McDONOUGH, S.K.; LARSEN, S.; BRODY, R.S.; STOCK, N.D., and HARDY, W.D., jr.: A transmissible feline fibrosarcoma of viral origin. Cancer Res. *31:* 953–956 (1971).

58 McGRATH, P.; LAI, P.K.; DEINHARDT, F.; WOLFE, L.G.; SCHUDEL, A., and FALK, L.A.: Unpublished data.

59 MELENDEZ, L.V.; DANIEL, M.D.; HUNT, R.D., and GARCIA, F.G.: An apparently new herpesvirus from primary kidney cultures of the squirrel monkey *(Saimiri sciureus)*. Lab. Anim. Care *18:* 374–381 (1968).

60 MELENDEZ, L.V.; DANIEL, M.D.; GARCIA, F.G.; FRASER, C.E.O.; HUNT, R.D., and KING, N.W.: *Herpesvirus saimiri*. I. Further characterization studies of a new virus from the squirrel monkey. Lab. Anim. Care *19:* 372–377 (1969).

61 Melendez, L.V.; Hunt, R.D.; Daniel, M.D.; Garcia, F.G., and Fraser, C.E.O.: Herpes saimiri. II. An experimentally-induced malignant lymphoma in primates. Lab. Anim. Care *19:* 378–386 (1969).

62 Melendez, L.V.; Hunt, R.D.; Daniel, M.D.; Fraser, C.E.O.; Barahona, H.H.; Garcia, F.G., and King, N.W.: Lymphoma viruses of monkeys. *Herpesvirus saimiri* and *Herpesvirus ateles*, the first oncogenic herpesviruses of primates – a review; in Oncogenesis and herpesviruses, pp. 451–461 (IARC, Lyon 1972).

63 Melendez, L.V.; Hunt, R.D.; Daniel, M.D.; Blake, B.J., and Garcia, F.G.: Acute lymphocytic leukemia in owl monkeys inoculated with *Herpesvirus saimiri*. Science *171:* 1161–1163 (1971).

64 Melendez, L.V.; Hunt, R.D.; Daniel, M.D.; Fraser, C.E.O.; Barahona, H.H.; King, N.W., and Garcia, F.G.: *Herpevirus saimiri* and *ateles* – their role in malignant lymphomas of monkeys. Fed. Proc. Fed. Am. Socs exp. Biol. *31:* 1643–1650 (1972).

65 Melendez, L.V.; Hunt, R.D.; Garcia, F.G.; Daniel, M.D., and Fraser, C.E.O.: Prevention of lymphoma in cotton-top marmosets by *Herpesvirus saimiri* antiserum. J. med. Primatol. *3:* 213–220 (1974).

66 Melendez, L.V.; Hunt, R.D.; King, N.W.; Barahona, H.H.; Daniel, M.D.; Fraser, C.E.O., and Garcia, F.G.: *Herpesvirus ateles*, a new lymphoma virus of monkeys. Nature new Biol. *235:* 182–184 (1972).

67 Miller, G.: The oncogenicity of Epstein-Barr virus. J. infect. Dis. *130:* 187–205 (1974).

68 Miller, G. and Coope, D.: Epstein-Barr viral nuclear antigen (EBNA) in tumor cell imprints of experimental lymphoma of marmosets. Trans. Ass. Am. Physns *87:* 205–218 (1974).

69 Miller, G.; Coope, D.; Niederman, J., and Pagano, J.: Biological properties and viral surface antigens of Burkitt lymphoma- and mononucleosis-derived strains of Epstein-Barr virus released from transformed marmoset cells. J. Virol. *18:* 1071–1080 (1976).

70 Miller, G. and Lipman, M.: Release of infectious Epstein-Barr virus by transformed marmoset leukocytes. Proc. natn. Acad. Sci. USA *70:* 190–194 (1973).

71 Miller, G.; Robinson, J.; Heston, L., and Lipman, M.: Differences between laboratory strains of Epstein-Barr virus based on immortalization, abortive infection and interference; in Oncogenesis and herpesvirus II. IARC Sci. Publ. No. 11, pp. 395–408 (Lyon 1975).

72 Miller, G.; Shope, T.; Lisco, H.; Stitt, D., and Lipman, M.: Epstein-Barr virus: transformation, cytopathic changes, and viral antigens in squirrel monkey and marmoset leukocytes. Proc. natn. Acad. Sci. USA *69:* 383–387 (1972).

73 Neubauer, R.H.; Wallen, W.C., and Rabin, H.: Inhibition of the mitogenic response of normal peripheral lymphocytes by extracts or supernatant fluids of a *Herpesvirus saimiri* lymphoid tumor cell line. Infec. Immunity *12:* 1021–1028 (1975).

74 Noyes, W.F.: Tumors induced in the South American marmoset monkey by Rous sarcoma virus. J. natn. Cancer Inst. *45:* 579–587 (1970).

75 Rabin, H.: Assay and pathogenesis of oncogenic viruses in nonhuman primates. Lab. Anim. Sci. *21:* 1032–1049 (1971).

76 Rabin, H.; Pearson, G.; Chopra, H.C.; Orr, T.; Ablashi, D.V., and Armstrong, G.R.: Characteristics of *Herpesvirus saimiri*-induced lymphoma cells in tissue culture. In Vitro *9:* 65–72 (1973).

77 Rabin, H.; Theilen, G.H.; Dungworth, D.L.; Sarma, P.S.; Nelson-Rees, W.A., and Cooper, R.W.: Continuing studies of feline sarcoma virus-induced tumors in nonhuman primates; in Unifying concepts of leukemia. Biblthca haemat., No. 39, pp. 244–250 (Karger, Basel 1973).

78 Rabson, A.S.; O'Conor, G.T.; Lorenz, D.E.; Kirschstein, R.L.; Legallais, F.Y., and Tralka, T.S.: Lymphoid cell-culture line derived from lymph node of marmoset infected with *Herpesvirus saimiri*. Preliminary report. J. natn. Cancer Inst. *46:* 1099–1109 (1971).

79 Rangan, S.R.S.; Wong, M.C.; Ueberhorst, P.J., and Ablashi, D.V.: Brief communication. Mixed culture cytopathogenicity induced by virus preparations derived from cultures infected by simian sarcoma virus. J. natn. Cancer Inst. *49:* 571–577 (1972).

80 Ruscetti, S.L. and Parks, W.P.: Natural immunity in cats to feline leukemia viral antigens. J. Immun. *117:* 2029–2035 (1976).

81 Schaffer, P.A.; Falk, L.A., and Deinhardt, F.: Brief communication. Attenuation of *Herpesvirus saimiri* for marmosets after successive passage in cell culture at 39 °C. J. natn. Cancer Inst. *55:* 1243–1246 (1975).

82 Schauf, V.; Falk, L., and Deinhardt, F.: Effect of Bacillus Calmette-Guérin immunization in marmosets infected experimentally with *Herpesvirus saimiri*. J. natn. Cancer Inst. *54:* 721–726 (1975).

83 Schauf, V.; Massey, R.; Deinhardt, F., and Kruse, R.: Immunization of Rous sarcoma virus-inoculated marmosets with BCG and transformed allogeneic cells. J. natn. Cancer Inst. *54:* 151–155 (1975).

84 Schudel, A.; Falk, L.A., and Deinhardt, F.: Unpublished data.

85 Scolnick, E.M. and Parks, W.P.: Isolation and characterization of mammalian cells transformed by a primate sarcoma virus: mechanism of rescue. Int. J. Cancer *12:* 138–147 (1973).

86 Scolnick, E.M.; Parks, W.; Kawakami, T.; Kohne, D.; Okabe, H.; Gilden, R., and Hatanaka, M.: Primate and murine type-C viral nucleic acid association kinetics: analysis of model systems and natural tissues. J. Virol. *13:* 363–369 (1974).

87 Seman, G.; Levy, B.M.; Panigel, M., and Dmochowski, L.: Brief communication. Type-C virus particles in placenta of the cottontop marmoset *(Saguinus oedipus)*. J. natn. Cancer Inst. *54:* 251–252 (1975).

88 Shope, T.C.: Prevention of Epstein-Barr virus (EBV)-induced infection in cotton-topped marmosets by human serum containing EBV neutralizing activity; in Oncogenesis and herpesviruses II. IARC Sci. Publ. No. 11, pp. 153–159 (Lyon 1975).

89 Shope, T.; Dechairo, D., and Miller, G.: Malignant lymphoma in cotton-top marmosets after inoculation with Epstein-Barr virus. Proc. natn. Acad. Sci. USA *70:* 2487–2491 (1973).

90 Snyder, S.P. and Theilen, G.H.: Transmissible felline fibrosarcoma. Nature, Lond. *221:* 1074–1075 (1969).

91 Theilen, G.H.; Gould, D.; Fowler, M., and Dungworth, D.L.: C-type virus in tumor tissue of a woolly monkey (*Lagothrix* spp.) with fibrosarcoma. J. natn. Cancer Inst. *47:* 881–889 (1971).

92 Theilen, G.H.; Snyder, S.P.; Wolfe, L.G., and Landon, J.C.: Biological studies with viral-induced fibrosarcomas in cats, dogs, rabbits, and non-human primates; in Comparative leukemia research 1969. Biblthca haemat., No. 36, pp. 393–400 (Karger, Basel 1970).

93 Theilen, G.H.; Wolfe, L.G.; Rabin, H.; Deinhardt, F.; Dungworth, D.L.; Fowler, M.E.; Gould, D., and Cooper, R.: Biological studies in four species of nonhuman primates with simian sarcoma virus; in Unifying concepts of leukemia. Biblthca haemat., No. 39, pp. 251–257 (Karger, Basel 1973).

94 Werner, J.; Henle, G.; Pinto, C.A.; Hoff, R.F., and Henle, W.: Establishment of continuous lymphoblast cultures from leukocytes of gibbons *(Hylobates lar.)* Int. J. Cancer *10:* 557–567 (1972).

95 Werner, J.; Pinto, C.A.; Hoff, R.F.; Henle, W., and Henle, G.: Responses of gibbons to inoculation of Epstein-Barr virus. J. infect Dis. *126:* 678–681 (1972).

96 Wolf, H.; Werner, J., and Zur Hausen, H.: EBV DNA in nonlymphoid cells of nasopharyngeal carcinomas and in malignant lymphoma obtained after inoculation of EBV into cottontop marmosets. Cold Spring Harb. Symp. quant. Biol. *39:* 791–796 (1974).

97 Wolfe, L.G. and Deinhardt, F.: Oncornaviruses associated with spontaneous and experimentally induced neoplasia in nonhuman primates; in Medical primatology 1972, part. III, pp. 176–196 (Karger, Basel 1972).

98 Wolfe, L.G.; Deinhardt, F.; Theilen, G.H.; Rabin, H.; Kawakami, T., and Bustad, L.K.: Induction of tumors in marmoset monkeys by simian sarcoma virus, type 1 *(Lagothrix)*. A preliminary report. J. natn. Cancer Inst. *47:* 1115–1120 (1971).

99 Wolfe, L.G.; Falk, L.A., and Deinhardt, F.: Oncogenicity of *Herpesvirus saimiri* in marmoset monkeys. J. natn. Cancer Inst. *47:* 1145–1162 (1971).

100 Wolfe, L.G.; Marczynska, B.; Rabin, H.; Smith, R.; Tischendorf, P.; Gavitt, F., and Deinhardt, F.: Viral oncogenesis in nonhuman primates; in Medical primatology 1970, pp. 671–682 (Karger, Basel 1971).

101 Wolfe, L.G.; Smith, R.K., and Deinhardt, F.: Simian sarcoma virus, type 1 *(Lagothrix):* focus assay and demonstration of non-transforming associated virus. J. natn. Cancer Inst. *48:* 1905–1908 (1972).

102 Wolfe, L.G.; Smith, R.D.; Hoekstra, J.; Marczynska, B.; Smith, R.K.; Mc Donald, R.; Northrop, R.L., and Deinhardt, F.: Oncogenicity of feline fibrosarcoma viruses in marmoset monkeys: pathologic, virologic and immunologic findings. J. natn. Cancer Inst. *49:* 519–539 (1972).

103 Wright, J.; Falk, L.A.; Wolfe, L.G.; Ogden, J., and Deinhardt, F.: Susceptibility of common marmosets *(Callithrix jacchus)* to oncogenic and attenuated strains of *Herpesvirus saimiri*. J. natn. Cancer Inst. *59:* 1475–1478 (1977).

104 Wright, J.; Falk, L.A.; Collins, D., and Deinhardt, F.: Brief communication. Mononuclear cell fraction carrying *Herpesvirus saimiri* in persistently infected squirrel monkeys. J. natn. Cancer Inst. *57:* 959–962 (1976).

105 Wright, J.; Falk, L.A.; Wolfe, L.G., and Deinhardt, F.W.: Lymphocyte populations of *Callithrix jacchus* marmosets. Cell. Immunol. (in press).

L.G. Wolfe, DVM PhD, Department of Microbiology, Rush-Presbyterian-St. Luke's Medical Center, 1753 West Congress Parkway, *Chicago, IL 60612* (USA)

Prim. Med., vol. 10, pp. 119–134 (Karger, Basel 1978)

Spontaneous Colonic Adenocarcinoma in Marmosets [1]

C.C. LUSHBAUGH, G.L. HUMASON, D.C. SWARTZENDRUBER, C.B. RICHTER and N. GENGOZIAN

Medical and Health Sciences Division, Oak Ridge Associated Universities, Oak Ridge, Tenn.

Introduction

The rarity of spontaneous neoplasms of any kind in nonhuman primates is well known [3, 4, 6, 7, 10, 15, 18, 20, 22, 24–26, 28–30, 40, 41, 46]. Although it has led some to believe that these species have a lower susceptibility to cancer than man and rodents, this low incidence may be apparent rather than real. It could stem from the young age at which most captive primate species die or are killed for experimental purposes [18, 21]. For instance, JUNGHERR [18] observed only one neoplasm (a malignant lymphoma) in 12,000 necropsies of rhesus monkeys *(Macaca mulatta);* an incidence (0.008%) which would be unbelievably low in human terms if it were not that all of his animals were prepubertal when killed and examined. Improvements in animal husbandry have decreased the numbers of early deaths of wild-caught primates, prolonged their life spans, and enabled breeding colonies of some rare and endangered primate species to be established successfully in recent years. Human cancer experience suggests that as more of these animals live to natural old age their true cancer incidence will be revealed, particularly as deaths from infectious and parasitic infestations are prevented. Most case reports of cancer in nonhuman primates show the well-known relationship of carcinoma with old age, and in keeping with these observations, the new successful attempts to produce malignant neoplasms experimentally in simians have required long exposure periods to chemical carcinogens [1, 9] or long survival periods after exposure to ionizing irradiation [19, 49]. However, as in man, spontaneous hematopoietic malignancies

1 Supported by the Energy Research and Development Administration.

in monkeys have not shown this phenomenon of age dependence [11, 17, 25, 39]. In experiments with *Herpesvirus saimiri* malignant lymphomas have been induced in marmosets that kill within 28 days of inoculation [17, 31–33, 35]. A short induction period has also been found for various sarcomas in marmosets injected with Rous sarcoma [23, 37] and feline sarcoma viruses [9, 11, 48]. Solid epithelial tumors have not been produced in primates by these viruses, and as spontaneous malignancies in animals, epithelial cancers are much rarer than connective tissue neoplasms [21, 28, 47].

Although colonic adenocarcinomas 'are among the most prevalent human cancers in the United States... they are infrequently seen as spontaneous cancers in animals' according to LINGEMAN and GARNER [24]. In a review of the animal literature, these authors found reports of 553 instances of *small* intestinal adenocarcinomas (434 of which were in Australian sheep), and only 199 instances of large intestinal adenocarcinomas. Among the latter, 71 originated in the rectum, and 78 were without specified sites; most were in dogs. Their literature survey covered about 60 years of scientific reporting and represents man's experience with many generations of domestic, experimental, and captive exotic animals, and demonstrates the unexplained rarity of intestinal carcinomas in animals even when they share man's environment and food. LINGEMAN and GARNER [24] found only three reported cases of large intestinal cancer in nonhuman primates. Two instances, one of which was rectal in origin was reported by PLENTL *et al.* [38], and the other by SNYDER and RATCLIFFE [42], emphasize this disparity between the incidence of spontaneous intestinal carcinomas in animals and man even further. This reported necropsy experience in primate pathology contrasts markedly with our observations over the last 14 years of the spontaneous occurrence of mucinous adenocarcinomas of the colon with lymph node metastases in 13 animals of two species of captive and laboratory-bred marmosets living in a large research colony in Oak Ridge where these New World primates are being studied immunologically by GENGOZIAN [12].

Most of the previously reported spontaneous and induced malignancies of primates have been reported in rhesus monkeys and related species [6, 18, 21, 25].

This apparent species' predilection for cancer may reflect the relatively large numbers of necropsies performed on this species, compared to the relatively few in other nonhuman primates. Although NOYES [37] did not find *any* spontaneous neoplasms among an unstated number of *Saguinus fuscicollis* and *Saguinus oedipus oedipus* in the Sloan-Kettering marmoset colony during a five-year observation period ending in 1970, the observations of ADAMSON *et al.* [1] suggest that marmosets like other primitive primates may have a high sensitivity to carcinogens and may not be as resistant as the higher species of monkeys, apes, and man. These latter observations suggest that an upward revision of the presently low cancer incidence in various primate species may soon be seen as the establishment of large breeding colonies of New World monkeys becomes more successful and many imported and colony-bred animals are able to survive for more than 10 years [3, 6, 9].

Our experience over the last 14 years seems to support this prediction and suggests that the primitive subhuman primate may have a high sensitivity to natural carcinogens.

Animals and Observational Methods

These observations which are reported here were made during the course of surveillance by necropsy for defining natural diseases occurring in a large colony of marmosets (average monthly census almost 500 animals). The purpose of the necropsies was to identify endemic diseases, thus aiding in their therapy and eradication and helping prevent epidemic disease. In previous studies [8, 36], the parasitic diseases of marmosets of this colony [36] were defined. In those studies, 506 necropsies of young adult, imported, hairy-faced

Table I. Fourteen-year incidence of colon cancer in various species of marmosets necropsied after living more than ten months in the colony from March 1963 to March 1977

Marmoset species	Necropsies[1]	Colon cancers
1 Bare-faced tamarins (imported)	81	12 (+2)[2]
2 Hairy-faced tamarins		
a) Imported	483	0
b) Colony-bred	223	(1)[3]
3 Goeldi's marmoset		
a) Imported	7	0
b) Colony-bred	2	1 (+1)[2]
Total all species	796	13
Average annual incidence all species	57	1
bare-faced	6	1

[1] Adults, in colony more than ten months.
[2] Suspected case.
[3] Small intestine.

Table II. Identity and pathological summary of intestinal cancers

Case No.	Date of death	Species	Sex	Residence time in colony, months	Number of primary sites	Involved intestinal region	Local invasion	Lymph node metastases (positive/total)
1	2–6–68	*C. goeldii*	M	45[1]	>1	colon	massive	5/9
2	3–19–70	*S. o. oedipus*	M	30	>1	colon	massive	1/14
3	7–3–71	*S. o. oedipus*	M	71	>1	cecum	massive	7/14
4	8–17–71	*S. o. oedipus*	F	25	2	cecum, asc. colon	early	6/16
5	12–20–71	*S. o. oedipus*	M[2]	13	2	cecum	early	11/14
6	6–1–72	*S. o. oedipus*	F	56	7ea	cecum, ileum 7	extensive	5/16
7	9–1–72	*S. o. oedipus*	M	49	1	asc. colon	early	1/10
8	7–6–73	*S. o. oedipus*	M	50	>6	cecum, asc. colon	extensive	5/15
9	3–1–74	*S. o. oedipus*	M	41	1	desc. colon	extensive	0/18[3]
10	9–17–74	*S. o. oedipus*	F	48	>8	desc. colon	extensive	13/27
11	5–16–75	*S. o. oedipus*	F	55	5	cecum	early	0/13[3]
12	5–19–75	*S. fuscicollis*	M	49[1]	2	mid-ileum	extensive	3/3
13	6–11–75	*S. o. oedipus*	F	57	1	asc. colon	extensive	11/16
14	12–2–76	*S. o. oedipus*	M	74	>1	asc. colon	massive	8/10

[1] Laboratory-bred animal.

[2] Animal with *Prosthenorchis elegans* infestation.

[3] Diagnosis based on invasion through external muscle coats.

tamarin marmosets (largely *S. fuscicollis* subspecies) were studied, and only one neoplasm, a malignant lymphoma, was found. In the ensuing 14 years, we have studied 796 additional necropsies of adult marmosets among which there were 225 colony-bred animals that died at greater than one year of age. The three categories of species, the numbers of each necropsied, and the number of colony-bred animals studied are listed in table I. The cases are identified in table II in order of their discovery, by date of death, sex, and length of residence in the colony (in lieu of actual age except in two instances, case 1 and 12, where the animals were colony-bred and their ages known).

At necropsy, a complete gross and histologic examination is made routinely. Paraffin embedded sections of all organs are stained for light microscopic examination after fixation in Gomori's fixative or in 10% formaldehyde. More recently, the difficulty in locating small primary sites accurately has led us to routinely examine longitudinal rolls of the entire colon cut into stylized segments. For light microscopy the hematoxylin-PAS stain, lightly counterstained with orange-G, has been used routinely [16, 27]. Without the brilliant PAS-positive staining of mucin, many of the primary and secondary sites of these cancers would have been missed. For electron microscopy, small pieces are fixed in sodium cacodylate buffered 3% glutaraldehyde and postfixed in 1% osmium tetroxide. Sometimes small pieces of formaldehyde-fixed paraffin-embedded tissue are excised after initial identification of a carcinoma, cleared of paraffin, postfixed in osmium, and embedded in Epon 812.

Observations

Tabulated in table I are the numbers of primary colon cancers found among 796 necropsies of adult marmosets dying of various causes after residing in our colony for more than ten months. Thirteen died with colon cancer as the primary cause of death. Twelve of the 13 were *S. o. oedipus* and one was a *Callimico goeldii.* Three other *in situ* colon cancers are listed in parentheses and noted as 'suspected cases' but excluded because they were not disseminated and had not yet contributed to the death of those three animals. No neoplasms of the large intestine were found among the hairy-faced tamarins (chiefly *S. fuscicollis illigeri)* but one ileal primary neoplasm was found that produced a fatal mid-ileal obstruction. The immediate cause of death of the 14 animals detailed in table II was intestinal obstruction except for cases 6, 9, 10, and 11. Death occurred during a phlebotomy being done for diagnostic purposes in case 6. The immediate cause of death was ulcerative colitis in case 9, ascites and severe anemia in case 10 and 11. Except in one instance the intestinal obstructions were due to submucosal fibrosis, cicatrization and smooth muscle hyperplasia which decreased the lumen. The exception was a death from paralytic ileus and dilation of the colon in case 7 where although only one local lymph node contained metastases, regional nonmyelinated mesenteric nerves were invaded.

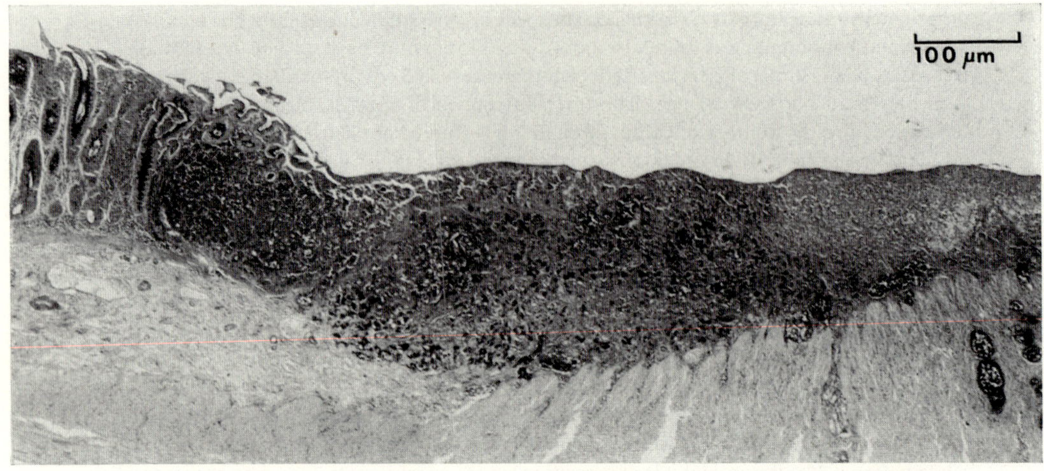

Fig. 1. Primary mucosal site of colon cancer in an *S. o. oedipus* (case 13, table II). Note hyperplastic but normal mucosa at left abruptly replaced by disorganized masses of PAS-positive cells that infiltrate the smooth muscle coat (at far right). PAS-HE.

Histopathology

Histologically, all of these cancers are quite similar. A typical primary focus is seen in figure 1 as a disruption of mucosal architecture by disorganized PAS-positive epithelial cells. These grow through the lamina propria and form small masses of cells without any orientation to a glandular lumen. Usually such foci are found among hyperplastic colonic glands that resemble adenomatous hyperplastic colitis of man. These hyperplastic glands in this commonly benign disease often penetrate the muscularis mucosa and form glands and cell-lined mucus-filled cysts in submucosal lymphoid follicles. The adenocarcinoma, however, was not associated with such pseudo-invasive extensions. Neither does it appear to arise in mucosal polyps since no polyps were found in 302 necropsy examinations by NELSON *et al.* [36] and us (this paper). Examination of many serial sections has often been required to find the minute primary sites and show where the adenocarcinoma cells penetrate through the base of the lamina propria into the submucosal connective tissue and lymphatics. Remarkably small areas of this kind can be associated with voluminous lymph node metastases and mesenteric nerve invasion. As seen on the right side of the photomicrograph (fig. 1), the

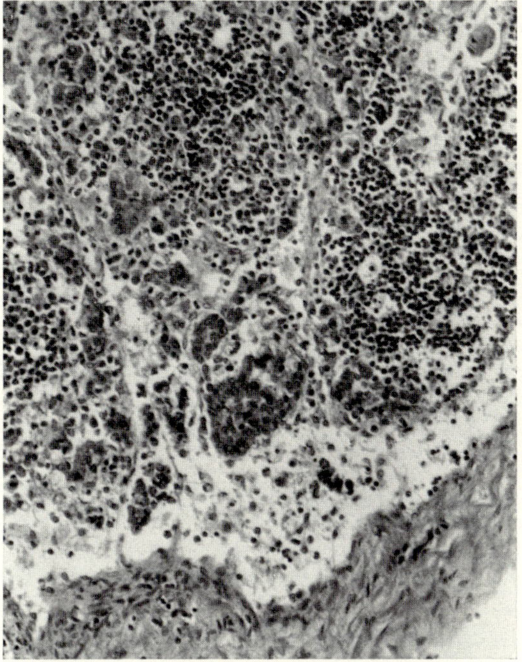

Fig. 2. Peripheral sinus of paracolonic lymph node that contains metastatic colon cancer cells (case 4, table II). PAS-HE. ×300.

adenocarcinoma can form large mucin-filled lakes as it penetrates the circular and longitudinal muscles of the colon to reach the peritoneal surface.

The paracolonic and mesenteric lymph nodes usually contain metastatic cancer cells as shown in figure 2 and tabulated in table II. In most cases the lymph node metastases do not elicit a connective tissue response but lymph node sclerosis from metastases does occur. Invasion via lymphatics occurs into other mesenteric and retroperitoneal structures – nonmyelinated nerves (fig. 3), ureters, renal pelves, pancreatic capsule and gastrocolic ligaments. The only blood vascular metastases we found have been pulmonary alveolar capillaries.

In primary sites, as well as metastatic foci, three kinds of adenocarcinoma cells can be identified. The most common one is a mucin-filled, granular appearing, PAS-positive cell which in plain HE staining is recognized as the so-called signet-ring cells of the colloid type of human colon cancers. The next most common is hematoxylinophilic and PAS-negative.

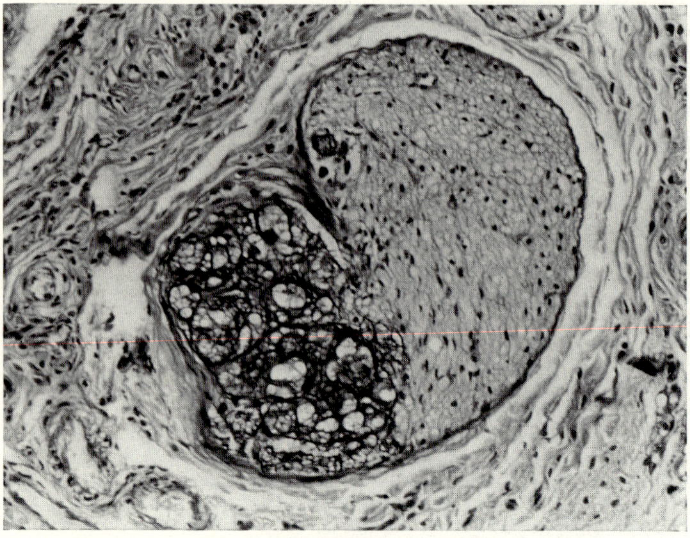

Fig. 3. Partial replacement of nonmyelinated mesenteric nerve by mucin-secreting colon cancer in a *C. goeldii* (case 1, table II). PAS-HE. ×300.

It has a large, irregular, dark nucleus and scanty cytoplasm. Among these cells, the third type can be found in small numbers. This one has a brilliant PAS-positive, almost crystalline body in the center of an intracytoplasmic space. As shown by electron microscopy, this intracytoplasmic cyst is lined by a cell membrane and has typical intestinal microvilli protruding into the cystic space shown in figure 4. By ultramicroscopic examination, we find that this cyst does not always contain a condensed mucin droplet but can be fluid-filled or filled with stainable material and minute vesicles that resemble those of an intestinal glycocalyx. These cells have proved to be extremely useful in drawing the light microscopist's attention to the presence of minute foci of adenocarcinoma cells that would otherwise be missed [45]. Surprisingly, all of these animals except one (case 5) were free of the ileal acanthocephalid, *Prosthenorchis elegans,* which has been the cause of death of many of our marmosets within ten months of importation. All, however, had various degrees of a chronic hyperplastic colitis that is commonly seen

---

Fig. 4. Electron micrograph of *S. o. oedipus* cell. The intracellular cyst lined by microvilli seen in the center of the photograph is also to be found in carcinogen-induced rodent colon cancers and those of man. Approximately ×10,000.

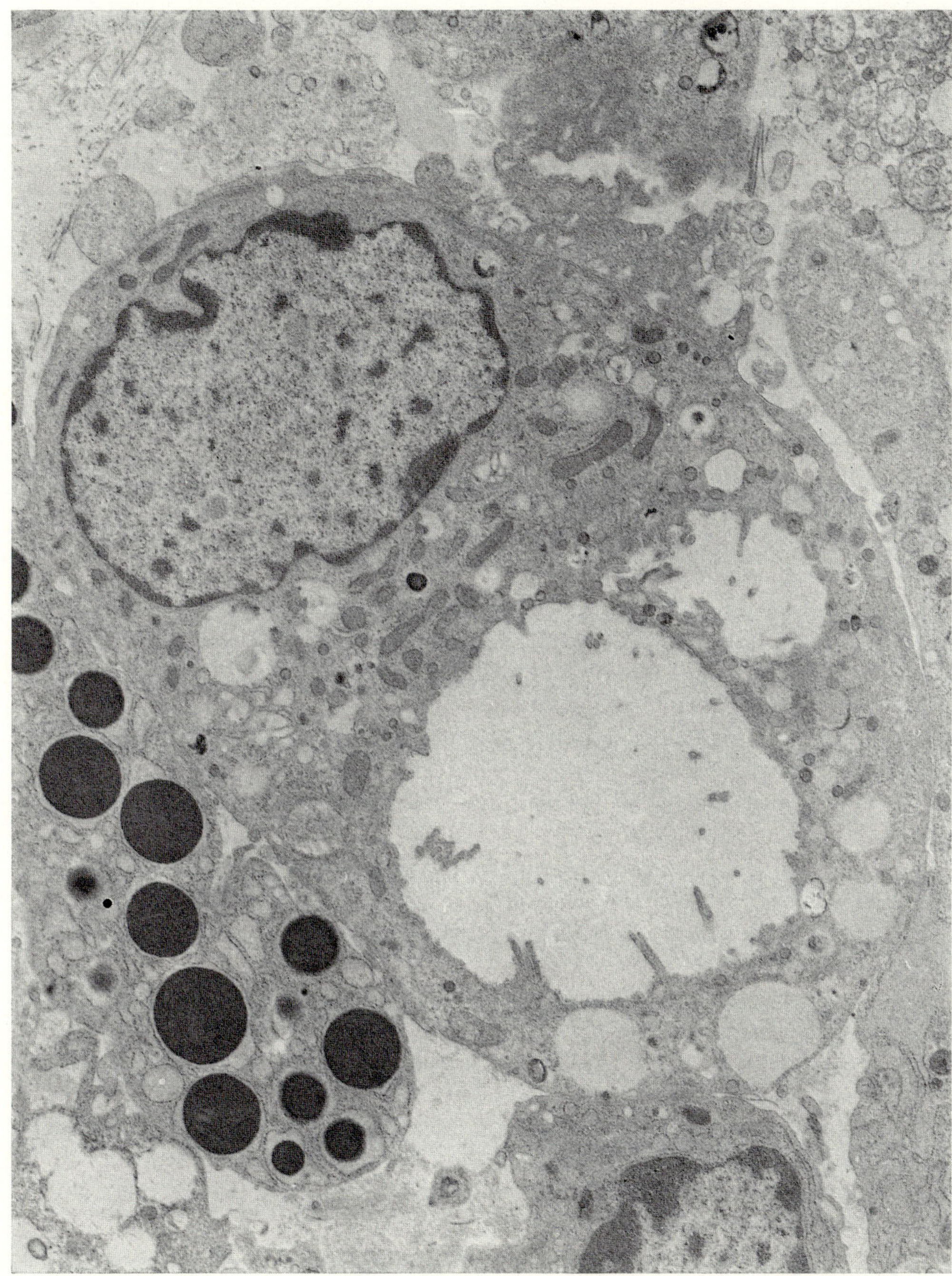

in various regions of the gastrointestinal tract of experimental animals and man [29, 34]. While this disease appears to be a nonspecific inflammatory reaction to a wide variety of irritating substances, carcinogens and infectious agents, some observers considers it intuitively a precancerous mucosal disease.

Other Pathologic Features

Additional unusual features of these cancers, summarized in table II, are the multiple primary sites of origin, the extensive invasion and the large number of lymph node metastases [2]. Because of the massive local invasion, reactive fibrosis, smooth muscle hyperplasia, mucosal ulceration, and postmortem autolysis in cases 1, 2, and 3, the actual number of independent primary sites of origin could not be determined. But, in retrospect, because of the multiple independent sites found in the other cases, these original three cases must also be considered as having had multiple primary sites that coalesced with time. In case 6 there appeared to be seven primary sites in the terminal ileum as well as in the cecum. These ileal sites, however, may represent intramucosal metastases, because this animal had extensive local submucosal lymphatic invasion and retrograde extension. Only two cases had primary sites in the descending colon. (The marmoset has a poorly defined transverse colon and no sigmoid colon.) Comparison of these two cases with the others reveals that in them the chronic hyperplastic colitis commonly found associated with these adenocarcinomas was present in the descending colon, whereas in the other cases it was confined to, or most severe in the cecum and ascending colon. In case 7, the solitary primary site was halfway between the ileocecal valve and the ascending/descending colon juncture. In contrast with this, in case 10, the entire ascending colon had multiple discrete primary mucosal sites all along its course. In only three cases local submucosal invasion occurred after mucosal lymphatic vascular invasion. In two of these (cases 4 and 5), however, mesenteric lymph node metastases were numerous and easily found. In case 7, a solitary mesenteric lymph node metastasis was found on microscopic examination which led the pathologist to the 1-mm diameter primary site in the colon that had been overlooked at gross necropsy even though a paralytic ileus was present. In only two cases (cases 9 and 11) was it necessary to abandon the criterion of incontrovertible proof of malignancy that is afforded by lymph node metastases. In these animals without lymph node metastases, Stewart's alternate criterion of malignancy was found – extension of cellular growth through the muscularis propria to the peritoneum [43, 44]. In all the other animals, at least one mesenteric lymph node was involved and in the majority 30–80% of the nodes along the mesenteric veins contained metastases. The unusual close proximity of the renal pelves, ureters, pancreas, ascending colon and their mesenteric attachments in the extremely narrow upper abdomen of the marmoset probably explains the unusually extensive metastases we found in the walls of the ureters, unmyelinated mesenteric nerves and peripheral portions of pancreas. In some instances, the large relatively cell-free, mucin-filled lakes that are common in human metastatic colloid cancers were found occupying numerous neural structures (fig. 3).

No metastases to liver, adrenals, kidneys, brain or bone were found among these animals, but in the process of reviewing these cases recently, the reviewer (C.R. and G.H.) found several instances of mucin-producing cancer cell emboli in the lungs of these animals. In cases 3 and 14, these cells had penetrated the pulmonary alveolar capillaries and were growing as bona fide pulmonary metastases.

Discussion

These observations suggest that among subhuman primates, *S. o. oedipus,* the cotton-topped tamarin is uniquely susceptible to colon carcinoma while the hairy-faced species like most other animals of laboratories and zoos are not. The question naturally arises how this phenomenon could have been overlooked in the numerous necropsies of these animals in the last several decades. The implication that the cases reported here of colonic adenocarcinomas are merely incidental findings in animals dying from some other primary cause cannot be accepted because ten of 14 animals had grossly apparent severe intestinal obstructions and another had an ileus secondary to carcinomatous invasion of regional intestinal nerves. The three other deaths from this neoplasm were not from obstruction but from severe inanition, anemia and wasting. All of these animal lost 25–50% of their best recorded weight before their deaths (average weight loss 38%). While the horny-headed ileal parasite, *P. elegans,* is the most frequent cause of intestinal obstruction and death within ten months of importation of feral animals, it was found in only case 5 of these 14 cases of intestinal cancer and then it was obviously incidental. Perhaps, however, common every-day experience with this intestinal parasitism has discouraged routine intestinal histologic examination in all animals, and thereby has obscured the presence of the neoplasm. The parasite may also have shortened the lives of many animals so colony residence times have not been sufficiently long for the neoplasm to develop. It should be mentioned again that the 14 cases reported here among 796 necropsies occurred over a 14-year observation period and at an average of only one colon neoplasm per 60 animals per year. On the basis of this percent incidence, this apparent frequency would be unusual in human epidemiologic experience even though colonic adenocarcinoma is the second most common cancer among men and women together [5]. The small number of total animals examined, however, precludes such comparison with the human incidence and mortality. By excluding the *in situ* cases which do not meet Stewart's criteria for the experimental diagnosis of malignancy (see number of cases in parentheses in table I), we have tried to equate incidence with mortality because in human epidemiology the true incidence of the disease is not known because many cases are cured, few people now are autopsied and human statistics are based on official death certificates.

Biologically, if not statistically, this colon cancer in marmosets resembles the 'signet-ring' colon carcinoma of humans where it also metastasizes early and widely, often before causing symptoms of intestinal obstruction. As in man, induction of this neoplasm in tamarins seems to be related to aging even though the 'ages' of the animals listed in table II are colony residence times except in two instances. The average colony residence time of all the animals with intestinal cancer was found to be 47.5 months (ranging from 13 to 74 months), indicating that about a four-year induction period is required for development of this disease in our colony. It also occurred in two adult animals born and raised in the colony. These two colony-bred animals with intestinal cancer that died at 45 and 49 months of age are, of course, the only animals whose age is known. Nine of the 14 animals resided in the colony from 40 to 60 months. The average colony residence time of the colon cancer-resistant hairy-faced marmosets figures out to be about 57 months, but some animals now coming to necropsy have resided in the colony for more than 11 years without developing it. If there is a 'minimum residence time' requirement for this disease to develop, both kinds of tamarins have had adequate periods of risk and sufficient time for induction of the tumor. The resistant *S. fuscicollis* species have actually had more individuals at risk for longer periods of time than the few cotton-topped tamarins we have seen. These 'residence time' data afford some insight into the increasing rate of this neoplastic disease with aging. A preliminary analysis by maximum likelihood analysis of our data from the variable numbers of animals at risk at any time, and the incidence of cancers found during this 14-year period, suggests that the rate at which the incidence of this tumor increases with age in cotton-topped tamarins has a doubling time of 340 days. The slope constant (K) is commonly computed in experimental carcinogenesis in mice and human oncology using the GOMPERTZ [13] model:

$$R_t = R_o e^{kt}$$

where R_t is the risk of developing a specific tumor at some age (t), R_0 is the theoretical or initial risk at zero age (birth?) and k (the hazard constant) is the exponent of the natural logarithm. k equals 0.693/T where T is the doubling time of R. The doubling time for risk of most neoplasms has been found to be closely related to the life span of the species under study. Our preliminary estimate of 340 days for this risk to double in the tamarin, compares logically with the estimate of 100 days for most mouse neoplasms and 2,700 days for human malignancies [14]. Such a doubling time (about a year) would be an encouraging factor in consideration of the tamarin as a surrogate primate model for experimental colon carcinogenesis. From the points of view of biology, clinical course and pathology, the tamarin colonic adenocarcinoma has numerous similarities with that of man which encourages further study.

Summary

We find that colonic adenocarcinoma, which is an extremely rare neoplasm of all animals except man and carcinogen-treated rodents, occurs spontaneously in some marmosets. The cotton-topped *Saguinus oedipus oedipus* is particularly prone to develop it, but we have found it also at necropsy in *Callimico goeldii* (Goeldi's marmoset). Numerous metastases to regional lymph nodes develop. The cancers arise *de novo* in the mucosa and early invade the submucosa and lymphatic apparatus and paracolonic lymph nodes. These findings and the continuing occurrence of this cancer in our colony suggests that the marmoset may be the long-sought primate model for experimental intestinal carcinogenesis.

References

1 ADAMSON, R.H.; COOPER, R.W., and O'GARA, R.W.: Carcinogen-induced tumors in primitive primates. J. natn. Cancer Inst. *45:* 555–559 (1970).

2 ALFORD, J.E. and FALSETTI, D.F.: Multiple cancers of the large bowel. Dis. Colon Rectum *13:* 316–329 (1970).

3 ALLEN, J.R.; HOUSER, W.D., and CARSTENS, L.A.: Multiple tumors in a *Macaca mulatta* monkey. Archs Path. *90:* 167–175 (1970).

4 AMBRUS, J.L.; STRANDSTROM, H.V., and KAWINSKI, W.: 'Spontaneous' occurrence of Yaba tumor in a monkey colony. Experientia *25:* 64–65 (1969).

5 BURKITT, D.P.: Epidemiology of cancer of the colon and rectum. Cancer *28:* 3–13 (1971).

6 CHAPMAN, W.L., jr.: Neoplasia in nonhuman primates. J. Am. vet. med. Ass. *153:* 872–878 (1968).

7 CHAPMAN, W.L., jr. and ALLEN, J.R.: Multiple neoplasia in a rhesus monkey, *Macaca mulatta.* Pathol. vet. *5:* 342–352 (1968).

8 COSGROVE, G.E.; NELSON, B.M., and SELF, J.T.: The pathology of pentastomid infection in primates. Lab. Anim. Care *20:* 354–360 (1970).

9 DEINHARDT, F.: Induction of tumors in marmoset monkeys with ST-feline fibrosarcoma virus. Bibl.thca haemat., No. 36, pp. 401–402 (Karger, Basel 1970).

10 DEINHARDT, J.B.; DEVINE, J.; PASSOVOY, M.; POHLMAN, R., and DEINHARDT, F.: Marmosets as laboratory animals. I. Care of marmosets in the laboratory, pathology and outline of statistical evaluation of data. Lab. Anim. Care *17:* 11–29 (1967).

11 DEINHARDT, F.; WOLFE, L.G.; THEILEN, G.H., and SNYDER, S.P.: ST-feline fibrosarcoma virus. Induction of tumors in marmoset monkeys. Science *167:* 881 (1970).

12 GENGOZIAN, N.: Marmosets: their potential in experimental medicine. Ann. N.Y. Acad. Sci. *162:* 336–362 (1969).

13 GOMPERTZ, B.: On the nature of the function expressive of the law of human mortality and on a new mode of determining the value of life contingencies. Phil. Trans. R. Soc. *1825:* 513–585.

14 GRAHN, D.: Biological effects of protracted low dose radiation exposure of man and animals; in Late effects of radiation. Proc. Coll. Chicago 1969, pp. 101–136 (Taylor & Francis, London 1970).

15 HENDERSON, J.D., jr.; WEBSTER, W.S.; BULLOCK, B.C.; LEHNER, N.D.M., and CLARKSON, T.B.: Naturally occurring lesions seen at necropsy in eight woolly monkeys (*Lagothrix* sp.). Lab. Anim. Care *20:* 1087–1097 (1970).

16 HUMASON, G.L.: Animal tissue techniques; 3rd ed., pp. 328–330 (Freeman, San Francisco 1972).

17 HUNT, R.D.; MELÉNDEZ, L.V.; KING, N.W.; GILMORE, C.E.; DANIEL, M.D.; WILLIAMSON, M.E., and JONES, T.C.: Morphology of a disease with features of malignant lymphoma in marmosets and owl monkeys inoculated with *Herpesvirus saimiri.* J. natn. Cancer Inst. *44:* 447–465 (1970).

18 JUNGHERR, E.: Tumors and tumor-like conditions in monkeys. Ann. N.Y. Acad. Sci. *108:* 777–792 (1963).

19 KENT, S.P.: Spontaneous and induced malignant neoplasm in monkeys. Ann. N.Y. Acad. Sci. *85:* 819–827 (1960).

20 KNAUER, K.W.; VICE, T.E.; KIM, C.S., and KALTER, S.S.: Metastatic adenocarcinoma in the baboon. Primates *10:* 285–293 (1969).

21 LAPIN, B.A. and YAKOVLEVA, L.A.: Comparative pathology in monkeys (Thomas, Springfield 1963).

22 LAU, D.T. and SPINELLI, J.S.: A spontaneous carcinoid tumor in a cynomolgus monkey, *Macaca fascicularis.* Lab. Anim. Care *20:* 1145–1148 (1970).

23 LEVY, B.M.; TAYLOR, A.C.; HAMPTON, S., and THOMAS, G.W.: Tumors of the marmoset produced by Rous sarcoma virus. Cancer Res. *29:* 2237–2248 (1969).

24 LINGEMAN, C.H. and GARNER, F.M.: Comparative study of intestinal adenocarcinomas of animals and man. J. natn. Cancer Inst. *48:* 325–346 (1972).

25 LINGEMAN, C.H.; REED, R.E., and GARNER, F.M.: Spontaneous hematopoietic neoplasms of nonhuman primates. Review, case report and comparative studies; in Comparative morphology of hematopoietic neoplasms. Natn. Cancer Inst. Monogr. *32:* 157–170 (1969).

26 LOMBARD, L.S. and WITTE, E.J.: Frequency and types of tumors in mammals and birds of the Philadelphia Zoological Garden. Cancer Res. *19:* 127–141 (1959).

27 LUNA, L.G.: Manual of histologic staining methods of the Armed Forces Institute of Pathology; 3rd ed., pp. 153–173 (McGraw-Hill, New York 1968).

28 LUND, J.E.; BURKHOLDER, C., and SOAVE, O.A.: Renal carcinoma in an owl monkey *(Aotus trivirgatus)*. Pathol. vet. *7:* 270–274 (1970).

29 LUSHBAUGH, C.C.: Infiltrating adenomatous lesions of the stomach, cecum, and rectum of monkeys similar to early human carcinoma and carcinoma *in situ*. Cancer Res. *9:* 385–394 (1949).

30 MARUFFO, C.A.: Spontaneous tumours in howler monkeys. Nature, Lond. *213:* 521 (1967).

31 MELÉNDEZ, L.V.; HUNT, R.D.; DANIEL, M.D.; GARCIA, F.G., and FRASER, C.E.O.: *Herpesvirus saimiri*. II. Experimentally induced malignant lymphoma in primates. Lab. Anim. Care *19:* 378–386 (1969).

32 MELÉNDEZ, L.V.; DANIEL, M.D.; HUNT, R.D.; FRASER, C.E.O.; GARCIA, F.G.; KING, N.W., and WILLIAMSON, M.E.: *Herpesvirus saimiri*. V. Further evidence to consider this virus as the etiological agent of a lethal disease in primates which resembles a malignant lymphoma. J. natn. Cancer Inst. *44:* 1175–1181 (1970).

33 MELÉNDEZ, L.V.; DANIEL, M.D., and HUNT, R.D.: *Herpesvirus saimiri* induced malignant lymphoma. Recovery of the viral agent from the fatally affected animals. Biblthca haemat., No. 36, pp. 751–753 (Karger, Basel 1970).

34 MORGAN, C.N.: Malignancy in inflammatory diseases of the large intestine. Cancer *28:* 41–44 (1971).

35 MORGAN, D.G.; EPSTEIN, M.A.; ACHONG, B.G., and MELÉNDEZ, L.V.: Morphological confirmation of the herpes nature of a carcinogenic virus of primates *(Herpes saimiri)*. Nature, Lond. *228:* 170–172 (1970).

36 NELSON, B.; COSGROVE, G.E., and GENGOZIAN, N.: Diseases of an imported primate, *Tamarinus nigricollis*. Lab. Anim. Care *16:* 255–275 (1966).

37 NOYES, W.F.: Tumors induced in the South American marmoset monkey by Rous sarcoma virus. J. natn. Cancer Inst. *45:* 579–587 (1970).

38 PLENTL, A.A.; DEDE, J.A., and GREY, R.M.: Adenocarcinoma of the large intestine in a pregnant rhesus monkey *(Macaca mulatta)*. Report of a case. Folia primatol. *8:* 307–313 (1968).

39 PRICE, R.A.A. and POWERS, R.D.: Reticulum cell sarcoma in a Sykes monkey *(Cercopithecus albogularis)*. Pathol. vet. *6:* 369–374 (1969).

40 RUCH, T.C.: Neoplasia; in Diseases of laboratory primates, chapter 14, pp. 529–567 (Saunders, Philadelphia 1959).

41 SCHILLER, A.L.; HUNT, R.D., and DIGIACOMO, R.: Basal cell tumour in a rhesus monkey *(Macaca mulatta)*. J. Path. Bact. *99:* 327–329 (1969).

42 SNYDER, R.L. and RATCLIFFE, H.L.: Factors in the frequency and types of cancer in

mammals and birds at the Philadelphia Zoo. Ann. N.Y. Acad. Sci. *108:* 793–804 (1963).

43 STEWART, H.L.: Experimental cancer of the alimentary tract; in HOMBURGER and FISHMAN The physiopathology of cancer, pp. 3–45 (Hoeber, New York 1953).

44 STEWART, H.L. and LORENZ, E.: Histopathology of induced precancerous lesions of the small intestine of mice. J. natn. Cancer Inst. *7:* 239–268 (1947).

45 SWARTZENDRUBER, D.C.; HUXFORD, V.G.; LUSHBAUGH, C.C.; HUMASON, G.L., and SELLE, E.B.: Comparative ultrastructure of cells containing intracellular cysts in colonic adenocarcinoma of man, marmoset, and rat (submitted for publication).

46 VALERIO, M.; LANDON, J.C., and INNES, J.R.M.: Neoplastic diseases in simians. J. natn. Cancer Inst. *40:* 751–756 (1968).

47 WILLIAMSON, M.E. and HUNT, R.D.: Adenocarcinoma of the thyroid in a marmoset *(Saguinus nigricollis)*. Lab. Anim. Care *20:* 1139–1141 (1970).

48 WOLFE, L.; MCDONALD, R.; DEINHARDT, F.; SNYDER, S., and THEILEN, G.: Transmission of feline fibrosarcoma virus to marmoset monkeys. (Abstract) Fed. Proc. Fed. Am. Socs exp. Biol. *29:* 371 (1970).

49 ZALUSKY, R.; GHIDONI, J.J.; MCKINLEY, J.; LEFFINGWELL, T.P., and MELVILLE, G. S.: Leukemia in the rhesus monkey *(Macaca mulatta)* exposed to whole-body neutron irradiation. Radiat. Res. *25:* 410–416 (1965).

C.C. LUSHBAUGH, MD, Medical and Health Sciences Division, Oak Ridge Associated Universities, *Oak Ridge, TN 37830* (USA)

Prim. Med., vol. 10, pp. 135–141 (Karger, Basel 1978)

Oncogenicity of the C-Type Virus HL-23V in Marmosets and Characterization of Virus Isolated from an HL-23V-Induced Marmoset Tumor: Comparison with Simian Sarcoma Virus Type 1 [1]

C. M. BERGHOLZ, L. G. WOLFE and F. DEINHARDT

Rush-Presbyterian-St. Luke's and University of Illinois Medical Centers, Chicago, Ill.

Introduction

A type C virus, HL-23V, isolated from cultured human acute myelogenous leukemia cells, has structural proteins and genetic sequences identical to simian sarcoma virus type 1 (SSV-1/SSAV-1) [1, 2, 5]. Postmortem tissue and uncultured leukemic cells from patient HL-23, as well as cell cultures infected with HL-23V *in vitro*, were also shown to contain genetic sequences homologous with baboon endogenous virus (BaEV) [7]. In this study we investigated the oncogenicity of HL-23V in marmosets and compared the biologic and immunologic properties of HL-23V and virus isolated from an HL-23V-induced marmoset tumor with SSV-1.

Materials and Methods

Cells. Normal marmoset skin (HF or MFS) and lung (1283) cells were maintained in RPMI 1640 (Gibco) with 10% heat-inactivated fetal calf serum (FCS). Canine thymus cells (FCf-2th-A7573, Naval Biomedical Research Laboratory) and horse skin cells (R-1042, supplied by Dr. J. RHIM) were maintained in Eagles' basal medium (BME) improved with nonessential amino acids and supplemented with 10%-FCS.

Virus. Filtered (0.45 μm) supernatants from HL-23V-infected canine thymus cells (A7573/HL-23V, provided by Dr. R. GALLO) and SSV-1-transformed marmoset fibro-

1 This study was supported by research contracts NOI-CP-33219 within the Virus Cancer Program of the National Cancer Institute, US Public Health Service.

blasts (HF/SSV-1) were sources of virus stocks. All possible precautions were taken to avoid cross-contamination; HL-23V infected cultures were handled and incubated in a laboratory remote from other oncornavirus work.

Animals. White-lipped marmosets *(Saguinus fuscicollis nigrifrons, S.f. illigeri)* were inoculated intramuscularly or intraperitoneally with $3\text{–}7 \times 10^7$ virus-producing cells when three to five days old. The marmosets were born in our breeding colony and hand-reared by established procedures. Total and differential leukocyte counts were monitored monthly and samples of plasma stored at –20 °C.

Focus assay. Normal marmoset fibroblast cells were seeded at 6×10^5 cells/25 cm^2 flask and the following day were treated with DEAE-dextran (20 µg/ml) for 1 h, washed with Hanks' balanced salt solution (HBSS) and infected with 0.5 ml of virus inoculum. Virus was absorbed 90 min at 37 °C. Foci were counted 14 days postinoculation (PI).

XC test. Nontransforming virus was assayed by the mixed culture cytopathogenicity test with XC cells, as described previously [6].

Neutralization tests. Serum or plasma was diluted twofold in HBSS with 2% agamma FCS in a final volume of 0.6 ml and incubated at 4 °C for 4 h with 0.6 ml virus diluted to yield 50–100 foci/flask. Neutralization of focus-forming virus was determined by focus assay.

Indirect immunofluorescence. For detection of cytoplasmic viral antigen, acetone-fixed cell monolayers were treated for 30 min at 37 °C with marmoset plasma diluted in 2.5% bovine serum albumin (BSA) in PBS. Following two washes with PBS, monolayers were reacted with fluorescein-conjugated goat anti-human IgG (Hyland). Viral-related cell-surface antigens were detected by the membrane immunofluorescence test. Trypsinized cells were incubated in medium on a rocker overnight. Following two washes with PBS containing 2.5% BSA, 4×10^5 cells in 0.1 ml were incubated at 4 °C for 30 min with 0.2 ml of marmoset or goat anti-SSV-1/SSAV-1 neutralizing plasma or serum. Cells were washed three times, and fluorescein-conjugated goat anti-human Ig (cross-reacting with marmoset Ig) or rabbit anti-goat IgG was added to the appropriate samples, which were then incubated at 37 °C for 30 min, washed again and examined for membrane fluorescence.

Complement-dependent serum cytotoxicity. Cytotoxicity tests using SSV-1/SSAV-1 serum from a goat inoculated with SSV-1 transformed autologous cells were done according to a micro-^{51}Cr release assay described by JOHNSON *et al.* [3].

Karyotype analysis. Chromosomes were prepared and analyzed as described by MOORHEAD and NOWELL [4].

Results

HL-23V from chronically infected canine thymus cells (A7573/HL-23V) was tested for focus-forming activity by inoculation of normal marmoset cells with cell-free supernatants as routinely performed for an SSV-1 focus assay. Foci of transformed cells indistinguishable from those induced by SSV-1 appeared in five to seven days. Transformation was observed in all marmoset cell lines tested (15 total), a horse skin cell line, rat kidney cells

Table I. Tumor induction and antibody response in HL-23V-inoculated marmosets[1]

Animal No.	Route of inoculation	Tumor induction	Antibody titer[2] months postinoculation					
			1	2	3	4	5	6
75-BF-1	i.m.	fibroma	32	512	4,096	2,048	512	1,024
75-CN-1	i.m.	fibroma	<4	256	128	128	64	32
75-CP-1	i.m.	none	<4	64	128	128	64	128
75-BC-1	i.p.	none	NT	64	32	128	256	128

[1] Three-to-five-day-old marmosets were inoculated with $3–7 \times 10^7$ virus-producing A7573/HL-23V cells.

[2] The titer is expressed as the reciprocal of the highest plasma dilution with which virus-specific cytoplasmic fluorescence was observed in HL-23V-infected cells; NT = not tested.

(NRK, Dr. R. TING, personal commun.) and a feline embryo fibroblast cell line (FEF, Dr. P. MARKHAM, personal commun.). The presence of 10- to 100-fold excess nontransforming associated virus was demonstrated by the mixed culture cytopathogenicity test with XC cells. Results indicated that A7573/HL-23V cells and HL-23V-transformed marmoset cells produced levels of transforming (10^2–10^4 FFU/ml) and nontransforming (10^5) virus comparable to SSV-1-infected cells.

In agreement with previous reports, no antigenic differences between HL-23V and SSV-1 were detected by immunofluorescence tests, by measuring inhibition of focus induction in interference tests or in neutralization tests utilizing marmoset anti-SSV-1, marmoset anti-gibbon ape lymphoma virus, or goat anti-SSV-1/SSAV-1 sera.

Oncogenicity of HL-23V was investigated by inoculation of four newborn marmosets with $3–7 \times 10^7$ A7573/HL-23V cells (table I). Two of three marmosets inoculated intramuscularly developed palpable tumors by four weeks PI; the tumor in one animal had regressed completely by six weeks and a regressing tumor in the other, 75-BF-1, was partially excised at six weeks PI. A small mass was still present 28 weeks PI when the animal died of acute pneumonia. Both tumor specimens from 75-BF-1 were fibromas composed of fibroblasts, abundant collagen and some lymphoid cells. Extracellular type C virus and budding virus particles were observed by electron microscopy both in tumor tissue and in cell cultures established from ex-

Table II. Comparison of anti-SSV-1 neutralization titers with SSV-1 and virus (BFV) isolated from an HL-23V-induced marmoset tumor

Neutralizing antiserum	Virus		Source of BFV	
	SSV-1	BFV	cell cultures	passage level
75-BF-1 (3 months PI)	8 192[1]	8,192	tumor biopsy	4
75-BF-1 (3 months PI)	8,192	512	tumor biopsy	15
75-BF-1 (3 months PI)	8,192	32	tumor autopsy	3
75-BF-1 (6 months PI)[2]	1,024	64	tumor autopsy	9
Goat anti-SSV-1	512	512	tumor biopsy	4
Goat anti-SSV-1	1,024	128	tumor biopsy	15
Goat anti-SSV-1	2,048	32–64	MFS/BFV[3]	

[1] The titer is the reciprocal of the highest serum dilution yielding $\geq$50% inhibition of focus formation.

[2] Plasma sample at time of death.

[3] Virus sequentially cloned three times by isolation of BFV-foci induced on normal marmoset cells (MFS).

planted tumor tissue. The behavior and morphology of HL-23V-induced marmoset tumors were similar or identical to SSV-1 induced tumors [9]. All four marmosets developed antibodies that reacted with HL-23V antigens in indirect immunofluorescence tests, with the highest titer, 1:4,096, in marmoset 75-BF-1. Total and differential leukocyte counts remained normal in all four animals throughout the 18-month observation period.

Cell cultures established from 75-BF-1 tumor tissue were identified as marmoset by karyotype analysis. The growth pattern and morphology of the cells in culture closely resembled marmoset cells transformed *in vitro* by HL-23V or SSV-1. Filtered supernatants collected from the biopsy and autopsy tumor cell cultures contained 10^4–10^6 FFU/ml of transforming virus. The antigenic properties of virus produced by the 75-BF-1 tumor cell cultures were compared to HL-23V and SSV-1 by neutralization tests (table II). At passage 4 the 75-BF-1 tumor cells produced virus indistinguishable from SSV-1 or HL-23V. However, transforming virus produced by 75-BF-1 tumor biopsy cells at higher passages and virus produced by autopsy tumor cells was neutralized only by low dilutions of anti-HL-23V or anti-SSV-1 serum (1:32–1:64) relative to SSV-1 (1:1,024–1:2,048). The transforming virus, distinct from HL-23V and SSV-1, was designated BFV.

A newborn marmoset inoculated with 75-BF-1 autopsy tumor cells developed a large fibroma which has persisted for 12 months. A cell culture established from tumor tissue excised from this animal produced virus indistinguishable from BFV. It was noted that like plasma from 75-BF-1, plasma from this marmoset contained neutralizing activity against SSV-1 (titer: 1:1,024) but relatively low (titer 1:64) reactivity against BFV, suggesting that BFV envelope antigens may be related to those of an endogenous marmoset virus to which the marmoset is tolerant.

BFV was sequentially cloned three times to test for the presence of pseudotypes resulting from phenotypic mixing, or heterozygotes. However, each of 40 focus-derived cell lines established during the cloning procedure produced virus neutralized only by low dilutions (1:32–1:64) of anti-SSV-1 serum relative to SSV-1 (1:1,024–1:2,048), suggesting that BFV represented an HL-23V recombinant. The neutralization titer vs. BFV was the same regardless of whether virus supernatants were harvested after 8 or 72 h. Further studies were aimed at characterization of the biological, immunological, and biochemical properties of BFV derived from a stock which had been sequentially cloned three times.

Unlike SSV-1 or HL-23V-infected cells, cells producing cloned BFV failed to induce syncytia with XC cells, suggesting an altered viral envelope. Likewise BFV-infected cells failed to react in a complement-dependent serum cytotoxicity test utilizing goat anti-SSV-1 serum at dilutions as low as 1:5; when titered against SSV-1-infected cells the same serum yielded 50% of maximum cytotoxicity (44%) at a 1:320 dilution. Radioimmunoassay and immunodiffusion tests demonstrated the presence of SSV-1 p30 in BFV-infected cells, but no feline, murine, or baboon endogenous virus (BaEV) p30. Neutralization and KB tests for BaEV were also negative. A difference between BFV and SSV-1 host-range was not identified. Both BFV and SSV-1 induced foci on normal marmoset and human foreskin fibroblast cell strains. Virus production was also detected in BFV-infected dog, mink, rat and a number of tumor-derived and normal human cell lines. The presence of at least a 100-fold excess of BFV nontransforming associated virus (BFAV) was demonstrated. BFAV-infected cells did not induce syncytia with XC cells, and no SSV-1-related cell-surface antigen was detected in a membrane immunofluorescence test with a 1:32 dilution of goat anti-SSV-1 serum which titered 1:1,024 against SSV-1 infected cells. Immunodiffusion tests with detergent-disrupted BFAV revealed a line of identity with SSV-1 p30. By an indirect immunofluorescence technique utilizing a neutralizing goat anti-SSV-1/SSAV-1 serum which reacts with the SSV-1 envelope proteins gp71, p12 and p15, but not with p30, cytoplasmic antigen was detected in BFAV-infected cells. As glycoprotein(s) of mammalian oncornaviruses is primarily responsible for virus neutralization, this finding suggests that BFAV contains at least one SSV-1 related minor envelope protein as well as an SSV-1 p30. When challenged with SSV-1 or GALV pseudotype of SSV-1, BFAV-infected cells exhibited complete interference to focus induction. It is possible that the presence of a minor SSV-1 envelope protein (p12 or p15) may account for BFAV interference with SSV-1 as well as the neutralization of BFV consistently observed at dilutions of 1:32–1:64 of anti-SSV-1 or anti-HL-23V sera which titer 1:1,024 or higher with SSV-1.

BFAV-infected cells grown in the presence of ^{3}H-uridine yielded labeled particles with a density in sucrose of 1.15 g/cm^{3}. When the optimum divalent cation preference of BFAV and SSV-1 reverse transcriptases was compared, incorporation of ^{3}H-TTP was maximum with both viruses at a Mn^{++} concentration of 0.1 mM in an assay utilizing synthetic poly rA·oligo dT; little or no incorporation was observed in the presence of Mg^{++}. Preliminary hybridization studies indicate that 75% of the BFAV genome is homologous with SSAV-1 and 25% unrelated (Drs. M. REITZ and N. MILLER, personal commun.). As it has been demonstrated that the gene coding for envelope glycoprotein of Rous sarcoma virus represents 20% of the Rous sarcoma virus genome [8], the preliminary BFAV hybridization data are in agreement with the immunologic and biologic evidence suggesting that BFV only differs from SSV-1 in envelope protein(s).

In summary the results indicate that the SSAV-1-related component of HL-23V recombined with another virus, possibly an endogenous marmoset virus, yielding a virus with a new envelope gene. Studies are in progress to determine the nature and origin of the BFV envelope.

Summary

Dog thymus cells chronically infected with HL-23V, a C-type virus isolated from human acute myelogenous leukemia cells, produced both transforming and nontransforming virus indistinguishable from simian sarcoma virus type 1 (SSV-1/SSAV-1) and induced fibromas in newborn marmosets. All inoculated marmosets developed anti-HL-23V antibodies. A cell line established from a tumor biopsy produced transforming virus identical to SSV-1 and HL-23V at early passages. However, at later passages the cell line and a cell line established from residual tumor tissue removed at autopsy, produced virus which was neutralized only at low dilutions of anti-SSV-1 serum (1:32) relative to SSV-1 (1:1,024). This virus (BFV) was also distinguished from SSV-1 and HL-23V by XC tests, and by membrane immunofluorescence and serum cytotoxicity tests.

References

1 CHAN, E.; PETERS, W.P.; SWEET, R.W.; OHNO, T.; KUFE, D.W.; SPIEGELMAN, S.; GALLO, R.S., and GALLAGHER, R.E.: Characterization of a virus (HL-23V) isolated from cultured acute myelogenous leukaemia cells. Nature, Lond. *260:* 266–268 (1976).

2 GALLAGHER, R.E. and GALLO, R.C.: Type C RNA tumor virus isolated from cultured human acute myelogenous leukemia cells. Science, N.Y. *187:* 350–351 (1975).

3 JOHNSON, T.R.; MASSEY, R.J., and DEINHARDT, F.: Lymphocyte and antibody cytotoxicity to tumor cells measured by a micro-^{51}chromium release assay. Immunol. Commun. *1:* 247–261 (1972).

4 MOORHEAD, P.A. and NOWELL, P.C.: Chromosome cytology; in Methods in medical research, vol. 10, pp. 313–315 (Year Book, Chicago 1964).

5 OKABE, H.; GILDEN, R.; HATANAKA, M.; STEPHENSON, J.R.; GALLAGHER, R.E.; GALLO, R.C.; TRONICK, S., and AARONSON, S.A.: Immunological and biochemical characterization of type C viruses isolated from cultured human AML cells. Nature, Lond. *260:* 264–266 (1976).

6 RANGAN, S.R.S.; UEBERHORST, P.J., and WONG, M.C.: Syncytial giant cell focus assay for viruses derived from feline leukemia and a simian sarcoma. Proc. Soc. exp. Biol. Med. *142:* 1077–1082 (1973).

7 REITZ, M.S.; MILLER, N.R.; WONG-STAAL, F.; GALLAGHER, R.E.; GALLO, R.C., and GILLESPIE, D.H.: Primate type C virus nucleic acid sequences (woolly monkey and baboon types) in tissues from a patient with acute myelogenous leukemia and in viruses isolated from cultured cells of the same patient. Proc. natn. Acad. Sci. USA *73:* 2113–2117 (1976).

8 WANG, L.-H.; GALEHOUSE, D.; MELLON, P.; DUESBERG, P.; MASON, W.S., and VOGT, P.K.: Mapping of oligonucleotides of Rous sarcoma virus RNA that segregate with polymerase and group-specific antigen markers in recombinants. Proc. natn. Acad. Sci. USA *73:* 3952–3956 (1976).

9 WOLFE, L.G.; DEINHARDT, F.; THEILEN, G.H.; RABIN, H.; KAWAKAMI, T., and BUSTAD, L.K.: Induction of tumors in marmoset monkeys by simian sarcoma virus, type 1 *(Lagothrix)*. A preliminary report. J. natn. Cancer Inst. *47:* 1115–1120 (1971).

C.M. BERGHOLZ, PhD, Department of Microbiology, Rush-Presbyterian-St. Luke's Medical Center, 1753 West Congress Parkway, *Chicago, IL 60612* (USA).

Prim. Med., vol. 10, pp. 142–148 (Karger, Basel 1978)

Studies with the Baboon Endogenous Virus and its Pseudotype of Murine Sarcoma Virus in Marmosets[1]

R.L. Heberling, S.S. Kalter, J.W. Eichberg and B.M. Levy

Southwest Foundation for Research and Education, San Antonio, Tex., and University of Texas (Dental Branch), Houston, Tex.

Introduction

Although oncornavirus particles have been observed in the placentas and other tissues of a number of different primate species, an endogenous xenotropic virus has only been isolated from baboons *(Papio cynocephalus, P. hamadryas, P. anubis, P. papio, Theropithecus gelada)* [1, 2, 4, 6–8, 12] and the squirrel monkey *(Saimiri sciureus)* [5]. The presence of type-C viral particles or proviral DNA in all baboon tissues leads to conjecture on their role in normal or pathologic processes. In order to determine the potential of the *P. cynocephalus* baboon endogenous virus (BaEV) for initiating disease, this virus was inoculated into beagle dogs and several primate species, including the chimpanzee (*Pan* sp.), baboon *(P. cynocephalus)*, cebus monkey *(Cebus apella)*, and marmosets *(Saguinus oedipus oedipus, S. fuscicollis, Callithrix jacchus)* [3]. This report concerns itself with inoculating the three marmoset species with BaEV and a BaEV pseudotype of Kirsten murine sarcoma virus (MSV[BaEV]).

Materials and Methods

Animals

The marmosets used in this study were purchased through commercial sources or derived from breeding colonies at Southwest Foundation for Research and Education (SFRE), San Antonio, Tex., and University of Texas (Dental Branch), Houston, Tex.

1 This study was supported by Contract Number N01-CP-43214, within the Virus Cancer Program, of the National Cancer Institute.

Virus

A BaEV placental isolate strain, M7 [1], was grown in fetal canine cells (Cf2Th). This BaEV infected cell culture was grown and maintained on Dulbecco's modification of Eagle's medium containing 10% fetal bovine serum. Cell-free culture fluids containing $10^{4.5}$ $TCID_{50}$/0.5 ml were used to inoculate marmosets intramuscularly (0.5 ml) and intraperitoneally (1.0 ml).

The Kirsten murine sarcoma-BaEV pseudotype was prepared by cocultivating 10^6 BaEV infected Cf2Th cells with 10^6 NP/KHOS cells. NP/KHOS is a human osteosarcoma cell line which had been transformed with Kirsten murine sarcoma virus and cloned to derive a nonproducer (S^+L^-) culture [11]. It was generously supplied by Dr. J.S. Rhim. A transforming virus (MSV[BaEV]) was recovered from the coculture and shown to possess a BaEV envelope by neutralization of focus-forming ability on mink lung cells (MvlLu) with anti-BaEV dog serum. Approximately 10^5 FFU of cell-free MSV(BaEV) virus derived by filtration of culture fluids from the persistently infected sarcoma pseudotype culture through a 0.22 μm pore size membrane or 5×10^6 MSV(BaEV) infected cells were used to inoculate marmosets subcutaneously.

Pathology

Following inoculation of BaEV or MSV(BaEV) virus or MSV(BaEV) virus infected cells, animals were frequently observed for obvious signs of clinical disease, including the development of tumors. Periodically, blood samples were collected for serum in order to determine serum neutralizing (SN) antibody titers. At the time of sacrifice or death a complete necropsy was performed at which time tissues were collected for histopathologic examination and attempted virus isolation.

Virus Isolation

Pieces of tissue mince were placed in plastic tissue culture flasks (T-75) containing Cf2Th of Mv1Lu cells for isolation of BaEV or MSV(BaEV), respectively. The Mv1Lu cocultures were observed for foci of transformed cells and the culture fluids of Cf2Th cocultures were tested for RNA-directed DNA polymerase activity by determining incorporation of ^{3}H-thymidine in the presence of the synthetic template poly(rA)·oligo$(dT)_{12-18}$ in order to detect the presence of virus.

Virus isolates were identified by neutralization with anti-BaEV dog serum. Cocultures were passaged at least five times before being considered negative.

Determination of SN Antibody

Sera were tested for BaEV antibodies by mixing 50–100 FFU of MSV(BaEV) with twofold serial serum dilutions. After 1 h at room temperature the serum-virus mixtures were inoculated onto Mv1Lu cells. After ten days the number of FFU were counted. A reduction in FFU of 80% by a serum dilution was considered positive for antibody [9]. Appropriate controls were included.

Results

Inoculation of Marmosets with BaEV

Following the combined intramuscular and intraperitoneal inoculation of BaEV, marmosets were sacrificed for attempted virus isolation and de-

Table I. Inoculation of marmosets with BaEV

Animal	Age at inoculation	SN antibody	Sacrificed (S)/death (D)
S. o. oedipus			
301–X12	3 weeks	≥1:16	8 months (D)
S. fuscicollis			
281–2357	adult	1:2	1 month (S)
281–1898	adult	<1:4	3 months (S)
C. jacchus			
57–2720	adult	<1:4	4 months (S)
57–3140	adult	<1:4	18 months (D)
57–8174	1 day	<1:4	7 months (D)
57–8162	1 day	ND	4 months (D)

ND = Not done.

tection of pathologic changes or observed until they died (table I). Although the one cotton-topped marmoset *(S. o. oedipus)* and three of the four common marmosets *(C. jacchus)* died spontaneously four to 18 months postinoculation, none of these animals nor those sacrificed (two *S. fuscicollis,* one *C. jacchus)* showed any gross or microscopic evidence of pathologic lesions. In addition, attempts to isolate BaEV from organ specimens obtained at necropsy were all negative. Organs tested included lung, liver, spleen, kidney, adrenal gland, and mesenteric lymph nodes.

When serum samples from the inoculated animals were tested for the presence of SN antibody, the *Callithrix* sera were found to be negative, one of the two white-lipped marmosets was positive at a titer of 1:2 and the cotton-topped marmoset had a titer of ≥1:16.

Inoculation of Marmosets with MSV(BaEV)

When inoculated with MSV(BaEV) pseudotype infected cells, the three marmoset species responded differently (table II). Cotton-topped marmosets developed nodules at the site of subcutaneous inoculation which were approximately 1 cm in diameter after one week. These became increasingly larger and had achieved a size of approximately 3 cm when the animals died two to three months postinoculation. In some instances the tumors had ulcerated and become necrotic with an obvious bacterial infection resulting. A bacteremia was confirmed on microscopic examination of the tissues. The

Table II. Inoculation of marmosets with MSV (BaEV)

Animal	Age of inoculation	inoculum	tumor	Time of death, months	SN antibody
S. o. oedipus					
301–MA60	adult	MSV(BaEV) cells	+(P)	2	≥1:32
301–X87	adult	MSV(BaEV) cells, NP/KHOS	+(P), –	2	≥1:32
301–3268	adult	MSV(BaEV) cells, virus	+(P), –	2	<1:4
301–3248	adult	MSV(BaEV) cells, virus	+(P), –	3	≥1:32
301–X171	1 day	MSV(BaEV) virus	–	10	ND
301–3242	adult	control	–	–	<1:4
S. fuscicollis					
281–X2	adult	MSV(BaEV) cells	+(R)	1	1:4
C. jacchus					
57–3080	adult	MSV(BaEV) cells, virus	+(R), –	–	<1:4
57–3070	adult	control	–	–	<1:4

P = Progressing, R = regressing, ND = not done.

white-lipped and common marmosets both developed nodules at the site of inoculation which regressed over a period of several weeks. The white-lipped marmoset died four weeks postinoculation with multifocal suppurative nephritis. At this time a 5-mm nodule was present at the inoculation site. The *C. jacchus* is still alive after eight months and showing no signs of infection.

Biopsies of the cotton-topped and white-lipped marmoset tumors showed them to be fibrosarcomas. In the white-lipped marmoset tumor, sheets of interwoven neoplastic fibroblasts with areas of dense and sparse cellularity were present in the deep dermis and subcutis. The underlying skeletal muscle was invaded by the proliferating fibroblasts and myocytes were frequently individualized among the neoplastic cells. Mitotic figures were more apparent in this invasion zone. Areas of hemorrhage, necrosis, and suppuration were present throughout the neoplasm. The tumors in cotton-topped marmosets were similar histologically, showing signs of liquefactive necrosis.

Histopathologic studies of necropsy specimens failed to show any signs of metastasis beyond the site of inoculation. Neither were tumors observed at the site of cell-free virus or NP/KHOS cell inoculation.

A transforming virus, identified as MSV(BaEV), was isolated from tumors of two cotton-topped marmosets (301–X87, 301–3268), but all other tissues tested were negative.

SN antibody determinations indicated that the cotton-topped marmosets produced a good humoral response to MSV(BaEV), the white-lipped marmoset developed a low level response and no antibody was detected in the common marmoset. Two control sentinel animals were also negative for antibody and tumor development.

Discussion

The existence of endogenous xenotropic primate viruses with the properties of viruses known to be oncogenic in several mammalian species raises the question of their involvement in neoplastic disease. We are testing this possibility by inoculating BaEV into various experimental animals. We have demonstrated that beagle dogs are susceptible to infection by BaEV, as evidenced by virus isolation and seroconversion, but no pathology resulted from this infection [3, 4]. Chimpanzees, baboons, and cebus monkeys failed to show any evidence of disease following inoculation with BaEV although seroconversion resulted. In the present study, three marmoset species inoculated with BaEV also failed to show signs of disease.

In an attempt to demonstrate the possible role of BaEV in neoplasia following the acquisition of sarcoma genes we have used the MSV(BaEV) pseudotype virus to inoculate beagles and primates. With this virus we were able to induce fibrosarcomas in beagles, chimpanzees, and cynomolgus monkeys, but not squirrel or cebus monkeys following subcutaneous inoculation. A baboon developed a metastatic (to the lungs) brain tumor only after intracerebral inoculation. Baboons inoculated subcutaneously only developed regressing granulomas. These findings generally reflect *in vitro* host range susceptibility [1, 4]. Our results with marmosets differ in that marmoset cell cultures are not susceptible to *in vitro* transformation, but fibrosarcoma development was noted following inoculation of MSV(BaEV) infected cells. This tumor development is clearly related to the presence of virus since NP/KHOS cells by themselves were not tumor producing. On the other hand, cell-free MSV(BaEV) virus also failed to cause tumor formation. It is quite possible that the growth of virus-infected cells *in vivo* was the result of virus-induced cell surface alterations which prevented rejection of the inoculated cells and is unrelated to the transformation and growth of marmoset cells. Unfortunately, tumor explants could not be established for karyotyping in order to substantiate this possibility due to bacterial contamination. In any event, the cotton-topped marmoset appears to be a good model for the study of tumor development by sarcoma virus infected cells.

The relative susceptibility of the three marmoset species studied is similar to that previously noted for Rous and feline sarcoma virus [10, 13, 14]. The cotton-topped marmoset developed the largest and most persistent tumors when compared to either of the other two species. At the same time, the cotton-topped marmoset also appeared to be able to mount a better humoral response to BaEV membrane antigens, perhaps due to the larger antigenic mass of the tumor formed. However, this does not explain the elevated antibody response of the cell-free BaEV inoculated cotton-topped marmoset. In any event, the data illustrate the relative ineffectiveness of SN antibody in controlling tumor growth and emphasize the importance of cell-mediated immunity in accomplishing this.

In summary, BaEV failed to show any pathologic potential for the cotton-topped, white-lipped and common marmoset, but cells infected with a BaEV pseudotype of Kirsten MSV were able to induce fibrosarcomas. In cotton-topped marmosets, these tumors persisted until the animal's death two to three months postinoculation. In the white-lipped and common marmoset, tumors developed which ultimately regressed. Cell-free MSV (BaEV) was not oncogenic. In contrast to tumor development, viral SN antibody developed best in the cotton-topped, to a limited extent in the white-lipped and not at all in the common marmoset. The results suggest the usefulness of certain marmoset species for elucidating the immunologic parameters responsible for the progressive and regressive growth of tumors resulting from virus-infected cells.

Summary

Three marmoset species *(Saguinus oedipus oedipus, S. fuscicollis, Callithrix jacchus)* failed to show evidence of infection or disease following inoculation of baboon endogenous type-C virus (BaEV). Cells infected with a Kirsten murine sarcoma-BaEV pseudotype (MSV[BaEV]) produced a progressive fibrosarcoma in *S. o. oedipus* and tumors in *S. fuscicollis* and *C. jacchus* which regressed. Tumor formation is believed to be due to growth of inoculated, virus-infected cells and not transformation of marmoset cells. *S. o. oedipus* showed the best serum neutralizing antibody response to BaEV and MSV(BaEV), *S. fuscicollis* a moderate response and *C. jacchus* no response.

Acknowledgements

The technical assistance of George Peacher and Judy Koger and histopathology support by Dr. D.M. Boenig is appreciated.

References

1 BENVENISTE, R.E.; LIEBER, M.M.; LIVINGSTON, D.M.; SHERR, C.J.; TODARO, G.J., and KALTER, S.S.: Infectious C-type virus isolated from a baboon placenta. Nature, Lond. *248:* 17–20 (1974).

2 GOLDBERG, R.J.; SCOLNICK, E.M.; PARKS, W.P.; YAKOVLEVA, L.A., and LAPIN, B.A.: Isolation of a primate type-C virus from a lymphomatous baboon. Int. J. Cancer *14:* 722–730 (1974).

3 HEBERLING, R.L.; KALTER, S.S.; BARKER, S.T.; SMITH, G.C., and HELLMAN, A.: Infection of beagles with an endogenous baboon C-type virus. Fed. Proc. Fed. Am. Socs exp. Biol. *34:* 974 (4279) (1975).

4 HEBERLING, R.L.; KALTER, S.S.; BARKER, S.T., and WEISLOW, O.S.: Isolation and biological properties of endogenous baboon *(Papio cynocephalus)* type-C viruses. Biblthca haemat., No. 43, pp. 158–160 (Karger, Basel 1976).

5 HEBERLING, R.L.; BARKER, S.T.; KALTER, S.S.; SMITH, G.C., and HELMKE, R.J.: Oncornavirus. Isolation from a squirrel monkey *(Saimiri sciureus)* lung culture. Science, N.Y. *195:* 289–292 (1977).

6 KALTER, S.S. and HEBERLING, R.L.: Isolation of C-type viruses from baboon *(Papio cynocephalus)* placental tissue. Abstr. Ann. Meet. Am. Soc. Microbiol., p. 233 (V194) (1974).

7 KALTER, S.S.; HEBERLING, R.L.; HELLMAN, A.; TODARO, G., and PANIGEL, M.: C-type particles in baboon placenta. Proc. R. Soc. Med. *68:* 135–140 (1975).

8 KALTER, S.S.; HELMKE, R.J.; PANIGEL, M.; HEBERLING, R.L.; FELSBURG, P.J., and AXELROD, L.R.: Observations of apparent C-type particles in baboon *(Papio cynocephalus)* placentas. Science, N.Y. *179:* 1332–1333 (1973).

9 PEEBLES, P.T.: An *in vitro* focus-induction assay for xenotropic murine leukemia virus, feline leukemia virus C, and the feline-primate viruses RD-114/CCC/M-7. Virology *67:* 288–291 (1975).

10 RABIN, H.: Assay and pathogenesis of oncogenic viruses in nonhuman primates. Lab. Anim. Sci. *21:* 1032–1049 (1971).

11 RHIM, J.S.; CHO, H.Y.; VERNON, M.L.; ARNSTEIN, P.; HUEBNER, R.J.; GILDEN, R.V., and NELSON-REES, W.A.: Characterization of non-producer human cells induced by Kirsten sarcoma virus. Int. J. Cancer *16:* 840–849 (1975).

12 TODARO, G.J.; SHERR, C.J., and BENVENISTE, R.E.: Baboons and their close relatives are unusual among primates in their ability to release nondefective endogenous type-C viruses. Virology *72:* 278–282 (1976).

13 WOLFE, L.G. and DEINHARDT, F.: Oncornaviruses associated with spontaneous and experimentally induced neoplasia in nonhuman primates, a review; in GOLDSMITH and MOOR-JANKOWSKI Medical primatology 1972, Part III, pp. 176–196 (Karger, Basel 1972).

14 WOLFE, L.G.; MARCZYNSKA, B.; RABIN, H.; SMITH, R.; TISCHENDORF, P.; GAVITT, F., and DEINHARDT, F.: Viral oncogenesis in nonhuman primates; in GOLDSMITH and MOOR-JANKOWSKI Medical primatology 1970, pp. 671–682 (Karger, Basel 1971).

R.L. HEBERLING, PhD, Southwest Foundation for Research and Education, Infectious Diseases and Microbiology, PO Box 28147, *San Antonio, TX 78284* (USA)

Prim. Med., vol. 10, pp. 149–155 (Karger, Basel 1978)

Differences in Expression of Surface Marker Characteristics on Epstein-Barr Virus-Transformed Human and Simian Lymphoid Cell Lines[1]

James E. Robinson, Warren A. Andiman, Earl Henderson and George Miller

Yale University School of Medicine, New Haven, Conn.

Introduction

Most human lymphoid cell lines which carry the Epstein-Barr virus (EBV) genome have characteristics of bone marrow derived (B) lymphocytes [6]. Considerable evidence has accumulated that EBV transforms human B lymphocytes *in vitro* [4–10]. EBV transforms peripheral blood leukocytes of certain non-human primates, although primate cells are generally less susceptible to transformation than are human cells [3]. We compared lymphocyte surface markers on cell lines from human umbilical cord blood leukocytes and from peripheral blood leukocytes of cotton-topped marmoset (CTM) *(Saguinus oedipus oedipus)* and woolly monkeys (WM) *(Lagothrix* sp.). Since marmoset cells transformed by EBV did not have detectable complement receptors, the hypothesis that these cell lines might arise from cells other than B lymphocytes was tested.

Materials and Methods

Cell culture. Mononuclear leukocytes from human, marmoset, and WM blood were transformed by the B95-8 strain of EBV and maintained in medium RPMI 1640 and 20% FCS as described previously [3, and Andiman and Miller, unpublished].

Leukocyte markers. Methods for identifying leukocyte surface markers are described in detail elsewhere [8]. T-lymphocytes were identified by their ability to bind neuraminidase-treated sheep erythrocytes (E-rosette forming cells, E-RFC). Receptors for IgG-Fc were

1 Supported by grants from the American Cancer Society (VC-107); US Public Health Service (CA 12055, CA16038, AI611, HD00177, CA05230).

identified by the binding of 7S antibody-coated sheep erythrocytes, EA(7S). Cells with receptors for a component of fixed complement were identified by the ability to bind EAC (sheep erythrocytes sensitized with 19S antibody and mouse complement). Phagocytic cells which ingested EAC-IgG (EAC treated with additional 7S antibody) were enumerated by examination of Giemsa-stained smears [1].

Leukocyte subpopulations. Human cord blood and CTM blood leukocyte subpopulations were derived by separating rosetted from nonrosetted cells on Ficoll-Hypaque gradients as described in detail elsewhere [8].

Transformed center assay. The number of virus-exposed cells in a population which were capable of giving rise to transformants was determined in a transformed center assay performed in microwell tissue culture plates as described in detail previously [5].

Results

Leukocyte Markers on Primary and EBV-Transformed Cells from Humans, Marmosets, and WM

Similar proportions of E- and EAC-RFC were found in fresh mononuclear leukocyte preparations from all three species (table I). However, WM leukocytes contained a larger fraction of cells which formed rosettes with EA(7S). EBV-transformed cells from each species showed different patterns of lymphocyte markers (table II). Human transformants bound EAC but not EA(7S). Transformed WM cells had a high frequency EA(7S) binding and a somewhat lower frequency of EAC binding. CTM transformants did not form rosettes with EA(7S) or EAC. None of the cell lines formed E-rosettes.

Transformation Efficiencies of Partially Purified Leukocyte Subpopulations of Human and Marmoset Leukocytes

The lack of detectable complement (EAC) receptors on marmoset transformants suggested that these cells might arise from cells which lacked complement receptors, or alternatively, that these cells had lost this receptor in the process of transformation. In order to determine the origin of these cell lines, we derived different subpopulations of human and CTM leukocytes and we determined the transformation efficiency of these cellular subsets. The results are shown in tables III–V. Human cord blood leukocytes (table III) enriched in EAC-RFC transformed with a frequency greater than or equal to the unseparated population, while removal of EAC-RFC or enrichment in E-RFC diminished the transformation efficiency tenfold.

Subpopulations of CTM leukocytes enriched in EAC-RFC (table IV) showed a four- to elevenfold increase in transformation frequency compared

Table I. Lymphocyte markers on primary mixed blood leukocytes of different species

Species	Determinations	Markers, % cells[1]		
		E	EAC	EA(7S)
Human	6	48[2] (39–60)	38 (24–44)	13 (0–28)
Woolly Monkey	5	56 (47–60)	21 (18–27)	39 (25–52)
Cotton-topped marmoset	8	43 (41–53)	37 (19–48)	9 (6–11)[3]

[1] E = Cells binding E; EAC = cells binding EAC; EA(7S) = cells binding (EA (7S).
[2] Results expressed as median (range).
[3] Three determinations.

Table II. Lymphocyte markers on leukocytes of different species transformed *in vitro* by the B95-8 strain of EBV

Species	Number of lines tested	Markers, % cells		
		E	EAC	EA(7S)
Human	6[1]	nil	89 (85–90)	rare
Woolly	7[1]	nil	27 (18–34)	63 (42–75)
Cotton-topped Marmoset	6[2]	nil	nil	nil

[1] From six individuals.
[2] From four individuals.

to the unseparated population while the population depleted of EAC-RFC failed to transform. Removal of phagocytic cells by incubation of leukocytes with carbonyl iron before separation on Ficoll-Hypaque did not affect the transformation efficiency (table V). Transformants arising from each subpopulation from both human and marmoset experiments were examined for

Table III. Transformation frequency of partially purified subpopulations of human cord blood leukocytes using transformed center assay

Population number	Leukocyte population[1]	Surface markers, %		Virus exposed[2] cells/well	Transformants[3] per attempts	TE[4]
		EAC	E			
I	mixed	42	47	100	5/24	2.0×10^{-3}
				50	3/24	2.5×10^{-3}
II	E depleted	70	4	100	12/24	5×10^{-3}
				50	5/24	4.1×10^{-3}
III	E enriched	18	81	500	2/24	1.7×10^{-4}
				100	0/24	$<1.0 \times 10^{-4}$
IV	EAC depleted	28	49	500	2/24	1.7×10^{-4}
				100	0/24	$<1.0 \times 10^{-4}$
V	EAC pellet enriched	64	14	500	16/24	1.3×10^{-3}
				100	5/24	2.0×10^{-3}
VI	E and EAC depleted	26	26	500	2/24	1.7×10^{-4}
				50	0/24	$<1.0 \times 10^{-4}$
VII	E and EAC enriched	32	57	500	16/24	1.3×10^{-3}
				100	5/24	2.0×10^{-3}

1 E = Cells binding E; EAC = cells binding EAC.
2 Multiplicity 50 particles per cell.
3 Autochthonous leukocytes used as feeder system.
4 Transformation efficiency = $\frac{\text{Number observed transformants}}{\text{Number virus exposed cells/well} \times \text{number wells plated.}}$

Table IV. Transformation frequency of partially purified subpopulations of cotton-topped marmoset leukocytes using transformed center assay

Leukocyte population	Surface markers, %			Calculated, % B cells[2]	Virus exposed cells/well	Transformants per attempts	TE[3]
	EAC	E	P[1]				
Experiment No. 1							
Mixed	37	44	25	12	2×10^4	1/48	1.04×10^{-6}
EAC depleted	14	31	8	6	2×10^4	0/48	$<1.0 \times 10^{-6}$
EAC enriched	73	5	30	43	2×10^4	4/48	4.16×10^{-6}
Experiment No. 2							
Mixed	19	43	ND	ND	2×10^4	1/48	1.04×10^{-6}
EAC depleted	1	66	ND	ND	2×10^4	0/48	$<1.0 \times 10^{-6}$
EAC enriched	50	23	ND	ND	2×10^4	11/48	1.15×10^{-6}

1 P = Phagocytic cells.
2 %B cells = % EAC −%P.
3 TE = Transformation efficiency calculated as in table III.

Table V. Effect of macrophage depletion on transformation frequency of mixed mononuclear marmoset leukocytes

Leukocyte population	Surface markers, %			Infected cells/well	Number transformed[2]
	EAC	E	P[1]		
Macrophages present	48	53	25	2×10^4	7/192[3]
Macrophages removed	39	50	3	2×10^4	7/192[3]

[1] P = Phagocytic cells.
[2] Fraction of wells showing transformation.
[3] Transformation efficiency in each case = 1.8×10^{-6}.

leukocyte markers. In every case human cells retained the ability to bind EAC but marmoset cells had no detectable markers.

We were unable to do comparable experiments on WM leukocytes, since transformation occurs at a very low and irregular rate in this species [ANDIMAN and MILLER, unpublished].

Discussion

It was of interest to find that CTM transformants lacked detectable complement receptors which are regularly found on human lymphoblastoid cell lines. This difference between human and marmoset cell lines was observed with lines derived from many individuals and following transformation with not only the B95-8 virus, but also with virus derived from two Burkitt lymphoma cell lines, Olare and Nyevu (data not shown). In addition JONDAL and KLEIN [6] could not demonstrate surface immunoglobulin with certainty on marmoset transformants. WM transformants showed a still different pattern of surface markers; namely, a high rate of binding of EA (7S) and a lower rate of EAC binding. These findings suggest a strong host cell influence on the expression of surface markers on EBV genome positive lymphoid cell lines.

Both human and CTM leukocyte subpopulations enriched in B lymphocytes which bind EAC transformed with an efficiency as great or greater than the mixed population. Conversely, removal of EAC-RFC led to a marked decrease in transformation efficiency in both species. Since the human transformants expressed complement receptors and all CTM transformants failed to do so, our data suggest that human transformants arise from B-cell precursors and retain B-cell characteristics, whereas marmoset transformants arise from B-cell precursors which bear receptors for EAC but this marker appears to be extinguished as a result of transformation. This conclusion points out the danger of inferring a cell's origin from its final characteristics. Our findings are consistent with the proposal of YEFENOF *et al.* [11] based on co-capping experiments that the EBV receptor is closely linked to the complement receptor on the B lymphocyte.

The loss of the ability to bind EAC by transformed marmoset cells may be related to the greater susceptibility to the oncogenic effects of EBV that this species exhibits [3]. Conceivably, the absence of this receptor on these cells might render them less susceptible to immune surveillance mechanisms. Alternatively, the loss of EAC binding ability may reflect a more general alteration of surface membrane properties which contributes to their oncogenic potential. At the very least, our data suggest that transformation of growth properties and transformation of cell surface properties are separate events. It is notable that several EBV genome-positive human lymphoid cell lines derived from Burkitt lymphoma also lack the ability to bind EAC [2]. This similarity to marmoset transformants may be important in further understanding how EBV might induce neoplasia in humans.

Summary

Human lymphoblastoid cell lines transformed *in vitro* by the Epstein-Barr virus (EBV) had receptors for fixed complement detectable by a rosette test. EBV transformed cells derived from cotton-topped marmoset leukocytes did not express this receptor. Evidence is presented that both human and marmoset cell lines arose from precursor cells which have complement receptors. Our findings suggest that transformation of marmoset leukocytes by EBV results in the loss of a differentiated surface marker.

References

1 EHLENBERGER, A.; MCWILLIAMS, M.; PHILLIPS-QUAGLIATA, J.; LAMM, M., and NUSSENSWEIG, V.: Immunoglobulin-bearing and complement-bearing lymphocytes constitute the same population in human peripheral blood. J. clin. Invest. *57:* 53–56 (1976).

2 EPSTEIN, A.; HENLE, W.; HENLE, G.; HEWETSON, J., and KAPLAN, H.: Surface marker characteristics and Epstein-Barr virus studies of two established North American Burkitt's lymphoma cell lines. Proc. natn. Acad. Sci. USA *73:* 228–232 (1976).

3 Frank, A.; Andiman, W., and Miller, G.: Epstein-Barr virus and nonhuman primates. Natural and experimental infection. Adv. Cancer Res. *23:* 171–201 (1976).

4 Greaves, M.F.; Brown, G., and Rickinson, A.B.: Epstein-Barr virus binding sites on lymphocyte subpopulations and the origin of lymphoblasts in cultured lymphoid cells and in the blood of patients with infectious mononucleosis. Clin. Immunol. Immunopath. *3:* 514–526 (1975).

5 Henderson, E.; Miller, G.; Robinson, J., and Heston, L.: Efficiency of transformation of lymphocytes by Epstein-Barr virus. Virology *76:* 152–163 (1977).

6 Jondal, M. and Klein, G.: Surface markers on human B and T lymphocytes. II. Presence of Epstein-Barr virus receptors on B lymphocytes. J. exp. Med. *138:* 1365–1378 (1973).

7 Pattengale, P.; Smith, R., and Gerber, P.: B-cell characteristics of human peripheral and cord blood lymphocytes transformed by Epstein-Barr virus. J. natn. Cancer Inst. *52:* 1081–1086 (1974).

8 Robinson, J.; Andiman, W.; Henderson, E., and Miller, G.: Host determined differences in expression of surface marker characteristics on Epstein-Barr virus-transformed human and simian lymphoblastoid cell lines. Proc. natn. Acad. Sci. USA *74:* 749–753 (1977).

9 Schneider, U. and Hausen, H. zur: Epstein-Barr virus induced transformation of human leukocytes after cell fractionation. Int. J. Cancer *15:* 59–66 (1975).

10 Yata, J.; Desgranges, G.; Nakagawa, T.; Favre, M., and Thé, G. de: Lymphoblastoid transformation and kinetics of appearance of viral nuclear antigen (EBNA) in cord-blood lymphocytes infected by Epstein-Barr virus (EBV) Int. J. Cancer *15:* 377–384 (1975).

11 Yefenof, E.; Jondal, M.; Klein, G., and Oldstone, M.: Surface markers on human B and T lymphocytes. IX. Two-color immunofluorescence studies on the association between EBV receptors and complement receptors on the surface of lymphoid cell lines. Int. J. Cancer *17:* 693–700 (1976).

J.E. Robinson, MD, Department of Pediatrics and Epidemiology and Public Health, Yale University School of Medicine, 333 Cedar Street, *New Haven, CT 06510* (USA)

Prim. Med., vol. 10, pp. 156–162 (Karger, Basel 1978)

Characteristics of Cell Lines Established from Epstein-Barr Virus Induced Marmoset Tumors

RUSSELL H. NEUBAUER, HARVEY RABIN, RALPH F. HOPKINS, III and BARNET M. LEVY

Viral Oncology Program. NCI Frederick Cancer Research Center, Frederick, Md., and Dental Science Institute, The University of Texas Health Science Center at Houston, Houston, Tex.

Introduction

The Epstein-Barr virus (EBV) has been strongly associated with Burkitt's lymphoma (BL) in Africa on the basis of serologic and biochemical evidence [7]. In addition, it has been shown that EBV can infect and induce lymphoid disease in New World primates, most notably in cotton-topped marmosets [2, 12, 13]. Animals inoculated with high-titered preparations of transforming virus tend to develop multiple tumors which in many respects resemble BL in man. Lymphoblastoid cell lines have been established from human tumors and much has been published concerning their relationship with EBV, their surface markers and their functional properties. In contrast, little information has been reported concerning the nature of cell lines established from EBV-induced tumors of non-human primates. In this paper we report on the establishment and characterization of cell lines derived from multiple tumors of an individual marmoset inoculated with the B95-8 strain of EBV.

Materials and Methods

Marmosets and virus inoculum. Adult, female cotton-topped marmosets *(Saguinus oedipus oedipus)* were commercially obtained and maintained at the Dental Science Institute, The University of Texas Health Science Center, Houston, Tex. for several years prior to inoculation. The monkeys were inoculated intraperitoneally with 1 ml of EBV (harvested from B95-8 cells) which contained 1.6×10^4 TD_{50} as determined by titration in human cord lymphocytes.

Virus transformation and detection of viral antigens. Serial tenfold dilutions of virus were assayed in triplicate for the ability to transform human cord lymphocytes using the procedure of MILLER and LIPMAN [10]. Cultures were refed at weekly intervals by the replacement of half volumes of fresh media and observed for evidence of transformation, i.e., increased cell growth, clumping and metabolic activity (fall in pH).

Sera positive for antibodies to EBV antigens were provided by Dr. PAUL LEVINE, National Cancer Institute. These antibodies were detected by indirect immunofluorescent microscopy and the EBV nuclear antigen (EBNA) was detected by the three-step procedure of HENLE *et al.* [4].

Cell culture and cloning. Lymphoblastoid cell lines and transformation assays were cultured in RPMI 1640 medium supplemented with 10% heat-inactivated (56 °C for 60 min) fetal bovine serum (FBS), 50 U/ml penicillin and 50 μg/ml streptomycin, and incubated at 37 °C in a humidified atmosphere at 5% CO_2 in air.

For cloning the general method of KEAY [6] was employed using RPMI 1640 medium supplemented with 20% FBS, 200 μg/ml sodium pyruvate, 2 μg/ml zinc sulfate, 2 mM L-glutamine, 50 U/ml penicillin and 50 μg/ml streptomycin.

Cytogenetics. For cytogenetic analysis, cultured cells were treated for 3 h at 37 °C with 0.05 μg/ml of colchicine. The cells were then washed, placed in a hypotonic solution and fixed in acetic methanol. Chromosome spreads were stained with Giemsa, and for Giemsa banding the general procedure of SUMNER *et al.* [14] was used. Chromosome counts were made by direct microscopic examination of 50 spreads.

Surface markers. The lymphoblastoid cell lines were examined for their ability to form nonspecific rosettes with sheep erythrocytes (SE) and for the presence of complement receptors which bind erythrocyte-antibody-complement as previously described [15], or with a combined detection assay using rosette formation with SE and zymosan activated complement C_3 complexes [9]. The percentage of markers was calculated from the rosettes observed on 300 viable cells. Receptors for the fragment crystallizable (Fc) fraction of immunoglobulin were detected using heat-aggregated (60 °C for 15 min) goat IgG and fluorescein-conjugated rabbit anti-goat IgG.

Functional products. Cells were tested for the presence of surface or released immunoglobulins as described [15]. Fluids were tested for the presence of interferon using a 50% Herpes simplex virus type 1 plaque reduction assay in vero cells. The presence of lymphotoxin was assessed by a loss of radioactively labelled cellular protein [15] from αL929 cells (obtained from Dr. GALE GRANGER, University of California, Irvine, Calif.). The cells were also examined for fibrinolytic activity, a property associated with transformed monolayer cells and expressed by many lymphoid cells [11] using a ^{3}H-fibrin plate technique.

Results

Animal Inoculation

Two female cotton-topped marmosets were inoculated intraperitoneally with approximately 1.6×10^4 TD_{50} of B95-8 strain of EBV. After 143 days one marmoset (animal No. 1605) was judged to be in a moribund condition. At necropsy, a large tumor mass was observed in the spleen, multiple tumor nodules were evident in the liver, and the mesenteric lymph nodes were enlarged. Microscopic examination confirmed the diagnosis of lymphosarcoma in all three sites. The tumors consisted of neoplastic lymphocytic cells having considerable variation in maturation. Numerous mitotic figures were evident. Some cells resembled mature lymphocytes but most were poorly differentiated anaplastic lymphoblastoid cells. Most of the tumor cells were large with a rounded nucleus and one or two large nucleoli. Nuclear chromatin was finely dispersed with course condensations along the nuclear membrane and the cytoplasm was basophilic and sparse.

Establishment of Tumor Cell Lines

Portions of tumor tissue from spleen, liver and mesenteric lymph nodes were mechanically teased into single cell suspensions and put into culture, by three and a half months suspension cell lines were established from liver and spleen tumors. The lymph node cultures grew more slowly, but after nine months in culture these cells grew as rapidly as those derived from liver and spleen tumors. A portion of kidney was also taken for culture and maintained in low passage by seeding the cells at low density and subculturing at confluency.

Cytogenetic Analysis

Karyotypic examination of the kidney culture showed the modal number (2n = 46) and distribution normal for this species [5]. Chromosome counts in the liver and spleen tumor cell lines gave a consistent hypodiploid modal number of 2n = 45, in addition, 8% of the metaphase counts of the liver cell line were in the tetraploid range. Karyotypes of cells of liver and spleen origin were indistinguishable from each other. Initial analysis of the lymph node cells showed only 23 of 50 spreads with chromosome counts of 45, however, at five and a half months 90% of these cells were 2n = 45. Karyotypes of cells from the lymph node tumors showed the loss of the same medium metacentric chromosome observed in spleen and liver tumors, in addition they consistently showed a difference in one homologue of a pair of small metacentric chromosomes. This chromosome appeared larger than its homologue and, in most metaphases showed an apparent extra band in the centromeric region.

Virus Expression

The tumor cells were tested at four months for the presence of virus-associated antigens and release of transforming virus. Each of the lines was

Table I. Characteristics of marmoset (1605) EBV-induced tumor cell lines

Feature		Liver	Spleen	Lymph node
EBV expression		+	+	+
Chromosomes				
Modal number		45	45	45
Karyotype		CL	CL	CL+M
Rosettes, % positive				
2 months	SE	<1	<1	1
	EAC	85	84	84
15 months	SE	<1	2	2
	EAC	<1	1	2
Surface Ig, % positive				
4 months	λ	40	40	40
	μ	40	10	30
	γ	30	20	20
16 montgs	λ	30	15	2
	μ	30	15	2
	γ	5	5	1
Fc receptors, % positive		100	100	100
Lymphotoxin, units/ml		>32	4	8
Interferon, units/ml		<1	<1	<1
Fibrinolytic activity % of total		76	75	79

CL = Common loss; CL+M = common loss+marker.

positive for EBV-associated internal antigens in 2–5% of the cells and >90% of the cells stained for EBNA. Culture fluids, concentrated 20-fold by ultracentrifugation, had transforming virus of $10^{1.5}$ to $10^{2.2}$ TD_{50} per 0.2 ml.

Surface Markers

The cell lines had complement receptors on approximately 85% of the cells and were essentially negative for SE receptors after two months in culture. After 15 months all three cell lines had only background levels of rosette formation (table I). Single cell clones did not have significant levels of surface receptors for either SE or activated complement. Fc receptors could be detected on virtually 100% of the cells, from each parent line, when assayed at 15 months.

Functional Products

Assays for surface immunoglobulins were first performed after four months in culture and all three lines showed high levels of surface lambda light chain and both mu and gamma heavy chains. At five months the cell lines were tested for the release of immunoglobulins by radial immunodiffusion and were found to release both types of immunoglobulins at low levels (<40 µg/ml). At 16 months the lymph node line had marked reductions in levels of lambda, mu and gamma chains. Cells of liver and spleen tumors, however, retained high levels of surface lambda and mu chains but had reduced gamma chain staining (table I). An examination of surface immunoglobulins on the clonal cell lines, likewise, showed the presence of lambda, mu and gamma chains. Upon continued cultivation, lymph node clones lost essentially all surface staining, while clones of liver and spleen tumors showed reductions in the level of surface gamma chains.

All three tumor lines were shown to have lymphotoxin activity which had the general physical-chemical properties of human lymphotoxin [8] in that it was nonsedimentable (100,000 *g* for 1 h), nondialyzable, resistant to RNase, DNase and trypsin, relatively stable at 56 °C for 30 min, but inactivated at 80 °C for 30 min. The activity was also resistant to pH 2.0 for 24 h. The cultures were not found to produce interferon. The cell lines all demonstrated fibrinolytic activity to approximately the same extent. This activity was dependent on the presence of plasminogen and was not released from cells. These results are summarized in table I.

Discussion

The inoculation of cotton-topped marmosets with high titered ($>10^4$ TD_{50}) B95-8 strain of EBV has regularly resulted in the production of lymphomas [2, 12, 13]. These lymphomas tended to be multifocal involving many organ systems. In the present report, tumors developed in the spleen, liver and in the mesenteric lymph nodes and cell lines were established from each of these sites. The cell lines were all remarkably similar to each other. EBV was present in all cell lines and each showed the same levels of virus expression. Each cell line was of a B-cell type and all cell lines expressed surface immunoglobulin, complement and Fc receptors. All cell lines were hypodiploid with 45 chromosomes and the same chromosme was lost from each line. Additionally, each cell line expressed the same type of functional products.

In man, BL is a multifocal B-cell lymphoma [16] which is nearly uniformly associated with EBV [7]. On the basis of glucose-6-phosphate dehydrogenase isozyme studies, BL is considered to be a uniclonal disease [3]. The karyotypes of EBV-induced tumors which arise in marmosets bear strong resemblances to each other, suggesting the presence of a tumor stem line which through divergence leads to similar but recognizably different karyotypes. This idea, for clonality along with supporting evidence, has been previously presented [1]. The proof of uniclonality by isozyme in marmosets is not currently possible due to the lack of information on enzyme heterozygosis in this species, and due to the limitations which must be placed on experimentation in this endangered species. It is however, apparent that EBV, which is associated with human tumors, produces a lymphoma in cotton-topped marmosets which has many of the characteristics of BL in man.

Summary

The inoculation of a cotton-topped marmoset with B95-8 strain of EBV resulted in the induction of a multifocal lymphoma and lymphoblastoid cell lines were established from liver, spleen and mesenteric lymph node tumors. The cell lines were remarkably similar to each other with respect to the presence of EBV and its expression, the surface properties of the cells and their stability, and the functional products of the cells. Karyotypic examination of the cell lines revealed the common loss of a single chromosome. The slight differences noted in karyotypes suggest some divergence from a tumor stem cell and imply a clonal origin. Thus, EBV-induced lymphomas in cotton-topped marmosets resemble Burkitt's lymphoma in man.

References

1 Atkin, N.B. and Baker, M.C.: Chromosome abnormalities as primary events in human malignant disease. Evidence from marker chromosomes. J. natn. Cancer Inst. *36:* 539–557 (1966).

2 Deinhardt, F.; Falk, L.; Wolfe, L.G.; Paciga, J., and Johnson, D.: Response of marmosets to experimental infection with Epstein-Barr virus; in de Thé, Epstein and zur Hausen Oncogenesis and herpesvirus II, part 2, pp. 161–168 (International Agency for Research on Cancer, Lyon 1975).

3 Fialkow, P.J.; Klein, G.; Gartler, S.M., and Clifford, P.: Clonal origin for individual Burkitt tumours. Lancet *i:* 384–386 (1970).

4 Henle, W.; Guerra, A., and Henle, G.: False negative and prozone reactions in tests for antibodies to Epstein-Barr virus-associated nuclear antigen. Int. J. Cancer *13:* 751–754 (1974).

5 HSU, T.C. and BENIRSCHKE, K.: In An atlas of mammalian chromosomes, vol. 4, folio 198 (Springer, New York 1970).

6 KEAY, L.: A method for cloning plasmacytoma cells by attachment to histone-coated plastic surfaces. Tissue Culture Ass. Manual *1:* 177–180 (1975).

7 KLEIN, G.: Immunological surveillance against neoplasia; in The Harvey Lectures, series 69, pp. 71–102 (Academic Press, New York 1975).

8 KOLB, W.P. and GRANGER, G.A.: Lymphocyte *in vitro* cytotoxicity. Characterization of human lymphotoxin. Proc. natn. Acad. Sci. USA *61:* 1250–1255 (1968).

9 MENDES, N.F.; MIKI, S.S., and PEIXINHOE, Z.F.: Combined determination of human T- and B-lymphocytes by rosette formation with sheep erythrocytes and zymosan-C3 complexes. J. Immun. *113:* 531–536 (1974).

10 MILLER, G. and LIPMAN, M.: Comparison of the yield of infectious virus from clones of human and simian lymphoblastoid lines transformed by Epstein-Barr virus. J. exp. Med. *138:* 1398–1412 (1973).

11 NEUBAUER, R.H.; ARMSTRONG, M.E.; LAMBERT, L., and RABIN, H.: Fibrinolytic activity associated with established primate lymphoblastoid cell lines and normal lymphocytes; in Proc. 67th Ann. Meet. Am. Ass. Cancer Res., vol. 17, p. 24 (Williams & Wilkins, Baltimore 1976).

12 RABIN, H.; PEARSON, G.R.; WALLEN, W.C.; NEUBAUER, R.H.; CICMANEC, J.L., and LEVY, B.: Comparative studies with dfferent strains of Epstein-Barr virus in owl monkeys and marmosets; in CLEMMESEN and YOHN Comparative leukemia research 1975. Biblthca haemat., No. 43, pp. 326–330 (Karger, Basel 1976).

13 SHOPE, T.; DECHAIRO, D., and MILLER, G.: Malignant lymphoma in cotton-top marmosets after inoculation with Epstein-Barr virus. Proc. natn. Acad. Sci. USA *70:* 2487–2491 (1973).

14 SUMNER, A.T.; EVANS, H.J., and BUCKLAND, R.A.: New techniques for distinguishing between human chromosomes. Nature new Biol. *232:* 31–32 (1971).

15 WALLEN, W.C.; NEUBAUER, R.H., and RABIN, H.: *In vitro* characteristics of lymphoid cells derived from owl monkeys infected with herpesvirus saimiri. J. med. Primatol. *3:* 41–53 (1974).

16 WRIGHT, D.H.: The gross and microscopic pathology of Burkitt's tumour; in BURCHENAL and BURKITT Treatment of Burkitt's tumour. UICC Monograph Series, vol. 8, pp. 14–23 (Springer, Berlin 1967).

R.H. NEUBAUER, PhD, Viral Oncology Program, Frederick Cancer Research Center, Post Office Box B, *Frederick, MD 21701* (USA)

Prim. Med., vol. 10, pp. 163–170 (Karger, Basel 1978)

Susceptibility of Marmosets to Epstein-Barr Virus-Like Baboon Herpesviruses

F. Deinhardt, L. Falk, L. G. Wolfe, A. Schudel, M. Nonoyama, P. Lai, B. Lapin and L. Yakovleva

Rush-Presbyterian-St. Luke's and University of Illinois Medical Centers, Chicago, Ill., and the Institute of Experimental Pathology and Therapy, Sukhumi

Introduction

An Epstein-Barr virus (EBV)-like herpesvirus (HVP) has recently been isolated by three laboratories from baboons with lymphoproliferative diseases [4–6, 9, 13] and also from normal baboons [4–6]. This virus appears to occur naturally in at least four different species of baboons *(Papio hamadryas, P. anubis, P. papio, P. cynocephalus)*, it has an early (EA) and viral capsid antigen (VCA) cross-reacting with EBV, but lacks a detectable EBNA-like antigen and shares only about 40% homology with EBV DNA. HVP transforms B-lymphocytes of various primate species, including at least two species of marmosets *(Saguinus oedipus* and *S. fuscicollis)*. Marmoset lymphocytes transformed by HVP *in vitro* grow as lymphoblastoid cell lines (LCL) like EVB transformed cells, they have some B-cell characteristics and are either producer or nonproducer lines carrying one to two HVP genome equivalents per cell [6]. We report now the results of inoculation of marmosets with lymphoblastoid cells transformed by HVP derived from lymphomatous or from normal baboons.

Material and Methods

Animals. Wild-caught and colony-born marmosets of three species *(S. oedipus oedipus, S. fuscicollis* and *Callithrix jacchus jacchus)* were maintained as previously described [1, 2, 14, 15]. Animals were inoculated intramuscularly in the lateral aspect of the thigh, kept in isolation cages and were monitored for tumor development, hematological values and development of antibody to HVP and EBV VCA, EA and EBV nuclear antigen (EBNA). Complete necropsies were performed on all animals which died during the study and tissues were harvested for various laboratory investigations.

Isolation of HVP, cell culture techniques, immunofluorescence assays, virus rescue and biochemical studies have all been described previously [5, 6].

Chromosomal analysis. Karyotypes were prepared and analyzed by the technique of Moorhead and Nowell [12] as modified by Marczynska *et al.* [11].

Results

Inoculation of Adult Marmosets with Lymphoblastoid Cells Producing HVP Derived from a Lymphomatous Baboon

Six adult marmosets (four *S. oedipus* and two *S. fuscicollis nigrifrons*) were inoculated intramuscularly with $1.3\text{–}5 \times 10^8$ cells of an established baboon LCL (13 CB-1) producing HVP derived from a lymphomatous baboon (2–5% VCA positive cells) (table I). All of the inoculated animals developed a moderate leukocytosis (up to 30,000 cells/mm^3) with a relative lymphocytosis beginning seven to 16 days postinoculation (PI) and with the appearance of atypical cells (Downy types 1 and 2 and occasionally also type 3) and lymphoblasts in the peripheral circulation. Three of the animals, two of which also had developed large tumors at the site of inoculation (the third had only a slight swelling at the inoculation site), developed a marked generalized lymphadenopathy and died 13–22 days PI. At necropsy necrotic tumors were present at the inoculation site and pronounced hepatosplenomegaly and generalized lymphadenopathy was found. The histopathological findings were similar in all three animals and can be summarized as follows.

Inoculation site: dense accumulations of lymphoid cells, replacing normal musculature; hemorrhage and necrosis. Spleen: lymphoid hyperplasia; massive hemorrhage and necrosis. Lymph nodes: lymphoid hyperplasia, of variable degrees of severity from node to node; basic cytoarchitecture retained. Tonsils: lymphoid hyperplasia. Thymus: lymphoproliferation in an atrophic thymus; hyperplasia or lymphoma. Liver, heart, lung, kidney, adrenal gland: mild infiltration of lymphoid cells. Brain: prominent lymphoid infiltrates in meninges, choroid plexus and surrounding parenchymal vessels as perivascular cuffs. Diagnosis: lymphoproliferative disease with cellular infiltrates widely disseminated in tissues throughout the body (primarily lymphoblastic response).

The surviving animals had developed only very slight swellings at the inoculation sites immediately after inoculation and very mild to borderline inguinal lymphadenopathy which persisted approximately from ten to 25 days PI. Low titer antibodies (1:2) to VCA were detected 19 and 22 days PI

Table I. Adult marmosets inoculated with lymphoblastoid cells transformed by HVP derived from a lymphomatous baboon

Animal				Number of cells inoculated	Days postinoculation				Remarks
No.	sex	age years	species[1]		tumor[2]	hematologic abnormalities	antibodies to VCA	survival	
Inoculated with xenogeneic HVP producing 13-CB-1 baboon cells									
KZ-1	M	8	CT	1.3×10^8	3–D	12–D	none	17	died with generalized lymphoproliferative disease (KZ-1, DF-3, 5833)
DF-3	F	5	CT	1.3×10^8	3–D	12–D	(19)	22	
5833	M	>5	CT	5×10^8	10–D	7–D	none	13	
4472	F	>5	CT	5×10^8	(10–25)[3]	7–>30	–7	>130	alive
5471	M	>5	WL	5×10^8	(10–25)	16–23	16	>130	alive
4261	F	>5	WL	5×10^8	(10–25)	7–23	16	>130	alive
Inoculated with autologous nonproducer cells transformed in vitro *by cocultivation with 13-CB-1 cells*									
70-K1	M	6	CT	3.2×10^8	7–40	9–38	?	>200	alive
4086	F	>5	CT	3.4×10^8	7–30	9–21	?	>200	alive
5751	M	>5	WL	3.9×10^8	7–16	16–21	?	200	died of intercurrent disease
5774	F	>5	WL	1.6×10^8	7–16	9–38	?	110	died from pneumonia

? = Equivocal results, see text.

[1] CT = Cotton-topped marmoset *(S. o. oedipus)*, WL = white-lipped marmoset *(S. f. nigrifrons* or *S. f. illigeri)*.

[2] Day tumor at site of inoculation and/or inguinal lymphadenopathy first observed – day last observed before complete regression. D = Day of death.

[3] Period during which minor inguinal lymphadenopathy was present without a palpable tumor at the site of inoculation.

in one of the animals which died. One of the surviving animals (4472) had preexisting EBV-VCA antibodies which increased after inoculation and the other two developed low titer VCA antibodies (1:4 to 1:8) 16 days PI.

Tumor tissue from the sites of inoculation and tissues from enlarged lymph nodes were tested for viral DNA sequences by DNA-DNA hybridization kinetics with EBV-DNA. These experiments gave negative results at the level of 0.1–1 viral genome equivalent per cell. Several LCL were established from animals KZ-1, DF-3, 5833 and 5471. All of the lines had marmoset karyotypes and some B cell characteristics (they formed EAC but not E rosettes) but none contained VCA, EA or EBNA. Preliminary data indicate that one line tested so far from animal KZ-1 contains at least one EBV-like viral genome equivalent per cell as determined by EBV cRNA-cell DNA hybridization.

Inoculation of Newborn Marmosets with Lymphoblastoid Cells Producing HVP Derived from a Lymphomatous Baboon

Eight newborn marmosets (two *S. o. oedipus,* four *S. f. nigrifrons* and two *C. j. jacchus*) were inoculated intramuscularly with $0.5–2.9 \times 10^8$ cells of the same cell line as used for inoculation of adult marmosets. These animals were observed for clinical signs of disease but because of the small size of newborn marmosets, no blood was obtained until 17–41 days PI. The animals developed only minimal to borderline inguinal lymph node swellings during the first two to three weeks PI, they developed normally and no hematological abnormalities or antibodies to VCA, EA or EBNA were observed in the few blood samples which were obtained 17–81 days PI.

Inoculation of Adult Marmosets with Autologous Lymphoblastoid Cells Transformed in vitro *with HVP Derived from a Lymphomatous Baboon*

Four adult marmosets (two *S. o. oedipus,* one *S. f. nigrifrons,* and one *S. f. illigeri*) were inoculated with $0.39–3.4 \times 10^8$ autologous lymphoblastoid cells transformed by cocultivation with X-irradiated 13 CB-1 cells from the same cell line as used for the first two experiments. The transformed marmoset cell lines had some B-cell characteristics (formation of EAC rosettes, failure to form E rosettes but no or only borderline detectable surface immunoglobulins), no VCA, EA or EBNA could be demonstrated by FA techniques but three contained EBV related DNA sequences estimated by EBV cRNA-cell DNA hybridization to be equivalent to approximately one to two EBV related viral genomes per cell; the fourth line has as yet not been tested. The cell lines were tested for viral reverse transcriptase and were found negative [6]. The animals developed small pea-sized to large diffuse tumors (inoculated thigh twice the size of the uninoculated thigh) and inguinal lymphadenopathy beginning at about 7 days PI. The tumors and lymphadenopathy persisted for two to four weeks and had regressed completely by about 40 days PI and no tumors or lymph node enlargements reappeared during the remainder of the observation period of 110 to more than 200 days PI. A slight leukocytosis with a relative lymphocytosis occurred in three of the four animals and Downy cells and lymphoblasts appeared in the peripheral circulation of all four which persisted for one to four weeks. Tests for antibodies to VCA gave only equivocal results in some sera obtained nine to 113 days PI and no cell lines could be established from mononuclear cells derived by Ficoll-Hypaque centrifugation of several blood samples obtained two to 20 weeks PI from any of the animals.

Inoculation of Adult Marmosets with HVP Producing Lymphoblastoid Cells Derived from a Healthy Baboon

Six adult marmosets (two *S. o. oedipus,* 2 *S. f. nigrifrons,* and 2 *C. j. jacchus* were inoculated with 1.2×10^9 HVP producing lymphoblastoid cells (5% VCA positive) derived spontaneously from a clinically healthy baboon. The animals did not develop any tumors or lymph node swellings, and no hematological abnormalities. One of the animals had preexisting VCA antibodies and showed an anamnestic response after inoculation and four of the five originally VCA negative animals developed antibodies to VCA 14 days PI. LCL could not be established from mononuclear cells obtained from several bleedings two to six weeks PI. These studies are still in progress and the animals remain healthy >80 days PI.

Discussion

These studies, which are still incomplete, allow the following preliminary interpretation which however will need further confirmation. The strain of HVP isolated from a lymphomatous baboon of the Sukhumi colony of Professor Lapin [5, 9, 13] induces a marked lymphoproliferative disease in adult marmosets of at least two species *(S. o. oedipus, S. f. nigrifrons)* but no measurable disease is induced with the same inocula in newborn marmosets of the same species or in newborn common marmosets *(C j. jacchus).* The acute phase of the lymphoproliferative disease in adult animals is not infrequently severe and some animals die at the height of this illness. In the surviving animals the disease is self-limited and so far no late-occurring relapses of lymphoproliferation have been observed. The exact histopathological classification of the disease is difficult; the lesions could be compared with a very strong reactive response, in less severely affected animals with infectious mononucleosis or immunoblastic lymphadenopathy [10], and in sections of some organs, lymphoma. It is interesting that newborn marmosets do not react at all or at least not in a measurable way to inoculation of similar or even larger HVP inocula as used for adult animals. This is comparable to the difference in reaction of young infants versus older children and adults to EBV. The much milder reaction of the animals inoculated with HVP transformed autologous but not virus-producing cells may be explained by the lack of virus production in the autologous cell lines, whereas the complete lack of reaction to inoculation of HVP derived from a healthy baboon (HVP-N) [4, 6] may indicate strain differences between various HVP isolates similar to differences observed between various EBV strains. The lack of reactivity to the inoculation of even greater numbers of lymphoblastoid cells transformed by and producing HVP-N as compared to inoculation of cells transformed with HVP-L, also indicates that the lymphoproliferative disease observed after inoculation of lymphoblastoid cells producing HVP-L is not only a reactive response to the inoculation of xenogeneic cells, but is more likely a virus-induced lymphoproliferation.

These results denote further the previously noted similarities between human EBV and simian HVP and it will be of interest to see if comparable biological similarities will be found also between other simian EBV-like agents [8] and human EBV. The susceptibility of marmosets to both EBV [3, 7] and HVP and the range of responses to infection with either virus, comparable to the responses of man to EBV, make the marmoset an ideal experimental host for a further elucidation of the pathogenesis of self-limited and malignant lymphoproliferative diseases associated with infection by EBV or EBV-like agents.

Summary

Epstein-Barr virus (EBV)-like herpesviruses, strains of *Herpesvirus papio* (HVP), have recently been isolated from lymphomatous (HVP-L) and from normal baboons (HVP-N). Both HVP isolates infect some species of marmosets, HVP-L inoculated adult animals develop a mild to severe and sometimes fatal lymphoproliferative disease whereas newborn marmosets of the same species do not develop a measurable disease after inoculation with HVP-L. HVP-N infects adult marmosets but does not cause disease.

Acknowledgments

This study was supported by Research contracts NOI-CP-33219 within the Virus Cancer Program of the National Cancer Institute, US Public Health Service, and Research Grant VC-185 (Natl) and 76–77 (Illinois Div.) from the American Cancer Society.

LAWRENCE FALK is a scholar of the Leukemia Society of America, Inc., ALEJANDRO SCHUDEL was a recipient of a Pan American Health Organization postdoctoral fellowship award and PATRICK LAI held a fellowship from WHO, International Agency for Research on Cancer, Lyon, France. The DNA-cRNA filter hybridizations were performed by one of us (LAWRENCE FALK) while visiting in the Laboratory of T. LINDAHL, Department of Chemistry, Karolinska Institute, Stockholm, Sweden.

We thank the Board of Health, City of Chicago, for housing most of our experimental animals and Dr. MARCZYNSKA for evaluating some of the karyotypes of the transformed cell lines.

References

1 DEINHARDT, F. and DEINHARDT, J.: The use of platyrrhine monkeys in medical research; in Some recent developments in comparative medicine. Symp. Zool. Soc. London, No. 17, pp. 127–152 (Academic Press, New York 1966).

2 DEINHARDT, J.B.; DEVINE, J.; PASSOVOY, M.; POHLMAN, R., and DEINHARDT, F.: Marmosets as laboratory animals. I. Care of marmosets in the laboratory, pathology and outline of statistical evaluation of data. Lab. Anim. Care *17:* 11–29 (1967).

3 DEINHARDT, F.; FALK, L.; WOLFE, L.G.; PACIGA, J., and JOHNSON, D.: Response of marmosets to experimental infection with Epstein-Barr virus; in Oncogenesis and herpesviruses II, vol. 1, pp. 161–168 (IARC, Lyon 1975).

4 DEINHARDT, F.; FALK, L.A.; NONOYAMA, M.; WOLFE, L.; BERGHOLZ, C.; LAPIN, B.; YAKOVLEVA, L.; AGRBA, V.; HENLE, G., and HENLE, W.: Baboon lymphotropic herpesvirus related to Epstein-Barr virus (EBV); in Abstracts of the Third Herpesvirus Workshop, p. 64 (Cold Spring Harbor Laboratory, New York 1976).

5 FALK, L.; DEINHARDT, F.; NONOYAMA, M.; WOLFE, L.G.; BERGHOLZ, C.; LAPIN, B.; YAKOVLEVA, L.; AGRBA, V.; HENLE, G., and HENLE, W.: Properties of a baboon lymphotropic herpesvirus related to Epstein-Barr virus. Int. J. Cancer *18:* 798–807 (1976).

6 FALK, L.; HENLE, G.; HENLE, W.; DEINHARDT, F., and SCHUDEL, A.: Transformation of lymphocytes by *Herpesvirus papio*. Int. J. Cancer *20:* 219–226 (1977).

7 FRANK, A.; ANDIMAN, W.A., and MILLER, G.: Epstein-Barr virus and nonhuman primates. Natural and experimental infection; in Advances in cancer research, vol.23, pp. 171–199 (Academic Press, New York 1976).

8 GERBER, P.; PRITCHETT, R.F., and KIEFF, E.D.: Antigens and DNA of a chimpanzee agent related to Epstein-Barr virus. J. Virol. *19:* 1090–1099 (1976).

9 LAPIN, G.A.; AGRBA, V.Z.; YAKOVLEVA, L.A.; SANGULIA, I.A.; TIMANOVSKAJA, V.V.; CHUVIROV, G.N., and KOKOSHA, L.V.: The establishment of lymphoblastoid suspension cell lines, containing herpes-like virus, from hemopoietic organs of hamadryas

baboons with malignant lymphoma. (In Russian.) Rep. USSR Acad. med. Sci. *222:* 244–246 (1975).

10 Lukes, R.J. and Tindle, B.H.: Immunoblastic lymphadenopathy. A hyperimmune entity resembling Hodgkin's disease. New Engl. J. Med. *1:* 292 (1975).

11 Marczynska, B.; Treu-Sarnat, G., and Deinhardt, F.: Characteristics of long-term marmoset cell cultures spontaneously altered or transformed by Rous sarcoma virus. J. natn. Cancer Inst. *44:* 545–572 (1970).

12 Moorhead, P.S. and Nowell, P.C.: Chromosome cytology; in Methods in medical research, vol. 10, pp. 313–315 (Year Book, Chicago 1964).

13 Rabin, H.; Neubauer, R.H.; Hopkins, R.F.; Dzhikidze, E.K.; Shevtsova, Z.V., and Lapin, B.A.: Transforming activity and antigenicity of an Epstein-Barr-like virus from lymphoblastoid cell lines of baboons with lymphoid disease. Intervirology *8:* 240–249 (1977).

14 Wolfe, L.G.; Ogden, J.D.; Deinhardt, J.B.; Fisher, L., and Deinhardt, F.: Breeding and hand-rearing marmosets for viral oncogenesis studies; in Breeding primates, pp. 145–157 (Karger, Basel 1972).

15 Wolfe, L.G.; Deinhardt, F.; Ogden, J.; Adams, M., and Fisher, L.: Reproduction of wild-caught and laboratory-born marmoset species used in biomedical research (*Saguinus* sp., *Callithrix jacchus*). Lab. Anim. Sci. *25:* 802–813 (1975).

F.W. Deinhardt, MD, Max v. Pettenkofer-Institut, Pettenkoferstr. 9a, *D-8000 München 2* (FRG)

Prim. Med., vol. 10, pp. 171–172 (Karger, Basel 1978)

Summary and Perspectives, Marmosets and Oncology

HARVEY RABIN

Primate Virus Immunobiology, Viral Oncology Program,
NCI Frederick Cancer Research Center, Frederick, Md.

That marmosets have been important in experimental oncology has been amply demonstrated by the presentations given in this session. Reports have covered spontaneous tumors and tumors induced by RNA and DNA tumor viruses. The remarkable susceptibility of these primates to RNA viruses of nonprimate and primate hosts has been illustrated by tumor induction studies with avian, feline, and simian sarcoma viruses. Studies with primate herpesviruses have led to the establishment of marmoset models for both T-cell and B-cell lymphomas. The studies with Epstein-Barr virus (EBV) are of particular interest due to the fact that this human virus, which is associated with neoplastic disease, produces a lymphoma in marmosets which in many respects resembles the human EBV-associated tumor, Burkitt's lymphoma.

Marmosets have been instrumental in two major areas of recent research in viral oncology. First, they have demonstrated the oncogenic potential of several viruses: simian sarcoma virus, EBV, *Herpesvirus saimiri*, and *Herpesvirus ateles*. Second, they have shown their usefulness and potential importance as experimental models for a variety of tumors. As other experimental hosts have done. they may provide new viruses, through tumor formation, by selection, induction, or recombination as may be the case with the studies of HL23 virus. Data presented here on colonic cancer indicate that these animals may represent a model for spontaneous tumors as well. Thus, there is no question as to the contributions that these animals have made.

At the present time, primate cancer virologists are faced with a dilemma concerning future work with these endangered yet valuable and, in some instances, possibly irreplaceable animals. The dual recognition of these animals as endangered species and as valuable assets to medical research forces us to face some practical and ethical considerations. The burden is to arrive at solutions which will ensure preservation of these species, avoid adverse ecologic and economic impacts, and guarantee the continuation of research.

I feel it is appropriate for all of us who are directly involved in the use of marmosets and other endangered species to adopt a policy which indicates our concern for these animals, and which pledges our commitment to use them as experimental subjects only after strict guidelines governing their use have been followed. Illustrations of such guidelines are: (a) to forbid the use of marmosets for classroom exercises, e.g., demonstration of the oncogenicity of EBV; (b) to limit the use of members of endangered species to studies only when critical to the experimental protocol, and only after careful consideration of possible alternative hosts; (c) to maximize (through careful planning) the use of every animal, e.g., through collaborative projects; perhaps this might be expedited through the establishment of an animal availability network among interested investigators; (d) to avoid the use of freshly captured animals by utilizing colony-reared animals or retired or failed breeders. There should be a commitment to long-term breeding colonies with appropriate long-term funding.

If we fail to recognize the concern of conservationist groups, with whom we sympathize, and show leadership in the intelligent utilization of these animals, we may face a time when the use of this critical resource will be prohibited. In this context, the adoption of a code of ethical use of these species is the very least we should do now. The foresight and efforts of many of our colleagues and the agencies which have supported them should be recognized. Well known among these individuals are Drs. FRIEDRICH DEINHARDT, BARNET LEVY, and N. GENGOZIAN, who were among the first to establish colonies for research purposes. Careful consideration and due regard to the interests of all those involved with and dedicated to primates and primate research should be incorporated into establishing an environment for both conservation of species, especially in the wild, and judicious research.

H. RABIN, PhD, Primate Virus Immunobiology, Viral Oncology Program, NCI Frederick Cancer Research Center, *Frederick, MD 21701* (USA)

Session IV: Immunology, Genetics, Physiology and Behavior

Chairman: N. GENGOZIAN, Oak Ridge, Tenn.

Prim. Med., vol. 10, pp. 173–183 (Karger, Basel 1978)

Immunology and Blood Chimerism of the Marmoset[1]

NAZARETH GENGOZIAN

Marmoset Research Center, Medical and Health Sciences Division,
Oak Ridge Associated Universities, Oak Ridge, Tenn.

Marmosets, small South American primates of the *Callitrichidae* family, are unique in that they show a high frequency of fraternal twinning. Twinning is characterized further by the development of placental vascular anastomoses between the fetuses, a condition that leads to hemopoietic chimerism. As demonstrated previously in our laboratory, this blood chimerism permits tissue transplantation between co-twins and to date suggests complete immunological tolerance of the host to antigens expressed by the chimeric elements. In our effort to understand and identify the factors promoting tolerance in this natural blood chimera, we have found it necessary to obtain more information on the chimerism phenomenon itself and basic data on histocompatibility antigens and immunologic responsiveness of this species. The present report summarizes the status of our knowledge in these areas and presents information suggesting that blood chimerism may play a role in the immune response capabilities of these animals.

Materials and Methods

The data were obtained from two marmoset species, *Saguinus fuscicollis* (ssp. *illigeri*) and *Saguinus oedipus* (ssp. *oedipus*). The majority of the work was done with *S.f. illigeri* and will be presented as such unless otherwise noted. The technical procedures utilized in the immunologic, cytogenetic and tissue culture systems have been given in detail else-

1 Under contract with the US Energy and Research Development Administration. Supported in part by United States Public Health Service Grant ROI AI 12007–10 from the Division of Allergy and Infectious Diseases and HL 16757–01 from Thrombosis and Hemorrhagic Diseases Branch, National Institutes of Health.

where. These may be referenced as follows: sex chromosome analysis [10], Jerne plaque technique [15], *in vitro* antibody formation [21], mitogen stimulation and mixed lymphocyte reaction (MLC) [1], and T and B lymphocyte identification [20].

Results and Discussion

Chimerism

Frequency of Occurrence

The extensive placental anastomoses between twin fetuses were described by HILL [16] and WISLOCKI [28] in an excellent series of anatomical studies on the placentation of marmosets. The suggestion by these authors that twinning was fraternal has been confirmed by several investigators who have compiled the sex ratios (1mm: 2mf: 1ff) among marmoset twins born in a colony environment. BENIRSCHKE *et al.* [4] demonstrated that the placental anastomoses were functional by finding chimerism in blood and bone marrow tissue through sex chromosome analyses of metaphase plates. These observations on chimerism were extended in our laboratory to heterosexual co-twins for all hemopoietic tissues including lymph node, spleen, thymus, bone marrow, megakaryocytes, blood lymphocytes and neutrophils [6, 9, 10, 24]. Utilization of specific reagents for red blood cell antigens [5, 13] and more recently leukocyte antigens in *S.f. illigeri* marmosets [11] has revealed chimerism in isosexual co-twins, offering additional evidence that fraternal twinning is the rule; monozygotic twinning, if it occurs, must be an infrequent event. The *consistent* development of a functional placental vascular anastomosis between the twins is predicated by our finding of chimerism in all (more than 100) heterosexual co-twin pairs examined by sex chromosome analysis of blood lymphocytes.

The finding of chimerism in single-born marmosets initially suggested that twinning invariably occurred but death of one fetus *after* establishment of hemopoiesis and vascular anastomosis led to chimerism in the viable young [10]. If, however, this were the mechanism by which all single-born marmosets were derived, we would expect to find heterosexual blood chimerism in 50% of such offspring; on the contrary, an examination of 51 single-born in our colony showed only 5 or 10% to be chimeric [7]. This latter data suggest that (a) two ova are fertilized but one occasionally fails to implant properly or dies prior to development of any vascular system through which chimerism could be established in the viable co-twin, or (b) indeed, in most cases, only a single ovum is fertilized and successfully implants. In either case, the absence of blood chimerism in single-born animals would appear to be highly probable. This was increased experimentally in our laboratory by performing unilateral Fallopian tube ligations of breeding females; prior to the operation, single births among these animals occurred with a 10% frequency (2/20 deliveries); after surgery, the frequency increased to better than 60% (28/44 deliveries) [7]. In none of the latter cases, using both sex chromosome and a red cell antigen as markers, could blood chimerism be found. Although the number of phenotypic or genetic markers are still too limited to completely rule out blood chimerism in this select group, for the moment unilateral ligations appear to be the best method to procure a marmoset lacking blood chimerism. There are, however, two natural situations in which one can be reasonably assured of identifying such animals: (a) the complete disappearance of chimerism in an animal known to have had two populations of cells, an observation already made among animals in our colony; and (b) in triplets, when there are two males (or females) and one female (or male) and cells of the opposite sex cannot be found in the latter animal, chimerism may be presumed to be absent.

Percentage Chimerism and Alterations with Time

We had previously noted an inverse relationship in the percentage chimerism between co-twins, i.e. if the male co-twin has 60% male cells, the female co-twin will have approximately 40% female cells [6, 10]. Viewed in another way, each animal is shown to have the same proportion of male and female cells, 60 and 40%, respectively. While this relationship is not always exact, a similarity in the proportion of male and female cells has occurred in approximately 70% of the heterosexual co-twin pairs examined [6]. The degree of chimerism among such animals has ranged from as low as 1% to as high as 99%, with these levels being maintained for periods of more than five years. As indicated above, however, significant changes in chimerism have been observed in a few animals, even to the extent of complete disappearance of the chimeric (co-twin's) cells. For example, one animal that had maintained 30–36% chimerism (by sex chromosome analyses of over 175 cells) over a period of one year, showed a gradual decline during the next two years and then a complete disappearance of the foreign cell population in the succeeding 12 months. Upon its death, cytogenetic analysis of 475 cells obtained from the blood, bone marrow (direct and indirect), spleen and lymph node tissues showed the absence of the chimeric cells. Such observations, however, have been limited to a few animals, while the majority show fluctuations ranging from 10 to 40% over a period of several years. Thus, the disappearance of blood chimerism or a consistent trend in one direction (increase or decrease) is an infrequent happening and a stable or fluctuating level of chimerism is the rule [6, 8].

Most of our studies on chimerism have focused on blood lymphocytes from heterosexual chimeras; *in vitro* stimulation with plant lectins provided cells for sex chromosome analysis. The percentage chimerism in blood lymphocytes is the same as that found in stimulated and nonstimulated bone marrow preparations [10]. These data suggested but did not prove that the percentage chimerism would be the same for all hemopoietic elements. With a few exceptions in the erythrocytic series, this has indeed been found to be the case [13]. A significant finding relating to the question of hemopoietic chimerism has been the demonstration of an equivalent percentage chimerism among the T (thymus-derived) and B (bone marrow-derived) lymphocyte populations. Using *in vitro* separation procedures on blood lymphocytes followed by appropriate mitogenic stimulation, we have shown that the percentage chimerism (by sex chromosome analysis) in these two cell populations is essentially identical [19]. This observation not only suggests a common origin for cells found in the various lymphoid organs but also supports the current concept in immunology that T lymphocytes are derived from a stem cell pool such as the bone marrow, and acquire T cell characteristics under influence of the thymic epithelium. Thus, chimerism in the peripheral T-cell population would not be expected if thymocytes originated *in situ* in the thymic gland.

Immunologic Responsiveness and Chimerism

In vitro *Studies*

Any question concerning the immunologic capabilities of the marmoset must take into consideration the dual lymphocyte populations in this animal and consequently the relative contributions each makes in any response situation. If cohabitation is mutually compatible, we may anticipate full expression by each cell type to any antigenic challenge. Whether the marked alterations in chimerism observed in a few animals is reflective of immunologic incompatibilities or is due to other unidentified physiologic stress factors is not known; the observation alone, however, dictates that we should not tacitly accept the thesis that complete immunologic tolerance exists in this chimeric model. In any event, it was felt that some answer to these questions may be forthcoming from an analysis of the relative involvement of the two populations in cell-mediated and humoral immune response studies.

The one-way mixed lymphocyte culture (MLC) reaction with cells from heterosexual chimeras was used to determine whether the animal's own or chimeric population of cells responded to mitomycin-treated allogeneic or xenogeneic white blood cells (WBC). Sex chromosome analyses of metaphase plates obtained from such cultures revealed the percentage chimerism among the responding cells to be no different from that of cells that had been stimulated with phytohemagglutinin or concanavalin A [18, 19]. Thus, when either T-enriched populations or unseparated WBC from a chimera were used in the MLC reaction, the blastogenesis obtained with both cell populations following stimulation with histocompatibility antigens was equivalent to that obtained with a nonspecific plant lectin. These data would suggest that the chimeric status at least does not interfere with the responsiveness of these cells to alien antigens; whether, however, this mitotic activity culminates in the development of cytotoxic cells and whether both cell populations would realize the same potential, is not known.

A similar question was posed for T and B cell cooperation leading to antibody formation by a chimeric population of cells. To identify the lymphocytes producing the antibody, however, it was necessary to first obtain a reagent capable of reacting with specific histocompatibility antigens of one of the cell types in the chimeric population. Extensive isoimmunizations within one subspecies of *S. fuscicollis, S.f. illigeri,* ultimately yielded an antiserum which, when combined with a Coombs reagent (anti-marmoset globulin) and complement, was 90–95% cytotoxic for WBC from some *S.f. illigeri* marmosets and completely nonreactive with others [11]. Extensive screening of animals revealed some to have two populations of cells, one sensitive to the isoimmune reagent, and the other viable after treatment. These observations and subsequent breeding data indicated that the reagent was specific for a leukocyte antigen (MLA-1) which was inherited in a simple Mendelian dominant fashion. More importantly, it would permit *in vitro* identification of antibody producing cells in a chimeric mixture. Blood lymphocytes (see below) from several animals known to have MLA-1 positive and MLA-1 negative cells were stimulated *in vitro* with sheep RBC and seven days later the cultures tested for plaque-forming cells. Treatment of two aliquots from each culture, one with normal serum and the other with the isoimmune reagent in the presence of complement, revealed rather divergent and unexpected results. Thus, while most showed both cell types of the chimera pool to be producing plaque-forming cells, in a few instances only the host or the chimeric population responded to the test antigen [11]. The explanation for this dichotomy of function in the cells from some of the chimeric animals is not readily apparent. Analogous studies performed in tetraparental mice having chimeric populations derived from low and high responder type mouse strains have shown that these animals may not form antibody to synthetic antigens reflective of the genetic origin of their lymphocytes [2, 3, 27]. To draw a relationship of these findings to that in the marmoset, however, would be highly speculative at this time. For the moment, at least, it would appear that cell-mediated immunity (i.e. T-cell response in the MLC reaction) and humoral antibody formation may not be functioning at equivalent levels with respect to participation of both cell types in the chimera pool.

A second observation pertaining to chimerism and immune responsiveness concerned the ability to obtain antibody formation by blood leukocytes of this species when cultured with an antigen. We had found that *in vitro* cultivation of leukocytes with sheep RBC for five to 11 days yielded highly significant numbers of plaque-forming cells [21]. The relative ease with which we were able to do this prompted us to speculate that natural blood chimerism of the marmoset may be playing a role in the response. Thus, the 'allogeneic effect' phenomenon, a heightened or synergistic type of immune response with artificial allogeneic cell mixtures, *in vitro* or *in vivo,* is well documented for other systems. A chimeric population of lymphocytes from the marmoset, albeit compatible in its *in vivo* environment, may under *in vitro* conditions yield a subtle yet unrecognizable allogeneic reaction which would facilitate any potential response to a foreign antigen such as sheep RBC. As one test of this hypothesis we attempted to stimulate cells from known nonchimeric marmosets. These consistently failed or gave only marginal responses. Surprisingly, however, we could demonstrate antibody formation only with leukocytes from chimeric *S. fuscicollis* marmosets, not *S.o. oedipus,* even though chimeric animals of this species were also used. The relative immune competence of the two marmoset species may bear on this point. *S.o. oedipus* marmosets have shown a consistently lower immune capacity relative to that of *S. fuscicollis* when challenged with a variety of antigens; this observation may explain the lack of any response by their blood leukocytes, chimerism notwithstanding. This species differential is further emphasized by failure of *S.o. oedipus* spleen cells, but not *S. fuscicollis* cells, to respond *in vitro* to sheep RBC antigen [22]. We suggest, therefore, that with respect to the blood leukocyte studies, the innate response potential of the lymphocyte populations is the major factor but chimerism provides a significant nonspecific catalyst for the development of cellular events necessary for antibody formation *in vitro*.

In vivo *Studies*

As yet we have no direct data bearing on the influence of blood chimerism on the immune response following *in vivo* challenges with various antigenic stimuli. Tolerance among co-twins to histocompatibility antigens expressed on reciprocally exchanged skin or kidney grafts was expected and observed in our laboratory [12, 14, 23]. The rejection of primary skin allografts among unrelated marmosets of the same species, however, was much slower than that observed in other mammalian and primate species, e.g. skin graft survival greater than 100 days in *S.o. oedipus* and beyond 60 days in *S. fuscicollis* [14]. Initially, two interdependent factors, biologic and geographic, were cited to account for the extended survival time. Biologically, the marmosets' natural blood chimerism provides each animal with two sets of transplantation antigens recognized as 'self': its own and that of its co-twin. Such ubiquitous chimerism in a random-bred population increases the possibility of shared major transplantation antigens and consequently a decrease in incompatibility among its members. Geographically, each marmoset species has evolved in a fairly restricted area and finite pooling of genes for histocompatibility antigens is likely. Although these observations may explain the long survival times of intraspecies grafts, it should be noted that even in interspecies situations, e.g. *S. fuscicollis* vs. *S.o. oedipus,* graft survival greater than 30 days has been obtained [14]. Whether this reflects (a) the influence of chimerism via 'closely related histocompatibility antigens' of the graft, (b) a more basic effect of chimerism on allograft responsiveness, or (c) merely the mode and tempo of allograft responsiveness characteristic of this species independent of any chimerism phenomenon, is not known.

Studies on the humoral immune response of *S. fuscicollis* and *S.o. oedipus* marmosets to three commonly used antigens, sheep RBC, *Salmonella typhi* flagella, and *Escherichia coli* lipopolysaccharide revealed marked differences between the two species. With each antigen, independent of the dose, route and number of injections, *S. fuscicollis* animals responded with significantly greater titers (fig. 1) [15, 17]. The lower immune response of *S.o. oedipus* marmosets is all the more striking when a comparison is made to data obtained in other more commonly used laboratory animals challenged with the same antigens. Although such species comparisons may not be justified and certainly cannot form a basis for classifying animals as 'immuno-

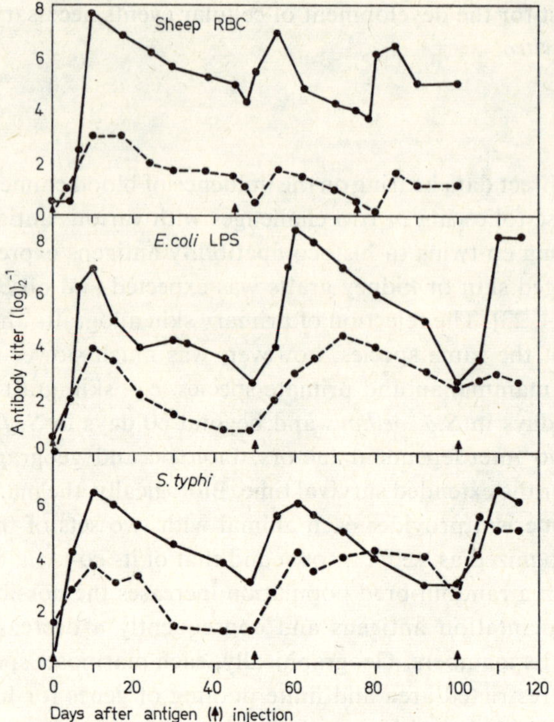

Fig. 1. Primary, secondary and tertiary antibody response of *S. fuscicollis* (–) and *S. oedipus* (---) marmosets to sheep red blood cells (RBC), *E. coli* lipopolysaccharide (LPS) and *S. typhi* flagella. Groups of four to five marmosets were utilized for each of the three antigens used. Data adapted from GENGOZIAN *et al.* [15] and KATELEY and GENGOZIAN [17].

incompetent', it is recognized that the antigens used in these studies are known to be highly immunogenic in other animals. That the low immune response profile of *S.o. oedipus* may not be due merely to *in vivo* processing of antigen, e.g. antigen distribution, retention, etc., is indicated by our *in vitro* studies with spleen cells of this species. Primary *in vitro* antibody formation to sheep RBC with such tissue has been essentially negative (see above), while under similar culture conditions, *S. fuscicollis* spleen cells responded as well (on a cell dose basis) as those of mice. Before one can conclude that *S.o. oedipus* is incompetent, if only with respect to *S. fuscicollis*, it must be noted that the blood lymphocyte response of these two species to a plant lectin (concanavalin A) and xenogeneic WBC stimuli (one-way

MLC) were almost identical. Thus, the mean stimulation indexes (SI) with concanavalin A and xenogeneic antigens for *S.o. oedipus* were 30.0 (SE± 10.8; n = 14) and 10.2 (SE±4.2; n = 15), respectively, and with the same stimuli, cells from *S. fuscicollis* marmosets showed SIs of 30.0 (SE±11.9; n = 14) and 11.4 (SE±4.6; n = 15), respectively [18]. The latter assays are indicative of a T-cell function, which is to be contrasted to the *in vivo* challenges for antibody formation (B cells) with T-dependent and T-independent antigens. It is conceivable, therefore, that B cell function or its collaborative steps with helper T cells in eliciting humoral antibody may be impaired in *S.o. oedipus* when compared to *S. fuscicollis;* alternatively, macrophage or accessory cell processing of antigens may be abnormal. These and other possible explanations are currently under investigation in our laboratory.

Excepting the demonstration of co-twin tolerance, in none of the *in vivo* studies cited can we state that chimerism is exerting a positive or negative effect on the immune mechanism of marmosets. In other chimeric animal models, however, the influence of chimerism is apparent. Most notable is the radiation allogeneic bone marrow chimera, in which T and/or B cell functions have been found to be abnormal; in some mouse strain combinations, the defect may be expressed greater in one cell type than the other [25, 26]. Admittedly, comparisons of the immune response capabilities of these animals to that of marmosets must be made with reservations since the methods of establishment and degree of chimerism are quite different. However, even in the tetraparental animal, the laboratory model which comes closest to the marmoset in mechanism of chimerism induction and degree of chimerism, there are data which suggest a subtle but definite alteration in the responses engendered by the chimeric populations of these animals. It is of interest to note here also that both the antigens and the genetic strain combinations used to create the tetraparental animal are important; thus, normal [3] and subnormal [2, 27] responses have been reported. These observations, therefore, dictate that we should not preclude the concept that chimerism *per se* may be a major factor in the marmosets' immune response profile.

References

1 Barnhart, D.D. and Gengozian, N.: An evaluation of the mixed lymphocyte culture (MLC) reaction in marmosets. Transplantation *20:* 107–115 (1975).

2 Bechtol, K.B. and McDevitt, H.O.: Antibody response of C3H↔(CKB × CWB)F_1

tetraparental mice to poly-L(Tyr,Glu)-poly-DL-Ala—poly-L-Lys immunization. J. exp. Med. *144:* 123–144 (1976).

3 BECHTOL, K.B.; WEGMANN, T.G.; FREED, J.H.; GRUMET, F.C.; CHESEBRO, B.W.; HERZENBERG, L.A., and MCDEVITT, H.O.: Genetic control of the immune response to (T,G)–A—L in C3H←→C57 tetraparental mice. Cell. Immunol. *13:* 264–277 (1974).

4 BENIRSCHKE, K.; ANDERSON, J.M., and BROWNHILL, L.E.: Marrow chimerism in the marmosets. Science *138:* 513–515 (1962).

5 GENGOZIAN, N.: A blood factor in the marmoset, *Saguinus fuscicollis.* Its detection, mode of inheritance and species specificity. J. med. Primatol. *1:* 276–290 (1972).

6 GENGOZIAN, N.: Male and female cell populations in the chimeric marmoset; in GOLDSMITH and MOOR-JANKOWSKI Proc. 2nd Conf. Exp. Med. Surg. in Primates, New York 1969, pp. 165–175 (Karger, Basel 1971).

7 GENGOZIAN, N. and BATSON, J.S.: Single-born marmosets without hemopoietic chimerism: naturally occurring and induced. J. med. Primatol. *4:* 252–261 (1975).

8 GENGOZIAN, N. and BATSON, J.S.: In preparation.

9 GENGOZIAN, N.; BATSON, J.S., and EIDE, P.: Hematologic and cytogenetic evidence for hematopoietic chimerism in the marmoset, *Tamarinus nigricollis.* Cytogenetics *3:* 384–393 (1964).

10 GENGOZIAN, N,; BATSON, J.S.; GREENE, C.T., and GOSSLEE, D.G.: Hemopoietic chimerism in imported and laboratory-bred marmosets. Transplantation *8:* 633–652 (1969).

11 GENGOZIAN, N.; BAZZELL, S.J., and NICKERSON, D.A.: Histocompatibility antigens and antibody-forming cells in the marmoset, a natural blood chimera. Abstract. Fed. Proc. Fed. Am. Socs exp. Biol. *35:* 353 (1976).

12 GENGOZIAN, N.; LEE, S., and PORTER, R.P.: In preparation.

13 GENGOZIAN, N. and PATTON, M.L.: Identification of three blood factors in the marmoset, *Saguinus fuscicollis* ssp. Medical Primatology 1972. Proc. 3rd Conf. Exp. Med. Surg. in Primates, Lyon, pp. 349–360 (Karger, Basel 1972).

14 GENGOZIAN, N. and PORTER, R.P.: Transplantation immunology in the marmoset; in GOLDSMITH and MOOR-JANKOWSKI Proc. 2nd Conf. Exp. Med. Surg. in Primates, New York 1969, pp. 165–175 (Karger, Basel 1971).

15 GENGOZIAN, N.; SALTER, B.L.; BASFORD, N.L., and KATELEY, J.R.: Characterization of the antibody response of the marmoset to sheep red blood cells. Clin. exp. Immunol. *23:* 525–536 (1976).

16 HILL, J.P.: II. Croonian Lecture. The developmental history of the primates. Phil. Trans. R. Soc. Lond. *221B:* 45–178 (1932).

17 KATELEY, J.R. and GENGOZIAN, N.: Marmoset species variations in the humoral antibody response to two bacterial antigens (in preparation).

18 KATELEY, J.R.; NICKERSON, D.A., and GENGOZIAN, N.: Blastogenic response of marmoset lymphocytes: cytokinetics and identification of responsive cells. Immunology (in press).

19 NIBLACK, G.D.; KATELEY, J.R., and GENGOZIAN, N.: T and B lymphocyte chimerism in the marmoset. Immunology *32:* 257 (1977).

20 NIBLACK, G.D. and GENGOZIAN, N.: T and B lymphocytes in the marmoset; a natural hemopoietic chimera. Clin. exp. Immunol. *23:* 536–543 (1976).

21 NICKERSON, D.A. and GENGOZIAN, N.: Primary *in vitro* antibody formation by blood leucocytes of a sub-human primate. Cell. Immunol. *27:* 171–176 (1976).

22 Nickerson, D.A. and Gengozian, N.: Unpublished.
23 Porter, R.P. and Gengozian, N.: Immunologic tolerance and rejection of skin allografts in the marmoset. Transplantation *8:* 653–665 (1969).
24 Porter, R.P. and Gengozian, N.: *In vitro* proliferation and differentiation of hemic precursor cells from marrow and blood of naturally chimeric marmosets. J. cell. Physiol. *79:* 27–41 (1972).
25 Urso, P. and Gengozian, N.: T-cell deficiency in mouse allogeneic radiation chimeras. J. Immun. *111:* 712–719 (1973).
26 Urso, P. and Gengozian, N.: Variation in T and B cell deficiency in different allogeneic radiation chimeras. J. Immun. *113:* 1770–1778 (1974).
27 Warner, C.M.; Fitzmaurice, M.; Maurer, P.H.; Merryman, C.F., and Schmerr, M.J.: The immune response of tetraparental mice to two synthetic amino aicd polymers: 'high-conjugation' 2,4-dinitrophenyl-glutamic acid^{57}-lysine^{38}-alanine^{5} (DNP-GLA^{5}) and glutamic acid^{60}-alanine^{30}-tyrosine^{10} (GAT^{10}). J. Immun. *111:* 1887–1893 (1973).
28 Wislocki, G.B.: Observations on twinning in marmosets. Am. J. Anat. *64:* 445–483 (1939).

N. Gengozian, Marmoset Research Center, Medical and Health Sciences Division, Oak Ridge Associated Universities, *Oak Ridge, TN 37830* (USA)

Prim. Med., vol. 10, pp. 184–192 (Karger, Basel 1978)

Lymphocyte Functions and Subpopulation Distribution in Marmosets

W.C. Wallen, A.P. Claysmith and J.L. Cicmanec

National Institute of Neurological and Communicative Disorders and Stroke, National Institutes of Health, Bethesda, Md., and Litton Bionetics, Inc., Kensington, Md.

Introduction

Marmosets have proven particularly useful in investigations of oncogenic viruses [14, 16] as well as in studies of immunologic parameters associated with mechanisms of transplant rejection and immunologic tolerance [11, 12]. They are capable of rejecting heterologous skin grafts [11], exhibiting delayed cutaneous hypersensitivity to DNFB [2], and participating in *in vitro,* cell-mediated immune responses [3, 6]. However, an extensive examination of marmoset lymphocyte function and subpopulations distributions has not been performed.

In this study, we examined five species of marmosets for lymphoproliferative responses to general mitogens, ability to function in direct cytotoxicity (T-cell) following mitogenic activation, and ability to react in the antibody-dependent lymphocyte cytotoxicity (ADLC) assay (K-cell). We also examined distribution of B- and T-lymphocyte subpopulations.

Materials and Methods

Animals

Sixteen marmosets including at least three primates from each of five species were employed in this study. Species studied were *Saguinus nigricollis* (4), *S. fuscicollis* (3), *S. oedipus* (3), *S. weddelli* (3) and *Callithrix jacchus* (3). Samples of 3 ml of heparinized blood were drawn biweekly from each monkey for these studies.

Lymphocyte Stimulation Assay

The lymphoproliferative response to general mitogens was determined by procedures previously described [15]. Briefly, lymphocytes were separated from peripheral blood placed in microtiter plate wells for 72 h at 2×10^5 cells/0.1 cm_3 (Linbro Scientific Co., Hamden, Conn.). Phytohemagglutinin (PHA; Burroughs-Welcome, Greenville N.C.) and Con A (Calbiochem, San Diego, Calif.) were used at 10 and 25 μg/ml. Pokeweed mitogen (PWM; Difco, Detroit, Mich.) was employed at dilutions of 1:20 and 1:80 of stock. Sixteen hours prior to termination, each culture received 1 μCi of tritiated thymidine (^{3}H-TdR; New England Nuclear, Boston Mass., sp. act. >6.0 Ci/mM). The data is presented as mean counts per minute (cpm) of triplicate samples from the mitogen dose giving maximum stimulation and as stimulation indexes (SI) which are determined as the ratio of the cpm in test cultures to the cpm in control cultures. Difference between the groups was determined by using the student's 't' test at 99% confidence. All means showed less than 10% standard error.

Direct Cytotoxicity

Activated marmoset lymphocytes (PHA, 10 μg/ml for 16 h) were washed and placed in flat-bottom microtiter plates in 0.1-cm^4 volumes at three attacker to target cell ratios (100:1, 20:1 and 4:1). Target cells (Raji cells) were labeled with ^{51}Cr (Amersham-Searle, Arlington eights, Ill., sp. act. >6.4 Ci/mM) for 30 min, washed (×3) and diluted to 2×10^4/ml in culture medium. Target cells (0.1 cm^3) were mixed with attacker cells and incubated at 37 °C for 4 h. The supernatant fluid was harvested and counted in a gamma spectrometer to determine the amount of radioactive ^{51}Cr released. Cytotoxicity was determined by the following formula:

$$\% \text{ Cytotoxicity} = \frac{\text{cpm (Ly)} - \text{cpm (spontaenous)}}{\text{cpm (total)} - \text{cpm (spontaneous)}} \times 100$$

where cpm (Ly) is cpm released by the target-lymphocyte mixture, cpm (spontaneous) is released by target cells alone over the 4-hour incubation period, and cpm (total) is total released ^{51}Cr following three cycles of freeze-thawing of the labeled target cells. Significant cytotoxicity was determined by comparing the means of ^{51}Cr released by PHA-treated lymphocytes against untreated lymphocytes at the same ratios by the student's 't' test with 99% confidence.

Antibody-Dependent Lymphocyte Cytotoxicity

This assay was performed as previously described [13]. Briefly, Raji cells, superinfected with Epstein-Barr virus (30% membrane antigen positive cells by immunofluorescence), were used as target cells and labeled with ^{51}Cr as described above. Convalescent serum from a patient with infectious mononucleosis (ADLC titer = 1:1,280) was employed at a dilution of 1:100. Three ratios of attacker to target cells were employed (100:1, 20:1, 4:1). Significant cytotoxicity was determined as previously described [13] utilizing student's 't' test at 99% confidence by comparing the amount of ^{51}Cr released from the target cells in the presence of serum plus lymphocytes compared to targets plus serum alone.

E-Rosettes

T-lymphocytes were quantitated by the procedure of Niblack and Gengozian [10]. Briefly, equal volumes (0.25 ml) of neuraminidase (20 U/ml) treated sheep red blood cells (SRBC) and lymphocytes (1×10^6 cells/ml) were incubated together at 37 °C for 5 min, gently centrifuged and held at 4 °C overnight. Viable cells binding three or more SRBC were considered as rosette-forming cells. At least 250 cells were counted from each sample.

EAC Rosettes

To detect the activated complement (C_3) receptor, bovine red blood cells (E), rabbit anti bovine E-IgM, prepared as described by KRAMMER *et al.* [8] and mouse serum were used to form the EAC complex. Equal volumes (0.25 ml) of EAC suspension and lymphocytes were mixed, gently centrifuged and incubated at 37 °C for 30 min. Rosettes were counted as above.

Membrane Immunofluorescence

To quantitate surface immunoglobulin (SIg) positive cells, the procedures of LOBO *et al.* [9] were used. They defined two populations of cells bearing easily detectable surface immunoglobulin – B-cells and L-cells. L-cells were defined as those containing labile IgG determinants while B-cells had stable immunoglobulin determinants. Lymphocytes were incubated at 4 or 37 °C for 30 min in media containing 1% pooled marmoset serum, washed (×3) at the same temperature, and stained at 4 °C with fluorescein-conjugated polyvalent antihuman γ-globulin (μ, γ, α specific, Hyland Laboratories, Costa Mesa, Calif.). Cells were counted with an AO fluorescent microscope. L-cell values were determined by subtracting the number of B-cells observed after incubation at 37 °C from the total number of fluorescent positive cells held at 4 °C. Monocytes were identified by latex particle phagocytosis and excluded from the counts.

Fc Receptor

The procedure of DICKLER [4] was used to quantitate Fc receptor cells. Cells were incubated at 37 °C for 30 min in medium containing 1% pooled marmoset serum, washed (×3) at 37 °C, and treated with 0.1 cm^3 of FITC-conjugated human IgG aggregates (10 mg/ml) and incubated at 37 °C for 30 min. The cells were washed (×2) and percent stained cells was estimated by counting 250 cells using an AO fluorescent microscope.

Table I. Response to general mitogens by lymphocytes from several marmoset species

		PHA		Con A		PWM	
		X̄cpm	index	X̄cpm	index	X̄cpm	index
Saguinus nigricollis							
Sn 949H	1	33,036	29.7	19,165	14.2	23,749	21.3
	2	33,132	100.6	35,775	108.6	12,960	39.4
Sn 979I	1	61,100	15.6	31,805	8.1	13,654	3.5
	2	183,224	422.2	98,873	227.8	43,786	100.9
Sn 125K	1	281,929	20.1	210,676	15.0	219,213	15.6
	2	115,881	243.7	62,527	131.5	33,329	70.1
Sn 590I	1	86,951	49.4	41,988	23.9	44,381	25.2
	2	154,320	22.7	183,049	26.9	160,931	23.7

Table I (continuation)

		PHA		Con A		PWM	
		X̄cpm	index	X̄cpm	index	X̄cpm	index
Saguinus fuscicollis							
Sf 957I	1	73,687	61.5	45,412	37.9	31,076	25.9
	2	225,024	35.4	190,060	24.9	94,007	17.3
	3	221,568	343.0	150,642	233.2	51,796	80.2
Sf 124K	1	44,703	15.0	55,477	16.3	39,527	14.5
	2	38,548	30.2	27,212	21.3	13,699	10.7
Sf 251I	1	184,165	42.2	119,125	27.3	158,580	36.4
	2	123,183	32.1	75,372	23.5	64,116	11.5
	3	167,312	502.4	130,619	392.2	55,730	167.3
Saguinus oedipus							
B 7110	1	131,709	43.1	73,631	27.6	135,993	44.3
	2	140,561	28.8	145,692	19.5	91,504	22.2
	3	235,933	55.5	254,910	59.9	156,981	30.5
B 7219	1	181,524	20.7	71,220	8.1	121,769	13.9
	2	191,372	59.1	157,586	50.5	140,328	46.1
	3	310,801	58.2	290,958	54.5	191,039	35.8
B 6919	1	45,339	21.3	31,870	15.0	29,700	14.0
	2	177,806	73.9	143,258	59.5	123,521	51.3
	3	145,350	44.9	159,507	49.0	62,980	19.3
Saguinus weddelli							
Sw 4821	1	50,002	56.4	31,332	35.4	30,602	34.5
	2	190,753	22.3	187,821	22.0	112,808	13.2
Sw 500J	1	141,892	40.0	36,410	8.0	110,092	24.2
	2	194,746	68.3	304,562	106.8	116,803	41.0
Sw 518J	1	116,605	39.8	272,755	93.3	217,090	74.2
	2	237,290	27.8	256,591	29.8	143,354	16.7
Callithrix jacchus							
B 7305	1	40,382	24.2	50,034	29.9	31,907	19.1
	2	329,265	26.8	252,788	20.6	261,012	21.3
	3	216,709	380.2	87,374	153.3	167,113	293.2
B 7281	1	76,692	42.8	41,824	24.4	24,267	13.6
	2	339,454	59.2	220,305	38.4	219,654	38.3
	3	75,228	200.9	10,304	27.5	25,678	68.6
Cj 369J	1	104,134	70.7	90,025	68.0	45,403	34.3
	2	144,004	30.8	120,784	26.9	173,612	35.8

Results

The lymphoproliferative responses of marmoset lymphocytes to general mitogens are presented in table I. Each animal gave a vigorous response to all three mitogens. In general, responses to PHA and Con A tended to be slightly greater than the response to PWM, and there were marked variations in the levels of responses from one sample to the next. On occasion, the untreated cultures exhibited markedly increased spontaneous DNA synthesis which significantly reduced the stimulation index value.

Table II. Cytotoxic response of marmoset lymphocytes against EBV superinfected Raji cells (ADLC) and following PHA activation against normal Raji cells (direct cytotoxicity)

	ADLC, %			Direct cytotoxicity, %[1]		
	100:1	20:1	4:1	100:1	20:1	4:1
Saguinus nigricollis						
Sn 949 H	3.0	6.2	1.8*	6.4	13.3	1.7*
Sn 979I	10.8	5.2	6.6	10.4	8.6	0*
Sn 125K	9.3	2.6*	2.9*	13.4	12.6	0.8*
Saguinus fuscicollis						
Sf 957I	4.2	0*	0*	7.7	12.5	4.0
Sf 124K	8.5	5.6	0*	0.3*	0*	0*
Sf 251I	3.3	3.3	0*	3.7	2.9	1.4*
Saguinus oedipus						
B 7110	9.4	4.5	6.1	12.1	1.8*	1.8*
B 7219	15.5	14.3	0	4.4	2.7*	2.4*
B 6919	10.2	13.4	8.7	12.3	2.3*	2.7*
Saguinus weddelli						
Sw 482J	12.7	7.5	0	6.3	0.7*	0*
Sw 500J	6.3	1.6*	0	14.8	7.2	0*
Sw 518J	9.0	27.1	0.3*	18.1	9.3	4.6
Callithrix jacchus						
B 7305	9.2	1.3*	0.1*	13.0	8.1	0.6*
B 7281	17.6	4.2	0.2*	19.8	10.8	0.1*
Cj 369J	8.1	7.6	7.2	5.2	4.3	1.9*

[1] Values are presented as % cytotoxicity of PHA activated lymphocytes minus spontaneous cytotoxicity of nontreated lymphocytes.
* Not significant at $p<0.01$.

Table III. Receptors for sheep erythrocytes (E), complement (EAC) and surface immunoglobulin (SIg) on marmoset lymphocyte subpopulations

Species	Number of samples	Mean % receptor positive cells (range)			
		E	EAC	SIg	
				stable	labile (L-cell)
S. nigricollis	9	66.6 (60.5–73.0)	14.4 (10.8–17.0)	12.3 (10.1–16.0)	26.9 (22.3–33.7)
S. fuscicollis	9	60.8 (54.0–68.0)	16.7 (14.2–18.3)	14.2 (10.8–16.2)	27.3 (18.4–31.6)
S. oedipus	9	60.2 (52.8–68.0)	19.0 (14.6–24.0)	16.4 (12.5–19.2)	36.8 (34.2–43.2)
S. weddelli	9	57.9 (52.9–61.0)	15.8 (12.0–19.6)	14.3 (10.7–20.0)	32.5 (30.5–37.4)
C. jacchus	8	46.4 (40.5–51.5)	10.3 (6.4–13.3)	10.6 (9.2–11.5)	37.5 (32.0–43.9)

Table IV. Fc receptors on marmoset lymphocytes

Species	Number of samples	Mean % receptor positive cells (range)		
		Fc	SIg	SIg^-Fc^+
S. nigricollis	3	23.7 (19.7–27.3)	12.9 (10.1–16.0)	10.8 (7.7–18.8)
S. fuscicollis	3	28.2 (24.2–31.4)	13.3 (11.0–14.8)	14.9 (10.1–20.4)
S. oedipus	3	28.6 (24.8–31.6)	15.2 (12.5–19.2)	13.4 (10.2–14.6)
S. weddelli	3	31.7 (30.6–33.4)	13.4 (12.0–15.7)	17.9 (14.9–21.0)
C. jacchus	3	24.9 (20.7–29.1)	10.4 (9.2–11.5)	14.2 (9.2–19.9)

All marmoset species exhibited both direct cytotoxicity following PHA stimulation and were able to participate in the ADLC response (table II). Both assays showed marked differences in the level of cytotoxicity among species and animal variation within species. The optimum ratio for obtaining maximum cytotoxicity frequently shifted between 100:1 and 20:1 among individuals.

The E receptor, which detects T-cells, was found to be most consistently high in *S. nigricollis* (66.6%) (table III). *S. oedipus* and *S. weddelli* had intermediate levels of circulating E receptor positive cells (60.2 and 57.9%, respectively). The receptor for modified complement (EAC), which detects B-cells, was present on 19.0% of *S. oedipus* lymphocytes and ranged to 10.3% of *C. jacchus* lymphocytes. SIg which was stable at 37°C and associated with B-lymphocytes was detected at similar levels to the EAC positive cells, i.e., *S. oedipus* had 16.4% SIg positive cells while *C. jacchus* had 10.6%

SIg cells. Interestingly, the two species with the highest percent T-cells, *S. nigricollis* and *S. fuscicollis* had the fewest circulating L-cells (26.9 and 27.3%, respetively).

Values for Fc receptor cells ranged from 23.7% for *S. nigricollis* to 31.7% for *S. weddelli* (table IV). It was apparent that not all Fc receptor positive cells also had SIg since the SIg positive cells were considerably fewer. These results demonstrated that marmosets have a large percentage of Fc^+SIg^- lymphocytes ranging from 10.8% for *S. nigricollis* to 17.9% for *S. weddelli* (table IV).

Discussion

Some previous studies have suggested that marmosets may be immunologically deficient [5, 6]. We have examined a variety of nonspecific lymphocyte functions and have shown that marmosets respond to general mitogens with a vigorous lymphoproliferative response, were able to mount a direct cytotoxic response following activation by PHA *in vitro,* and were able to participate in the ADLC response. In contrast to the finding of Harvey *et al.* [6] regarding an inferior lymphocyte stimulation potential, we found that marmosets were showing at times as much as 100- to 400-fold increase in DNA synthesis over control unstimulated cultures. These levels of DNA synthesis do not suggest that marmosets are inferior in this parameter of immunocompetence. The PHA induced cytotoxicity (T-cell) was not pronounced and may suggest that marmosets have fewer circulating T-killer cells, or may reflect the difficulty in killing these cells. Ability to participate in the ADLC response showed that marmosets also have functional K-cell cytotoxic lymphocytes which can function through an antibody intermediate.

Distribution of T- and B-lymphocytes appeared to be in the normal range (57–67% T-cells; 10–19% EAC cells) for that reported for humans [7] and comparable to those previously reported for marmosets [9] except for one species, *C. jacchus,* which had a markedly lower T-cell population (46.4%) and the lowest EAC positive population (10.3%). Marmosets also exhibit normal levels of stable surface membrane globulin (SIg) positive cells (11–16%) and normal levels of L-cells (27–38%) when compared to humans [7]. Fc receptor positive cells among all species ranged from 24 to 32%, greatly overlapping the stable SIg positive cells. This finding suggests that there is a significant proportion of marmoset cells which are SIg^-Fc^+ as was demonstrated by Abo *et al.* [1] for human lymphocytes. These cells may function to interact with antigen-antibody complexes and may play a role in the ADLC response.

Summary

Marmoset lymphocytes were highly reactive to general mitogens, participate in direct cytotoxicity (purportedly mediated by T-cells) and participate in an antibody dependent lymphocyte cytotoxicity assay (K-cell). The distribution of lymphocyte subpopulations among various species showed that the percent T-cells ranged from 46.4 to 66.6% in different species while complement receptor cells ranged from 10.3 to 19.0%. Surface immunoglobulin (SIg) stable B-cells ranged from 10.6 to 16.4% while the SIg labile L-cells ranged from 26.9 to 37.5%. A fourth receptor, Fc, was demonstrated on 23.7 to 31.7% of the marmoset lymphocytes.

References

1 Abo, T.; Yamaguchi, T.; Shimizer, F., and Kumagai, K.: Studies of surface immunoglobulins on human B-lymphocytes. J. Immun. *117:* 1781–1787 (1976).

2 Baer, H. and Lorenz, D.: The induction of immediate and delayed sensitivity in the marmoset. Proc. Soc. exp. Biol. Med. *134:* 410–412 (1970).

3 Barnhart, D.D. and Gengozian, N.: An evaluation of the mixed lymphocyte culture reaction in marmosets. Transplantation *20:* 107–115 (1975).

4 Dickler, H.B.: Studies of the human lymphocyte receptor for heat-aggregated or antigen-complexed immunoglobulin. J. exp. Med. *140:* 503–522 (1974).

5 Deinhardt, F.: Induction of neoplasms in marmosets. J. med. Primatol. *1:* 29–37 (1972).

6 Harvey, J.S.; Felsburg, P.J.; Herbling, R.L.; Kniker, W.T., and Kalter, S.S.: Immunological competence in nonhuman primates: differences observed in four species. Clin. exp. Immunol. *16:* 267–278 (1974).

7 Jondal, M.; Holm, G., and Wigzell, H.: Surface markers on human T- and B-lymphocytes. I. A large population of lymphocytes forming nonimmune rosettes with sheep red blood cells. J. exp. Med. *136:* 207–212 (1972).

8 Krammer, P.H.; Hudson, L., and Sprent, J.: Fc receptors, Ia antigens and immunoglobulin on normal and activated mouse T-lymphocytes. J. exp. Med. *142:* 1403–1409 (1975).

9 Lobo, P.I.; Westervelt, V.B., and Horwitz, D.A.: Identification of two populations of immunoglobulin-bearing lymphocytes in man. J. Immun. *114:* 116–119 (1975).

10 Niblack, G.B. and Gengozian, N.: T- and B-lymphocytes in the marmoset: a natural haemopoietic chimera. Clin. exp. Immunol. *23:* 536–543 (1976).

11 Porter, R.P. and Gengozian, N.: Immunological tolerance and rejection of skin allografts in the marmoset. Transplantation *8:* 653–665 (1969).

12 Porter, R.P. and Gengozian, N.: Immunological responsiveness and tolerance of marmoset lymphoid tissue *in vitro*. Transplantation *15:* 221–230 (1972).

13 PREVOST, J.M.; ORR, T.W., and PEARSON, G.R.: Augmentation of lymphocyte cytotoxicity by antibodies to *Herpesvirus saimiri* associated antigens. Proc. natn. Acad. Sci. USA *72:* 1671–1675 (1975).

14 SHOPE, T.; DECHAIRO, D., and MILLER, G.: Malignant lymphoma in cotton-top marmosets after inoculation with Epstein-Barr virus. Proc. natn. Acad. Sci. USA *70:* 2487–2491 (1973).

15 WALLEN, W.C.; RABIN, H.; NEUBAUER, R.H., and CICMANEC, J.L.: Depression in lymphocyte response to general mitogens by owl monkeys infected with *Herpesvirus saimiri*. J. natn. Cancer Inst. *54:* 679–685 (1975).

16 WOLFE, L.G.; FALK, L.A., and DEINHARDT, F.: Oncogenicity of herpesvirus saimiri in marmoset monkeys. J. natn. Cancer Inst. *47:* 1145–1162 (1971).

W.C. WALLEN, National Institute of Neurological and Communicative Disorders and Stroke, National Institutes of Health, Bldg. 36, Room 5C-11, *Bethesda, MD 20014* (USA)

Prim. Med., vol. 10, pp. 193–198 (Karger, Basel 1978)

Structure and Synthesis of Hemoglobin in Marmosets *(Saguinus fuscicollis)*

J.E. Fuhr and K.D. Lin

University of Tennessee Memorial Research Center,
Center for the Health Sciences, Knoxville, Tenn.

Introduction

Perhaps no molecule has been as well studied as the hemoglobin molecule. The intent of this paper, however, is to demonstrate that despite the vast literature detailing the production, structure and function of hemoglobin, there is need for further studies which can pertinently be performed in marmosets; and that the marmoset is a suitable animal model for the study of primate erythropoiesis.

Results and Discussion

Marmoset hemoglobin is composed of two pairs of polypeptide chains and a prosthetic group, heme, consisting of protoporphyrin IX and iron. Sequence analysis of the α- and β-globin polypeptides indicates a high degree of homology between marmoset and human globins [7]. There are five differences between the α-chains, and six differences between the β-chains of marmoset and human hemoglobin. All of the differences would be characterized as conservative since they do not affect the points of interaction between α- and β-polypeptides, or the point of heme attachment to the apoprotein (table I).

Biosynthetic studies in the bone marrow and reticulocyte have established that there is balanced synthesis of the α- and β-polypeptides. Moreover, studies with isonicotinic acid hydrazide (INH), a drug commonly used in the treatment of tuberculosis, have indicated that the synthesis of heme and globin are coordinated [2]. Thus, when INH blocked the heme pathway, mainly by interferring with pyridoxal phosphate activation of glycine, globin synthesis was also inhibited. Additional studies in marmoset reticulocytes

Table I. Differences in amino acid composition of human and marmoset hemoglobins

	α-Chain sequence					
	8	19	23	68	71	116
Human	Thr	Ala	Glu	Asn	Ala	Glu
Marmoset	Ser	Gly	Asp	Val	–	Asp
	β-Chain sequence					
	5	13	21	50	87	125
Human	Pro	Ala	Asp	Thr	Thr	Pro
Marmoset	Gly	Thr	Glu	Ser	Gln	Gln

on the relationship between the heme and globin pathways have indicated that cobalt may be capable of replacing heme in cases of iron deficiency [6].

More recently, interest has focused on the control of fetal hemoglobin synthesis. The marmoset has proved particularly valuable for these studies, since like the human, it has a fetal hemoglobin which is its principal hemoglobin during fetal life, and shortly before birth switches to adult hemoglobin production. Although human cord red blood cells may contain both fetal and adult hemoglobin in the same cell, it is believed that the almost 100% fetal hemoglobin producing cells arise from a different stem cell than the adult hemoglobin producing cells. Because of the potential medical benefit to individuals with hemoglobinopathies, such as sickle cell anemia and β-thalassemia, extensive studies by many investigators have been directed toward uncovering the mechanism or agents which may influence fetal hemoglobin synthesis specifically [8].

My own laboratory has focused on the role of thyroid hormone in the regulation of fetal hemoglobin synthesis in bone marrow from marmoset abortuses [3].

Figure 1 demonstrates that globin synthesis by fetal bone marrow was stimulated by the addition of L-thyroxine to the short-term (180 min) suspension culture. The globin polypeptides were separated and purified on carboxymethyl cellulose columns in 8.0 M urea, and specific activity was determined on both the α- and the non-α-globin polypeptides. As can be seen, normal, physiologic levels of the hormone stimulated the synthesis of both globin polypeptides. Increased levels of the hormone increased the

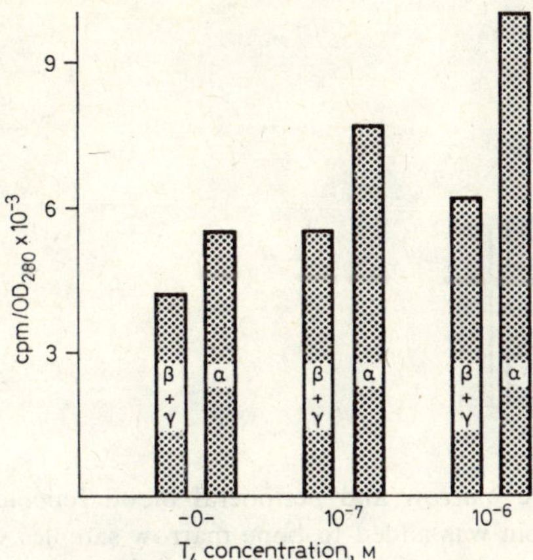

Fig. 1. Effect of different concentrations of L-thyroxine on globin synthesis by marmoset abortus bone marrow. Reactions contained 4×10^6 cells/ml of NCTC 135 in suspension culture and lasted 180 min at 37 °C. The hormone and the ^{3}H-leucine were present from the start of the incubation. After the addition of carrier hemoglobin, globin was prepared and the globin separated into individual polypeptides by chromatography on CMC in 8.0 M urea. Specific activity was determined on the individual polypeptides.

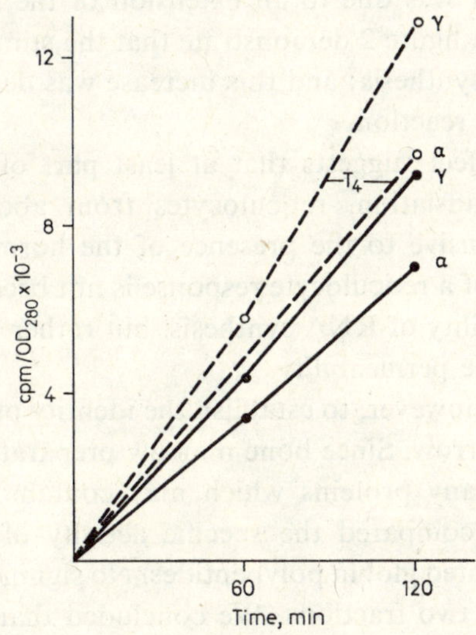

Fig. 2. Marmoset abortus bone marrow response to 0.001 mM L-thyroxine. The reaction contained 5×10^6 cells/ml and lasted for the times indicated. Dashed lines represent specific activity of globin prepared from cells incubated in the presence of thyroxine. Solid lines represent specific activity of globin from control incubations.

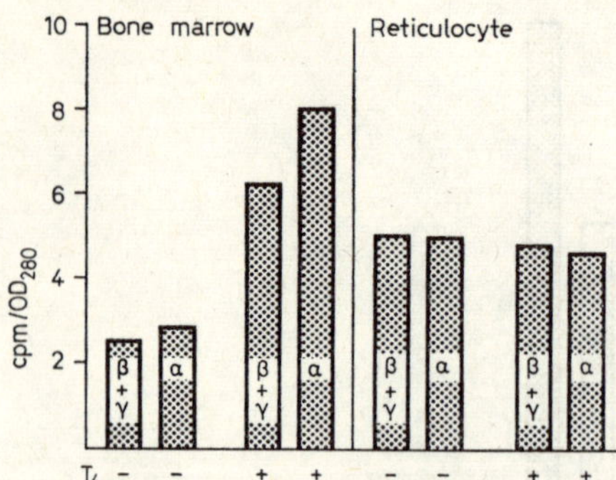

Fig. 3. Comparison of abortus bone marrow and peripheral blood reticulocyte response to L-thyroxine. Carrier hemoglobin was added to bone marrow samples when cells were lysed at the end of the incubation. Reactions lasted 180 min at 37 °C.

percent stimulation to 82 and 42% for the α- and non-α-globin polypeptides, respectively.

Because bone marrow suspension cultures very often lose a significant degree of activity shortly after the start of the incubation, it was necessary to determine whether the stimulation observed at 180 min represented an increased rate of globin synthesis, or was due to an extension of the time of linear incorporation. The results in figure 2 demonstrate that the stimulation was due to an increased rate of synthesis; and this increase was detectable within 60 min of the start of the reaction.

Although the rapidity of the effect suggests that at least part of the stimulation was at the level of translation, reticulocytes from abortus marmosets were completely unresponsive to the presence of the hormone (fig. 3). We believe the total absence of a reticulocyte response is not because the cell has no nucleus and no capability of RNA synthesis, but rather may be explained by changes in membrane permeability.

We were concerned at this time, however, to establish the identity of the globin synthesized by the abortus marrow. Since bone marrow preparations have been reported to synthesize many proteins which may contaminate purified globin polypeptides [1] we compared the specific activity of the leading and trailing edges of the separated globin polypeptides. No significant difference was observed between the two fractions. We concluded that the globin polypeptides, as isolated, were homogeneous and free of significant contamination.

Although the adult β-globin and the fetal γ-globin of marmosets co-chromatograph, we concluded that fetal globin synthesis was stimulated on

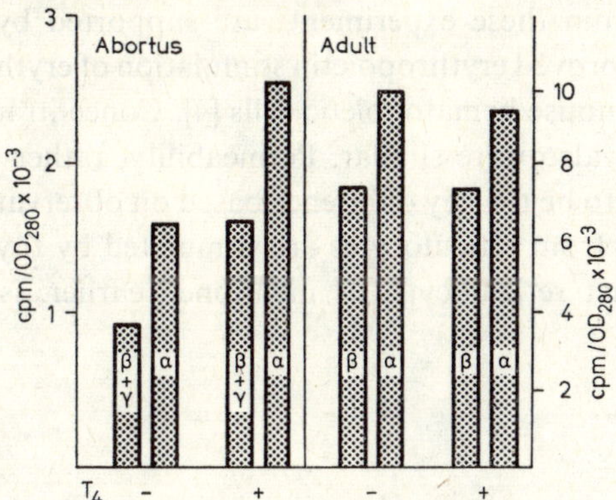

Fig. 4. Comparison of the responsiveness of adult and abortus bone marrow to L-thyroxine. Both bone marrow preparations contained 4×10^6 cells/ml which were incubated in sealed tubes for 180 min in the presence or absence of 0.001 mM L-thyroxine.

the basis of the following observations: (1) although only 1% of the hemoglobin from adult marmosets was resistant to denaturation by alkali, more than 80% of the abortus hemoglobin was resistant; (2) the molar extinction coefficient of the γ-globin from the abortus was apparently considerably lower than that of the adult β-globin; (3) examination of the peripheral smear from abortuses by means of the acid elution technique to visualize fetal hemoglobin-containing cells, revealed virtually 100% fetal cells; (4) synthesis of hemoglobin by abortus reticulocytes was significantly inhibited by the isoleucine analog, *o*-methylthreonine, whereas adult reticulocytes were not affected; (5) amino acid analysis of the fetal globin confirmed the presence of isoleucine residues which are not present in the adult hemoglobin.

Beyond the observation that fetal hemoglobin synthesis was stimulated by presence of thyroid hormone, there is perhaps a more significant observation, summarized in figure 4. While fetal bone marrow was consistently stimulated by thyroid hormone, bone marrow from adult marmosets was unresponsive, even at concentrations as high as 1 mM. These findings suggest that although fetal hemoglobin synthesis is a convenient marker for the response of fetal tissue, a more significant interpretation may be that erythroid precursor cells from abortus and adult marmosets differ not only in the type of hemoglobin synthesized, but also, and more importantly, in the permeability of their membranes, which dictates their responsiveness to hormonal stimuli.

The conclusions drawn from these experiments are supported by the report that thyroid hormone improved erythropoietin stimulation of erythroid precursors in colonies of fetal mouse hematopoietic cells [4]. Concentrations of hormone in the two studies also were similar. Permeability, rather than membrane receptor, is thought to be the key difference based on observations that cell-free systems prepared from reticulocytes are stimulated by thyroid hormone [5], whereas the intact reticulocyte, as mentioned earlier, is unresponsive to L-thyroxine.

Summary

Marmoset reticulocytes and bone marrow provide a valuable tissue resource for the study of primate hemoglobin structure and synthesis. The amino acid sequence of the marmoset hemoglobin has been determined, and the close homology between it and that of the human suggest it may be a valuable model for study of the effects of chemical modification on hemoglobin function. Hemoglobin synthesis by fetal bone marrow has been reported to be stimulated by physiologic levels of L-thyroxine. Differences in hormone responsiveness between fetal and adult red cell precursors have tentatively been ascribed to differences in permeability of the fetal and adult clones of cells.

References

1 CHALEVELAKIS, G.; CLEGG, J.B., and WEATHERALL, D.J.: Globin synthesis in normal human bone marrow. Br. J. Haemat. *34:* 535–557 (1976).

2 FUHR, J.E. and GENGOZIAN, N.: Coordination of heme and globin synthesis in primate reticulocytes. Biochim. biophys. Acta *320:* 53–58 (1973).

3 FUHR, J.E.; GENGOZIAN, N., and OVERTON, M.: *In vitro* stimulation of primate hemoglobin synthesis by L-thyroxine. Blood *49:* 407–413 (1977).

4 GOLDE, D.W.; BERSCH, N.; CHOPRA, I.J., and CLIN, M.J.: Potentiation of erythropoiesis by thyroid hormones. Clin. Res. *24:* 309a (1976).

5 KRAUSE, R.L. and SOKOLOFF, L.: Thyroxine stimulation of amino acid incorporation into protein chains of hemoglobin *in vitro*. Biochim. biophys. Acta *108:* 165–168 (1965).

6 LEISY, M.; OVERTON, M., and FUHR, J.E.: The effect of cobalt on haemoglobin synthesis in primate reticulocytes. Cytobios *12:* 109–114 (1975).

7 LIN, K.D.; KIM, Y.K., and CHERNOFF, A.I.: Primary structure of the marmoset *(Saguinus fuscicollis)* hemoglobin. I. Use of tryptic maleylated peptides in the solubilization and sequence elucidation of the α- and β-chains. Biochem. Genet. *14:* 427–440 (1976).

8 WEATHERALL, D.J.: Fetal haemoglobin synthesis; in Congenital disorders of erythropoiesis. Ciba Found. Symp. 37, pp. 307–323 (Elsevier, Amsterdam 1976).

J.E. FUHR, PhD, University of Tennessee Memorial Research Center, Center for the Health Sciences, *Knoxville, TN 37920* (USA)

Prim. Med., vol. 10, pp. 199–204 (Karger, Basel 1978)

Radiation-Induced Chromosome Aberrations in Somatic and Germ Cells of the Male Marmoset[1]

J. G. BREWEN and R. J. PRESTON

Biology Division, Oak Ridge National Laboratory, Oak Ridge, Tenn.

Introduction

One of the basic problems in radiation genetics is the establishment of reliable criteria for estimating human risk from various forms of exposure. Since chromosome aberrations constitute a large class of radiation-induced genetic effects, a great deal of information has been accumulated on aberration induction in human cells treated *in vitro*. The use of these data for estimating human genetic risk was restricted by two major problems. These were, (1) the uncertainty that the *in vitro* response was quantitatively similar to the *in vivo* response, and (2) what correlation existed between quantitative effects in somatic cells and in the genetically important germ cells.

The solution to these problems was to treat the same cell type, from the same organism, both *in vitro* and *in vivo* and compare the effects, and to analyze comparable aberration types in both the somatic cell of choice and the germ cells. The male marmoset offered an unique opportunity to make these comparisons. It is a primate and karyologically is similar to man thus providing an apparently excellent cytogenetic model. The experiments reported in this paper deal with the study of these problems and the use of the marmoset as a cytogenetic model for man.

1 Research sponsored by the Energy Research and Development Administration under contract with the Union Carbide Corporation.

Materials and Methods

The somatic cell of choice was the circulating peripheral leukocyte. Since the experiments were numerous and varied the reader is referred to the original article for details of the radiation regimes and culture procedures [2]. The blood samples were irradiated prior to phytohemagglutinin stimulation and metaphase figures collected between 50 and 52 h after culture initiation.

In the studies on spermatogonial stem cells mature male marmosets *(Saguinus fuscicollis illigeri* and *Saguinus oedipus oedipus)* were testicularly irradiated with 250 kV X-rays at 50 R/min. At various intervals, depending on dose, after irradiation either an entire testis, or a biopsy was taken. Preparations of diplotene-diakinesis were made by a modification of the technique of EVANS *et al.* [5]. A maximum of 200 cells were analyzed for multivalent associations from each sample. The doses employed in this study were 10, 25, 50, 75, 100, 150, 200, and 300 R and a minimum of two animals were used per dose point.

Results

Tables I and II summarize the data obtained in the peripheral leukocyte experiments. In table I the *in vivo* data were generated from blood samples taken immediately, and 24 h, after the exposure. Since there was no significant difference the data were pooled. The apparent excess of cells at the 100 and 200 R doses is the result of two animals being used at those doses. It can be seen that there is excellent agreement, both qualitatively and quantitatively, in the response of the leukocyte chromosomes to radiation whether the exposure was given to the live animal or freshly drawn blood.

Table II summarizes the data obtained when whole blood from marmosets and human volunteers was irradiated with 250 kV X-rays at 100 R/min. The aberration yields for the marmoset presented in table II are higher than those in table I. However, this can be readily explained by the fact that the data obtained in table I was for a low dose rate (3.7 R/min) and with ^{60}Co γ-rays which has an RBE of about 0.8 compared to 250 kV X-rays. The aberration yields in the marmoset are consistently higher than those in man although the dose response kinetics are very similar.

The data from the spermatogonial stem cell irradiation are summarized in table III and figure 1. It is seen that there is a linear increase with dose up to 100–150 R where a peak yield is obtained, followed by a slight decrease at higher doses. The average rate of induction over the linear portion of the dose response curve is 7.7×10^{-4} translocations per cell per R.

Table I. Aberration yields observed in marmoset leukocytes following *in vivo* or *in vitro* irradiation with ^{60}Co-γ-rays at 3.7 R/min

Dose R	In vitro			In vivo		
	number of cells	rings + dic. %	acentric frag. %	number of cells	rings + dic. %	acentric frag. %
0	300	0	0	1,200	0	0.07+0.07
100	300	13.3±2.1	8.0±1.6	1,000	13.7±1.2	7.9 ±0.9
200	300	33.3±3.3	16.0±2.3	950	40.2±2.1	12.7 ±1.2
300	300	68.7±4.8	31.7±3.2	300	70.7±4.9	23.3 ±2.8
400	300	98.3±5.7	49.7±4.1	250	94.9±6.1	33.6 ±3.7

Table II. Aberration yields in marmoset and human leukocytes following acute irradiation with 250 kV X-rays

Dose R	Marmoset			Human		
	number of cells	rings + dic. %	acentric frag. %	number of cells	rings + dic. %	acentric frag. %
0	300	0	0	400	0	0
50	200	6.5±1.8	5.0±1.6	400	5.0±1.1	3.5±0.9
100	300	16.7±2.4	11.0±1.9	400	13.5±1.8	8.0±1.4
150	300	40.0±3.7	24.3±2.8	–	–	–
200	300	59.0±4.4	32.0±3.3	400	41.8±3.2	20.0±2.2
300	300	99.0±5.7	53.7±4.2	400	75.5±4.3	41.0±3.2
400	–	–	–	400	125.5±5.6	87.3±4.7

Table III. Reciprocal translocation yields in marmoset primary spermatocytes after irradiation of spermatogonial stem cells with acute X-rays

Dose R	Number of cells scored	Translocations, %	Translocations cell/R × 10^4
0	600	0	0
10	600	0.67±0.33	6.7
25	1,000	2.4±0.5	9.6
50	600	3.3±0.8	6.7
75	675	4.7±0.8	6.3
100	600	7.8±0.8	7.8
150	200	5.5±1.7	3.7
200	200	7.5±1.9	3.8
300	600	6.2±1.0	2.1

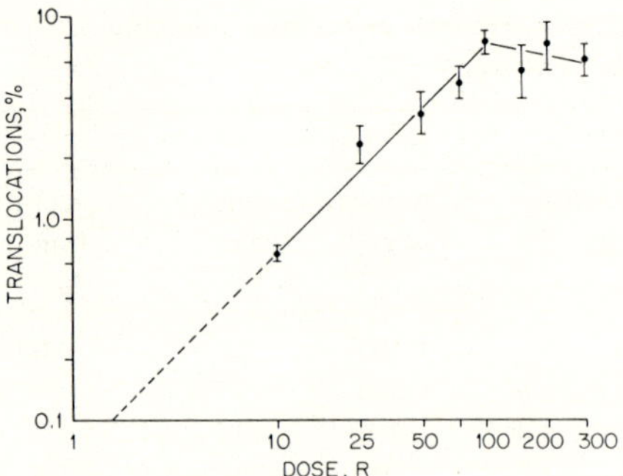

Fig. 1. A log versus log plot of the frequency of translocations at various acute X-ray doses. The slope of the line between 10 and 100 R is 1.

Discussion

The experiments on the irradiation, *in vitro* and *in vivo,* of marmoset peripheral leukocytes clearly show that the quantitative effect is similar in both situations. This observation has subsequently been extended to include the rabbit [1, 4] and Chinese hamster [7]. Thus the extensive data collected on *in vitro* treatment of human leukocytes can be assumed to reflect the cytogenetic effect that occurs when humans are exposed to ionizing radiations.

When comparing the quantitative effects on somatic cells and the spermatogonial stem cells several factors must be considered. First, the translocation events analyzed in the two cell types are different. In one instance (leukocytes) asymmetrical translocations are scored and in the other symmetrical translocations are quantitated. HEDDLE [6] has presented evidence that the two events occur at equal frequencies in *Vicia faba* and in human leukocytes [BUCKTON, personal commun.]. Thus we assume that the dicentric frequency in leukocytes corresponds to the reciprocal translocation frequency in the same cell population.

Second, the dose response kinetics are different for the two effects under study, translocations in the germ cells having linear kinetics and dicentric production in the leukocytes containing a large dose-square component. Hence, the ratios of the two events will vary over the dose range studied (table IV). The principle concern for human risk, however, is at low doses where the linear component predominates in the dose-response curve of dicentric induction, thus making it reasonable to assume that ratios of

Table IV. The ratio of dicentrics in marmoset leukocytes to translocation recovered from spermatogonial stem cells

Dose R	Dicentrics[1] cell/R × 10^4	Translocations cell/R × 10^4	Ratio
10	10.9	6.7	1.6:1
25	11.8	9.6	1.3:1
50	13.3	6.7	2:1
75	14.7	6.3	2.3:1
100	16.2	7.8	2.1:1

[1] Calculated from the coefficients for the best fitting regression $Y = 1.16 \times 10^{-3}D + 7.69 \times 10^{-6} D^2$.

translocations in somatic to germ cells derived at 50 R and below may be meaningful.

Third, in the leukocyte experiments the cells are analyzed for aberrations at the first cell division after treatment whereas in the spermatogonial experiments many cell divisions intervene between treatment and analysis of the primary spermatocyte. This latter fact offers the opportunity for selection to occur against translocation bearing cells. If it is assumed that the translocations are distributed randomly among both cells and chromosomes the degree of selection should not vary at low doses and the observed ratios thus can be used for predictive purposes.

The data presented here suggest that the ratio of recovered reciprocal translocations induced in spermatogonial stem cells to dicentrics induced in peripheral leukocytes is approximately 1:1.5 at low doses. We suggest, therefore, that existing data on dicentric induction in human leukocytes can be used to predict expected translocation yields from spermatogonial stem cell irradiation, with the proviso that at very low doses, e.g. less than 5 R, the ratio may be 1:1.

Summary

The induction of chromosome aberrations by low LET radiations was studied in peripheral lymphocytes and spermatogonial stem cells of the male marmoset. The data showed that there was no significant difference in the sensitivity of the lymphocytes whether they were irradiated *in vitro* or *in vivo*, but the frequency of heritable translocations recovered in the primary spermatocytes was considerably lower than that calculated to occur in the lymphocytes. The data are used to make estimates of human genetic risk from radiation based on limited interspecific comparisons.

References

1 Bajerska, A. and Liniecki, J.: The yield of chromosomal aberrations in rabbit lymphocytes after irradiation *in vitro* and *in vivo*. Mutation Res. *27:* 271–284 (1975).

2 Brewen, J.G. and Gengozian, N.: Radiation-induced human chromosome aberrations. II. Human *in vitro* irradiation compared to *in vitro* and *in vivo* irradiation of marmoset leukocytes. Mutation Res. *13:* 383–391 (1971).

3 Brewen, J.G.; Preston, R.J., and Gengozian, N.: Analysis of X-ray-induced chromosomal translocations in human and marmoset spermatogonial stem cells. Nature, Lond. *253:* 468–470 (1975).

4 Clemenger, J.F.P. and Scott, D.: A comparison of chromosome aberration yields in rabbit blood lymphocytes irradiated *in vitro* and *in vivo*. Int. J. Radiat. Biol. *24:* 487–496 (1973).

5 Evans, E.P.; Breckon, G., and Ford, C.E.: An air-drying method for meiotic preparations from mammalian testes. Cytogenetics *3:* 289–294 (1964).

6 Heddle, J.A.: Randomness in the formation of radiation-induced chromosome aberrations. Genetics *52:* 1329–1334 (1965).

7 Preston, R.J.; Brewen, J.G., and Jones, K.P.: Radiation-induced chromosome aberrations in Chinese hamster leukocytes. A comparison of *in vivo* and *in vitro* exposures. Int. J. Radiat. Biol. *21:* 397–400 (1972).

J.G. Brewen, Biology Division, Oak Ridge National Laboratory, *Oak Ridge, TN 37830* (USA)

Prim. Med., vol. 10, pp. 205–214 (Karger, Basel 1978)

Pathogenesis of Adjuvant-Induced Destructive Periodontitis in Marmosets [1]

Barnet M. Levy, Donald C. Nelms, Samuel Dreizen and Sol Bernick

University of Texas Dental Branch, Dental Science Institute, Houston, Tex., and Department of Anatomy, School of Medicine, University of Southern California, Los Angeles, Calif.

Introduction

Intragingival injection of complete Freund's adjuvant (CFA) in marmosets induces extensive destructive periodontitis within 21 days [5, 6]. This reaction has provided a time-shortened model for the study of the destructive aspects of periodontal disease. The purpose of the present report is to detail chronologically the sequence of histopathologic changes that culminate in a CFA-precipitated loss of periodontal integrity.

Materials and Methods

The interdental gingivae between the upper and lower right third premolar and first molar teeth of five pairs of cotton-top marmosets *(Saguinus oedipus)* were injected with 0.02 cm^3 CFA (Difco H37Ra). Corresponding areas on the left side were injected with 0.02 cm^3 incomplete Freund's adjuvant (IFA). One pair of animals was sacrificed on each of days 2, 4, 7, 11 and 14 after inoculation. The heads were removed and fixed in alcohol-formalin-acetic acid, or 10% neutral formalin for celloidin and paraffin sectioning, respectively. Selected gingival injection sites were fixed in 3% gluteraldehyde for epon embedding. Alternate serial sections were stained with hematoxylin and eosin, periodic acid-Schiff and hematoxylin, Masson's trichrome and Giemsa stains.

1 This study was supported in part by USPHS Grant DE-02232 from the National Institute of Dental Research, National Institutes of Health, Bethesda, Md.

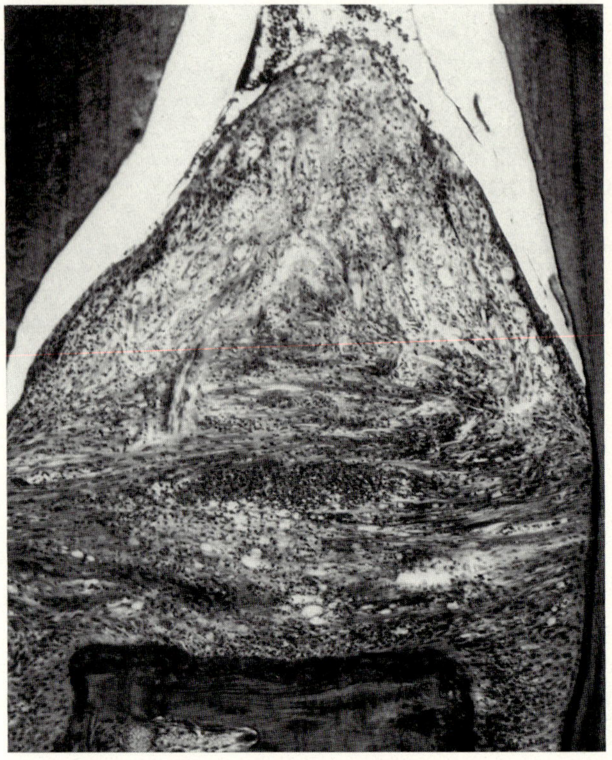

Fig. 1. Photomicrograph of interdental gingiva 2 days following injection of CFA. There is a moderate leukocytic infiltrate, mostly neutrophils, throughout the lamina propria. HE. ×75.

Results

2 days following injection of CFA into the interproximal gingiva, the lamina propria and epithelial layers were infiltrated by polymorphonuclear leukocytes (fig. 1, 2). Occasional small mononuclear cells (lymphocytes) were scattered throughout the infiltrate. In the areas of inflammatory cell concentration, the collagen fibers were lysed and fragmented and the polymorphonuclear leukocytes appeared to be entrapped in a fibrin network. Osteoclasts were present on the marrow side of the crestal bone. The periodontal ligament contained a perivascular infiltrate of polymorphonuclear leukocytes (fig. 3). Animals injected with IFA had similar acute inflammatory

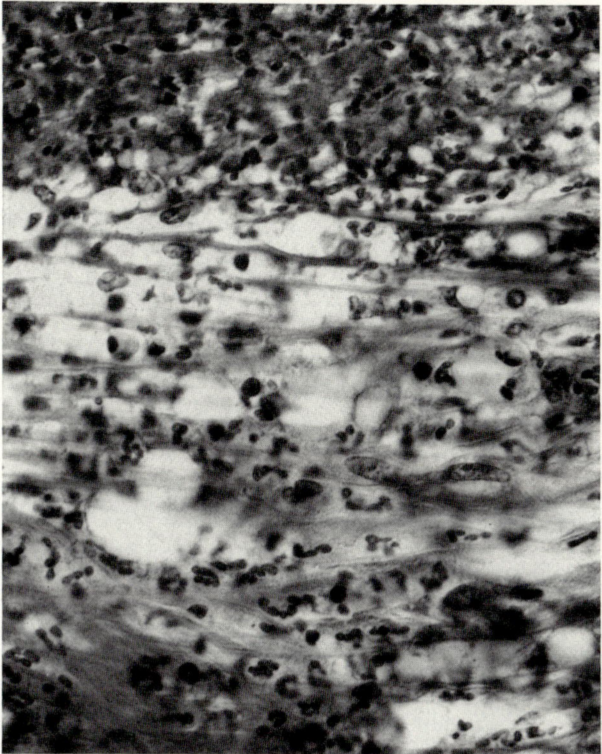

Fig. 2. High power photomicrograph of lamina propria of figure 1 illustrating the polymorphonuclear infiltrate entrapped in fibrin network. HE. ×500.

changes without an osteoclastic response. In contrast, the uninjected gingivae of the adjacent teeth were free of inflammatory cells.

4 days after the injection of CFA, the inflammatory infiltrate was still composed largely of polymorphonuclear leukocytes in various stages of degeneration. There was an increase in large and small mononuclear cells and a proliferation of endothelial cells and small capillaries, that gave the lesion an early granulomatous appearance. IFA-injected animals also showed a granulomatous reaction comprised of focal areas of polymorphonuclear leukocyte infiltration, endothelial proliferation and occasional collections of macrophages, many of which contained foamy cytoplasm.

7 days subsequent to the injection of CFA, the gingiva was infiltrated

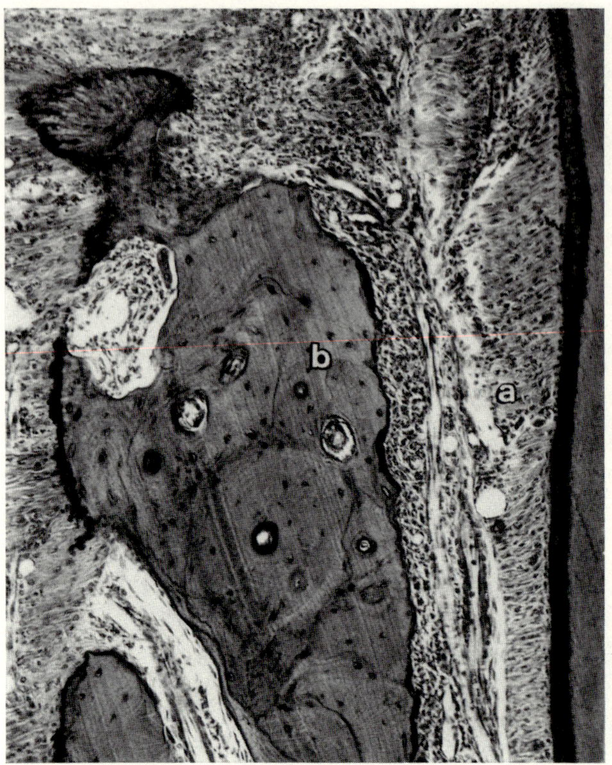

Fig. 3. Photomicrograph of periodontium of marmoset 2 days following intragingival injection of CFA. There is a considerable round cell infiltrate in the periodontal ligament (a). b = Periodontal bone. HE. ×125.

by mononuclear cells (mostly macrophages), endothelial cells, fibroblasts and large lymphocytes. Some of the large mononuclear cells could not be definitely classified. Both the gingiva near the crestal border and the periodontal ligament contained microabscesses. The abscesses were circumscribed by fibroblasts and collagen fibers (fig. 4). Many of the collagen fibers of the gingiva and periodontal ligament were destroyed or fragmented. The crest of the interseptal alveolar bone was characterized by large areas of focal resorption. Multinucleated osteoclasts were present in the resorption lacunae. The marrow was hyperplastic and consisted primarily of mononuclear cells and fibroblasts. In isolated areas, there was complete destruction of the alveolar bone, resulting in invasion of the periodontal ligament by marrow

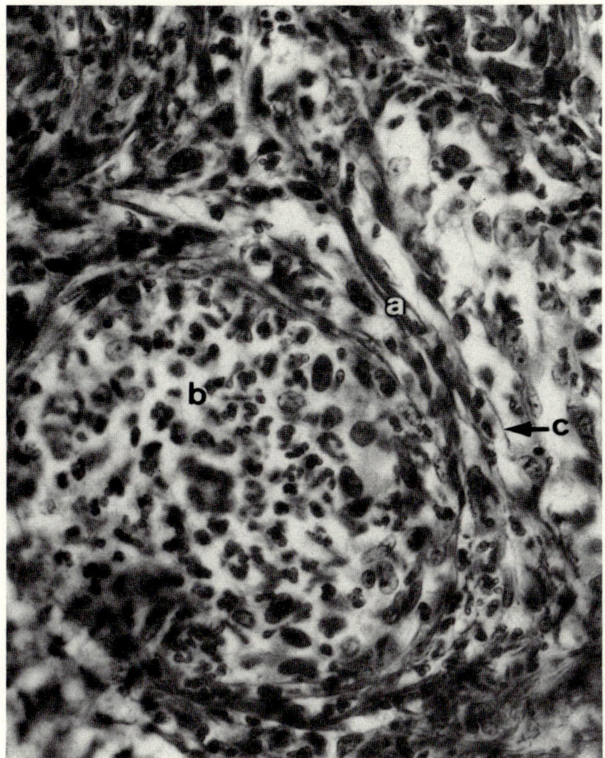

Fig. 4. Photomicrograph of gingival microabscess. The abscess is circumscribed by collagenous fibers (a), fragmented and lysed collagen fibers (c). The cellular infiltrate consists of degenerating neutrophils and macrophages (b). Fe-H.-picrofuchsin. ×500.

cells. The invading cells were mostly macrophages, fibroblasts, small mononuclear cells and bone marrow elements. A few polymorphonuclear leukocytes were also present. The principal fibers were so fragmented that only the cemental origin was demonstrable. The gingiva of IFA-injected animals was infiltrated by mononuclear and polymorphonuclear leukocytes. Although fibroblastic and endothelial proliferation was abundant, there was no evidence of bone destruction.

At 11 days following gingival injection of CFA, there was progressive destruction of the entire periodontium. The gingival epithelium, infiltrated by polymorphonuclear leukocytes, was thinner than normal and there were microareas of ulceration. The gingival connective tissues were infiltrated by

Fig. 5. Photomicrograph illustrating the gingival area injected 11 days previously with CFA. Note the intense lymphocytic and mononuclear cell infiltrate, ulceration of surface epithelium and bone resorption. HE. ×75.

large and small mononuclear cells, macrophages and lymphocytes. Scattered throughout the mononuclear cell infiltrate were polymorphonuclear leukocytes in various stages of degeneration. Microabscesses, consisting of small aggregates of degenerating polymorphonuclear leukocytes surrounded by epithelioid cells and histiocytes, were common. The principal attachment fibers of the periodontal ligament were lost and were replaced by a chronic inflammatory reaction consisting of proliferating fibroblasts and mononuclear cells. The crest of the interseptal alveolar bone was destroyed and replaced by granulation tissue. Numerous osteoclasts occupied the resorption lacunae (fig. 5).

By the 14th day after injection, there was complete loss of interseptal

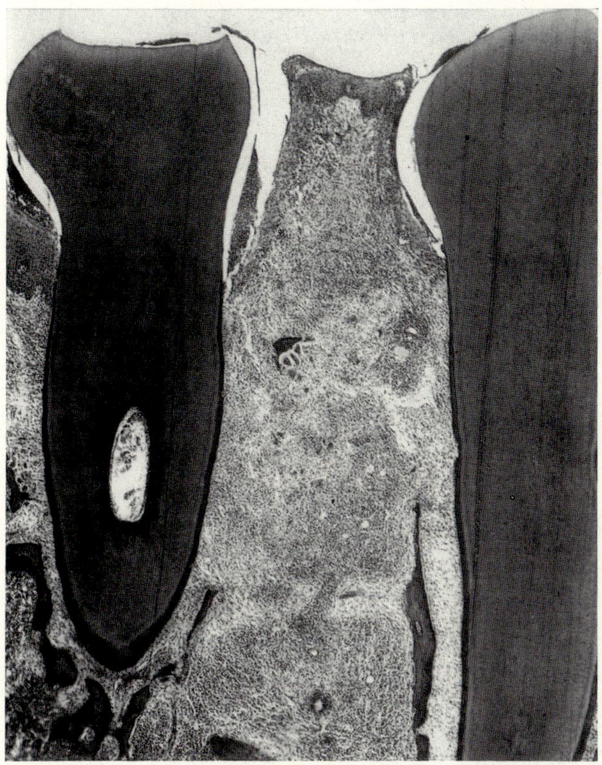

Fig. 6. Photomicrograph of periodontium 14 days after the injection of CFA into the gingiva. The alveolar bone is completely destroyed and replaced by granulomas and granulation tissue. HE. ×35.

bone (fig. 6). While the CFA-injected area was completely filled with granulomatous tissue, the adjacent noninjected interdental space was unaffected. In the injected areas the residual marrow spaces contained diffuse granulation tissue composed of fibroblasts, macrophages and large and small lymphocytes. Small well-organized granulomas were present in the periodontium and in areas previously occupied by alveolar bone. Some of the granulomas contained a core of polymorphonuclear leukocytes surrounded by epithelioid cells, lymphocytes and fibroblasts (fig. 7). There was no bone destruction in the IFA-injected areas and only an infiltrate of polymorphonuclear leukocytes, mononuclear cells, fibroblasts and endothelial cells marked the injection site.

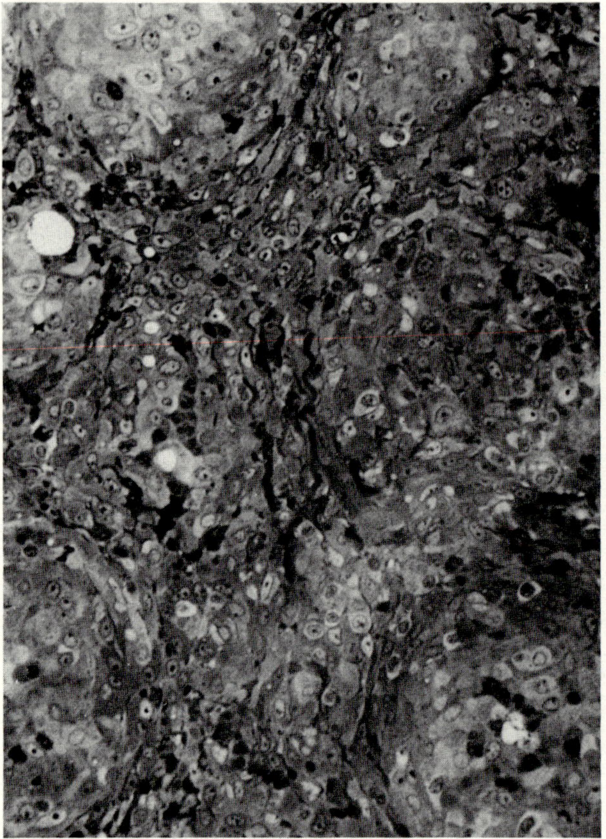

Fig. 7. Photomicrograph of epon-embedded tissue, illustrating typical, well-organized epithelioid granulomas. Toluidine blue. ×100.

Discussion

The present findings provide a strong indication that the process of destructive adjuvant-induced periodontitis is immunologically mediated since the IFA-induced granulomas did not induce bone destruction. During the first 4 days of the reaction to injected CFA or IFA, the cellular responses were similar with little difference in the basic cellular infiltrate. On day 7, however, the CFA-injected tissues were infiltrated primarily by mononuclear cells, mostly macrophages, and with large and small lymphocytes. Precise identification of some of the mononuclear cells could not be made. There is evidence that tuberculin sensitivity develops on or about the 4th day of immunization. SHAW *et al.* [7] found that tuberculin sensitivity became manifest 4 days after injection of CFA into guinea pigs and MACKANESS [2] detected delayed type hypersensitivity on the 4th day following intravenous injection of L-monocytogenes in mice. BOYDEN [1] noted delayed hypersensitivity to BCG in guinea pigs 11 days following vaccination.

Many of the unidentifiable cells in the 7-day lesions appeared to be lymphocytic in origin, young macrophages derived by quick succession of cell division [4]. MACKANESS [3] found that during strong stimulation of hypersensitive mice, large numbers of unclassifiable cells appeared in the peritoneal cavity in association with an intense mitotic response. When these cells were placed in tissue culture, they differentiated into typical macrophages.

The formation of well-organized granulomas composed of epithelioid cells, macrophages and occasional lymphocytes did not occur until 2 weeks after injection of CFA, whereas bone resorption was evident after only 1 week. The granulomas formed in areas where bone had previously been destroyed. While the changes produced by the injection of CFA into nonsensitized marmosets result in prompt and massive destruction of bone and other periodontal tissues, the mediating mechanism cannot be deduced from the morphologic studies. Investigation of the mediators produced by the infiltrating lymphocytes and other mononuclear cells in response to CFA may clarify the mechanism of bone and other connective tissue destruction.

Summary

The histologic reaction of the marmoset periodontium to a single intragingival injection of CFA was followed sequentially over a 2-week period. The earliest lesions were nonspecific and similar to IFA-induced responses. After 4 days the CFA-produced lesions were markedly destructive of periodontal bone, whereas the IFA reaction was not, indicating that the alveolar bone destruction that stemmed from the CFA injection may have been immunologically mediated.

References

1 BOYDEN, S.V.: The effect of previous injections of tuberculo-protein on the development of tuberculin sensitivity following BCG vaccination on guinea pigs. Br. J. exp. Path. *38:* 611–617 (1957).

2 MACKANESS, G.B.: Cellular resistance to infection. J. exp. Med. *116:* 381–406 (1962).
3 MACKANESS, G.B.: The behavior of microbial parasites in relation to phagocytic cells *in vitro* and *in vivo;* in Symp. Soc. Gen. Microbiol., vol. 14, p. 213 (1964).
4 MACKANESS, G.B. and BLADEN, R.V.: Cellular immunity. Prog. Allergy, vol. 11, pp. 89–140 (Karger, Basel 1967).
5 LEVY, B.; NELMS, D.C.; DREIZEN, S., and BERNICK, S.: Delayed hypersensitivity reaction in the periodontium. A preliminary report; in Medical primatology 1972. Proc. 3rd Conf. Exp. Med. Surg. Primates, Lyon 1972, part II, pp. 191–197 (Karger, Basel 1972).
6 LEVY, B.; ROBERTSON, P.B.; DREIZEN, S.; MACKLER, B.F., and BERNICK, S.: Adjuvant induced destructive periodontitis in nonhuman primates. A comparative study. J. periodont. Res. *11:* 54–60 (1976).
7 SHAW, C.; ALVORD, E.C., jr.; KAKU, J., and KIES, M.W.: Correlation of experimental allergic encephalomyelitis with delayed-type skin sensitivity to specific homologous encephalitogen. Ann. N.Y. Acad. Sci. *122:* 318 (1965).

B.M. LEVY, DDS, The University of Texas Dental Branch, PO Box 20068, *Houston, TX 77025* (USA)

Prim. Med., vol. 10, pp. 215–224 (Karger, Basel 1978)

The Use of Marmosets *(Callithrix jacchus)* in Teratological and Toxicological Research

ROBERT A. SIDDALL

ICI Pharmaceuticals Division Ltd., Alderley Park, Cheshire

Introduction

The choice of laboratory species for use in safety evaluation studies depends upon a number of criteria which include a background knowledge of histopathology, haematology, clinical chemistry and reproductive physiology and, where appropriate, a metabolic similarity to man for the drug under investigation. The availability of the species will also affect its selection as a laboratory animal. The species commonly used in toxicology are mice, rats, dogs and, for certain studies, rhesus monkeys. In teratology the most used species are mice, rats and rabbits.

In order to provide sub-human primates for use in safety evaluation studies – as an alternative to the dog and as a third species for teratology – it was decided to establish a breeding colony at ICI. An important factor in this decision was the need to support the conservation of primate species by breeding our total research requirement in such a colony. From our experience with a range of sub-human primates and following reports that *Callithrix jacchus,* the common cotton-eared marmoset, had been successfully bred and used as a laboratory animal, it was decided to evaluate this species further.

The first colony was established in the mid 1960s, although we had initially used marmosets in 1959, and these animals were evaluated in toxicology, teratology, reproductive physiology and behavioural research. HIDDLESTON [5] has described the establishment of the present breeding unit which has a potential output of over 1,000 home-bred marmosets a year.

Table I. Mean values for some clinical chemistry parameters in man, marmoset, dog and rat

	Sugar mmol/l	Urea mmol/l	Total protein g/l	Albumin g/l	Total bilirubin g/l	ALP U/l	ALT U/l	AST U/l	Sodium Na^+ mmol/l	Potassium K^+ mmol/l	n
Man	5.1	4.8	76.0	42.0	12.0	69.0	13.0	18.0	141.0	4.0	–
Marmoset	6.7	6.2	71.5	41.6	12.0	199.4	9.4	119.0	153.8	5.3	50
Dog	5.7	5.2	61.7	30.0	2.7	48.2	30.0	25.6	146.7	4.2	175
Rat	9.7	8.6	71.1	36.2	2.6	121.4	40.0	122.0	142.1	5.5	400

Table II. Mean values for some haematology parameters in man, marmoset, dog and rat

	Haemoglobin g/dl	Packed cell volume %	Total WBC $\times 10^9$/l	Total RBC $\times 10^{12}$/l	Platelets $\times 10^9$/l	Lymphocytes $\times 10^9$/l	Neutrophils $\times 10^9$/l
Man (males)	15.5 ±2.5	0.47 ±0.07	7.5 ±3.5	5.5 ±1.0	275 (150–400)	2.75 (1.5–4.0)	4.75 (2.0–7.5)
Marmoset	15.5 (12.3–18.1)	0.48 (0.41–0.56)	10.1 (5.5–16.6)	6.7 (5.2–8.1)	397.5 (183–693)	6.83 (2.69–13.13)	3.87 (1.08–9.73)
Dog	16.2 (12.2–19.6)	0.47 (0.37–0.38)	15.7 (9.5–26.0)	6.1 (4.5–7.8)	301.5 (135–568)	4.63 (2.3–7.53)	9.19 (5.13–14.57)
Rat	15.8 (14.2–17.5)	0.41 (0.41–0.52)	10.2 (4.1–20.3)	6.7 (4.9–8.3)	724 (450–1,065)	8.87 (3.45–15.65)	2.23 (0.62–4.80)

Toxicology

In order to determine whether the marmoset was a suitable species for toxicology a number of studies were performed and comparisons made between marmosets and the commonly used species, particularly the dog.

An important part of any toxicity study is the histopathological examination of post mortem specimens and a sound knowledge of anatomy, morphology of the organs, normal variation and possible artefacts are a prerequisite for a meaningful study. Tissues from 350–400 full necropsies have been examined as part of the validation of this species.

Observations have been made on a range of clinical chemistry parameters, as shown in table I, and on various haematological values, as shown in table II. The incidence of natural disease has been studied and a series of ophthalmological investigations have been completed. Metabolic studies with a variety of compounds have been performed and the findings will be the subject of a separate communication.

Oral dosing may be achieved by gavage, using a plastic catheter, or by drench – unpalatable compounds being mixed with a fruit syrup. Blood samples may be taken from the femoral vein without using mechanical restraining devices [4], or by using specially constructed stocks.

The results of these experiments indicate that the marmoset is a valuable, and sometimes preferable, alternative to the dog or rhesus – e.g. in the investigation of the toxicity of non-steroidal anti-inflammatory agents, the dog is known to be highly sensitive to these drugs.

They have a number of advantages over the larger sub-human primates such as ease of handling and housing and they have a small drug requirement. They are less likely to carry dangerous diseases and the hazard to the experimenter is consequently reduced.

Teratology

Benirschke [1] and Visek [13] recommended that the marmoset would make a useful species for the evaluation of teratogenicity during a conference held in 1963 and this view was restated in a WHO report published 4 years later [11]. One reason for these recommendations was that there are placental similarities between man and marmosets [6, 12]. Benirschke was of the opinion that marmosets were probably second only to chimpanzees in their potential usefulness in assessing the safety of chemical agents in pregnant women. Subsequent published findings of teratology experiments with the marmoset support this view [7, 8, 10]. POSWILLO *et al.* studied the effects of thalidomide, X-irradiation [7] and 'blighted' potato [8] and have concluded that the marmoset should command the attention of all concerned with the study of developmental pathology and teratology. After 10 years experience with this species we feel that it has an important role to play in the determination of teratogenic potential.

Before a laboratory species can be used to evaluate the teratogenicity of novel chemical agents it is essential to be able to diagnose and time pregnancy. It is necessary to define the organogenic period, determine the response to known teratogens and to be able to obtain fetuses and placentas with the minimum of difficulty.

Pregnancy Diagnosis

In a large colony the most reliable method of early pregnancy diagnosis is uterine palpation, whilst hormone assay [3] is a more accurate technique it is time-consuming and unsuitable for the numbers of females to be used in a teratology investigation. The timing of pregnancy based upon measurements of uterine width has been described previously [2, 9] and we have also used this method, i.e. the width of the uterine fundus was measured using dial callipers and the reading compared with a standard curve that had been previously prepared from 1,500 observations made on normal animals in the colony. The early onset of pregnancy, i.e. 21 days post coitus, can be detected with little difficulty, the uterus being palpated as a hard, spherical nodule low down in the pelvic region – the non-pregnant uterus is felt as a hard cylindrical organ. It is also possible to palpate the ovaries and, with experience, the presence of corpora lutea can be detected. Diagnosis of pregnancy by the use of commercially available test kits (e.g. Gravindex and Prepurex), which reveal the presence of chorionic gonadotrophin in the urine, is not reliable enough to be able to use as the sole means of determining pregnancy. The accuracy of timing pregnancy by uterine palpation can be enhanced by a knowledge of the post parturition mating period. The litter to litter interval in our colony has been found to be 156 days, the length of gestation 138 days and the median time from parturition to coitus to be 18 days, these results have been obtained from 700 normal pregnancies. Thus by the use of palpation and knowing the normal period from parturition to coitus it is possible to date pregnancy to within 1 day and compounds may be administered at any desired point in pregnancy. In this way the thalidomide-sensitive period has been determined as between days 30 and 35 of pregnancy.

Table III. Mean values showing neonatal growth in the marmoset

	Day 125 of pregnancy	Day 0 post partum	Day 7 post partum	Day 21 post partum
Body weight, g	21.51 ± 0.33	28.32 ± 0.40	32.13 ± 0.38	40.55 ± 0.41
Crownrump length, mm	78.68 ± 0.18	81.92 ± 0.86	87.48 ± 0.69	97.79 ± 1.06
Cranial width, mm	19.62 ± 0.07	22.62 ± 0.19	23.72 ± 0.17	25.69 ± 0.21
Cranial length, mm	29.69 ± 0.05	34.33 ± 0.28	35.68 ± 0.26	37.91 ± 0.25
Placental weight, g	3.03 ± 0.06	–	–	–

Table IV. The effect of aspirin and vitamin A on the fetus when administered during pregnancy to marmosets

Group	Dose	Dosing period	Body weight, g	Comments
Control	0	days 25–50	21.22±1.73 (n = 18)	no abnormalities detected
Aspirin	300 mg/kg	days 21–35	23.88±0.83 (n = 16)	no abnormalities detected, 2 females aborted and 1 died
Vitamin A	2×10^5 IU/kg	days 17–50	22.73±1.51 (n = 17)	no abnormalities detected

Hysterotomy

Fetuses and placentas are obtained by means of a hysterotomy with subsequent recovery of the female. The hysterotomy is performed under aseptic conditions using halothane for both induction and maintenance anaesthesia. A concentration of 2.0–2.5% halothane in a mixture of oxygen (95%) and carbon dioxide (5%) is delivered at a rate of 2 litres/min using a Fluotec atomiser. Overnight starvation is not required prior to anaesthesia, which can be useful if emergency surgery is required.

Table V. The effect of methotrexate on the fetus when administered during pregnancy to marmosets

Dose of methotrexate mg/kg	Days of pregnancy when dosed	Mean body weight, g	Comments
1.0	18–62	15.90	normal
(5 females)	20–43	22.49	normal
	24–59 (2 females)	20.41	normal
	43–64	34.00	normal (littered)
10.0	12–32	–	aborted[1]
(9 females)	20–30	28.30	normal
	23–33 (2 females)	–	2 aborted
	26–36 (2 females)	22.39	normal
	28–38 (2 females)	25.44	1 normal 1 aborted[1]
	30–40	–	aborted

[1] Clinical evidence of folic acid deficiency.

The fetuses and placentas are removed via a ventral midline incision in the abdominal wall and a median hysterotomy. Ampicillin powder is lightly dusted into the abdominal cavity prior to suturing. Great care is taken when suturing to preserve the breeding potential of the female. An intramuscular injection of ampicillin is given (15 mg/kg) daily for 3 days following the operation. The female is returned to her mate a few hours after recovery from the anaesthetic and a glucose/saline drink offered from a syringe to aid recovery. The skin sutures are removed 10 days later.

150 hysterotomies have been performed in this way and some females have received three consecutive hysterotomies without impairment of fertility. The operation is straightforward, taking about 30 min, and one female has been observed mating on the same day that surgery had taken place. Females usually become pregnant within 50 days of the hysterotomy, compared with 18 days following normal parturition. In the majority of our studies the hysterotomy was performed on day 125 of pregnancy.

All fetuses, whether born normally or delivered by hysterotomy, were weighed and crown-rump length, cranial length and cranial width measured; the results are shown in table III. Skeletal examinations were performed on all offspring following either radiology or alizarin processing. Placental

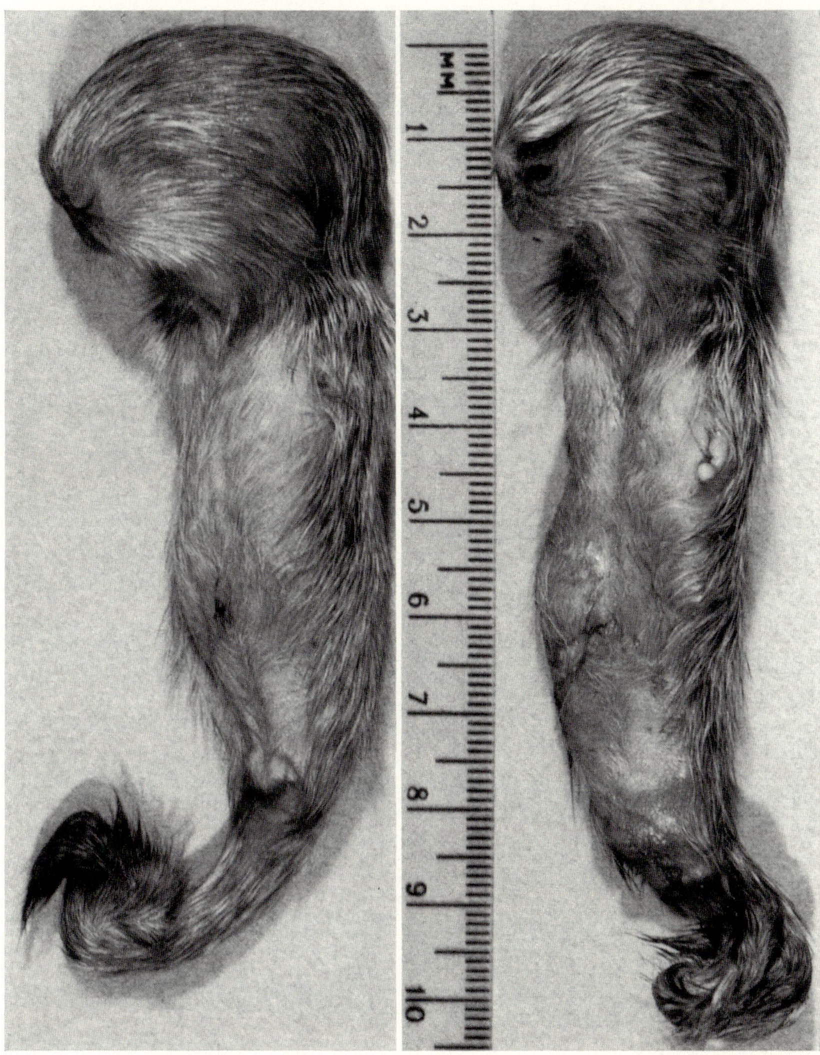

Fig. 1. Two fetuses from a marmoset that had been dosed during pregnancy with thalidomide (50 mg/kg days 25–50).

Table VI. The effect of ICI 72,222 on the fetus when administered during pregnancy to marmosets

Dose of ICI 72,222 mg/kg, days 25–50	Number of dams/group	Mean body weight, g	Mean crown-rump length, mm	Mean cranial width, mm	Mean cranial length, mm	Comments
0	21	27.26 ± 0.59	76.49 ± 0.91	21.00 ± 0.32	31.83 ± 0.36	no abnormalities detected
5	11	27.95 ± 0.86	80.93 ± 0.51	21.70 ± 0.25	32.83 ± 0.34	no abnormalities detected
25	11	25.93 ± 0.80	78.63 ± 0.56	21.76 ± 0.31	32.13 ± 0.31	no abnormalities detected
100	11	24.98 ± 0.74	77.78 ± 0.57	21.73 ± 0.14	32.35 ± 0.34	no abnormalities detected

Note: these females were subjected to hysterotomy on day 130 of pregnancy.

weights were recorded and the placentas kept for later hormone extraction. 650 offspring have been examined in this way.

The Effect of Drugs on the Fetus

A number of compounds have been administered to pregnant marmosets including thalidomide, vitamin A, aspirin, methotrexate, methallibure and several ICI compounds as part of a safety evaluation programme.

Thalidomide induced a severe embryopathy which was similar to the human syndrome, i.e. amelia, oto-mandibular defects and some soft tissue changes (see fig. 1), when administered between days 30 and 35 of pregnancy at a total dose of 50 mg/kg.

Hypervitaminosis A (doses up to 2×10^5) and aspirin (300 mg/kg) failed to induce malformations (table IV). Methotrexate (1 and 10 mg/kg) did not induce fetal malformations but at 10 mg/kg caused abortion and clinical evidence of folic acid deficiency (table V). An example of the results obtained from a routine safety evaluation study, with ICI 72,222, is shown in table VI.

Thus the marmoset may be used as a third species in teratology studies designed to evaluate the effect of new chemical agents in pregnant women. They could eventually be used as an alternative to the rabbit as a non-rodent species which is sensitive to thalidomide.

In conclusion our experiences with *C. jacchus* in toxicology and teratology indicate that this small, easily handled sub-human primate has a useful part to play in the safety evaluation studies that are essential to the development of new pharmaceutical agents.

Acknowledgments

The author wishes to thank ANDREW RUSSELL, DIANNE NICKEAS and FREDRICK DEAN for their technical assistance.

References

1 BENIRSCHKE, K.: Report to the Commission on Drug Safety; in Conference on prenatal effects of drugs, pp. 8–11 (Commission on Drug Safety, Chicago 1963).

2 GENGOZIAN, N.; SMITH, T.A., and GOSSLEE, D.G.: External uterine palpation to identify stages of pregnancy in the marmoset. *Saguinus fuscicollis* ssp. J. med. Primatol. *3:* 236–243 (1974).

3 HEARN, J.P. and LUNN, S.F.: The reproductive biology of the marmoset monkey,

Callithrix jacchus; in PERKINS and O'DONOGHUE Breeding simians for developmental biology. Laboratory animal handbooks, vol. 6, pp. 191–202 (Laboratory Animals Ltd., London 1975).

4 HEARN, J.P.; LUNN, S.F.; BURDEN, F.J., and PILCHER, M.M.: Management of marmosets for biomedical research. Lab. Anim. *9:* 125–134 (1975).

5 HIDDLESTON, W.A.: Large scale production of a small laboratory primate – *Callithrix jacchus;* in ANTIKATZIDES, ERICHSEN and SPIEGEL The laboratory animal in the study of reproduction. 6th ICLA Symposium, Thessaloniki, pp. 51–57 (Fischer, Stuttgart 1976).

6 JOLLIE, W.P.: Fine structural changes in the placental membranes of the marmoset with increasing gestational age. Rec. *176:* 307–320 (1973).

7 POSWILLO, D.E.; HAMILTON, W.J., and SOPHER, D.: The marmoset as an animal model for teratological research. Nature, Lond. *239:* 460–462 (1972).

8 POSWILLO, D.E.; SOPHER, D., and MITCHELL, S.: Experimental induction of foetal malformation with 'blighted' potato: a preliminary report. Nature, Lond. *239:* 462–464 (1972).

9 PHILLIPS, I.R. and GRIST, S.M.: The use of transabdominal palpation to determine the course of pregnancy in the marmoset *(Callithrix jacchus)*. J. Reprod. Fert. *43:* 103–108 (1975).

10 PHILLIPS, I.R.: Macaque and marmoset monkeys as animal models for the study of birth defects; in PERKINS and O'DONOGHUE Breeding simians for developmental biology. Laboratory animal handbooks, vol. 6, pp. 293–302 (Laboratory Animals Ltd., London 1975).

11 World Health Organisation: Technical Report, Series No. 364, p. 36 (1967).

12 WISLOCKI, G.B.: Placentation of the marmoset *(Oedipomidas geoffroyi)* with remarks on twinning in monkeys. Anat. Rec. *52:* 381–399 (1932).

13 VISEK, W.J.: Report to the Commission on Drug Safety; in Conference on prenatal effects of drugs, pp. 42–45 (Commission on Drug Safety, Chicago 1963).

R.A. SIDDALL, ICI Pharmaceuticals Division Ltd., *Alderley Park, Cheshire* (England)

Prim. Med., vol. 10, pp. 225–231 (Karger, Basel 1978)

Marmosets as Models of Speech and Language Disorders

Charles T. Snowdon

Department of Psychology, University of Wisconsin, Madison, Wisc.

The study of speech and language disorders has been greatly limited by the failure to develop appropriate animal models. Quite possibly the belief that speech and language are uniquely human phenomena has led to less development of research in this area than in other areas of experimental medicine.

Up to now the most successful animal models for speech and language have been developed in birds. In several species of birds researchers have uncovered auditory 'templates' for the recognition of certain vocal signals. These templates must be combined with appropriate environmental stimuli in order for normal song to develop [9]. Other evidence suggests a lateralization of neurological control over singing in birds that is inflexible after a certain period in development, similar to the permanence of lateralization of neurological control over speech in humans [12].

However, primates are phylogenetically more similar to humans than birds, and, because of this similarity, primates should be prime candidates for the search for appropriate animal analogs of speech and language phenomena. Unfortunately the search to date has been inconclusive. Dewson *et al.* [3] demonstrated an effect of temporal cortex ablation on the response of primates to human speech sounds and several studies have investigated the response of squirrel monkey auditory cortex to squirrel monkey vocalizations [14]. A few attempts have been made to elicit vocalizations in monkeys through electrical stimulation of the brain [10]. Several investigations have tried to demonstrate hemispheric lateralization in monkeys as a possible analogy to the lateralization of control over speech in humans. The data are conflicting on whether monkeys do demonstrate lateralization or not [2, 5, 21]. It seems unlikely that simple lateralization of function *per se* will be a sufficient model for speech and language studies.

Little work has been attempted on the ontogeny of vocal communication in primates. One study has revealed that monkeys reared in social isolation show vocal pathologies [13], one showed that squirrel monkeys deafened as adults displayed minor vocal pathology [20] and a few studies have indicated some difficulty in conditioning vocalizations in primates [19, 22]. Probably the most popularized primate model for language development has been the chimpanzee. There are now several demonstrations indicating that chimpanzees have the cognitive ability to develop symbol usage similar to our manipulation of signals in a language [4, 16, 17]. These studies, however, will not be very useful in developing analogs to human speech and language disorders. The investment required to train one chimpanzee on an artificial language is immense and time-consuming, whereas if one can work with an animal's natural communication system, such an investment in training need not be made.

In attempting to develop primate models of speech and language disorders research should focus on analogous but species-specific phenomena. It is unlikely that exact homologies to human speech and language will be found in any non-human species. There has been a long history of unsuccessful attempts to train chimpanzees in human speech [6]. The data is now quite convincing that chimpanzees and other primates do not have the anatomical structures necessary to be able to mimic most speech sounds [7].

In our search for analogous species-specific phenomena, we have successfully developed analogies to certain speech phenomena in our work with the contact vocalizations (trills) of the pygmy marmoset *(Cebuella pygmaea)*, which might contribute to models of speech and language disorders. Some parallels to humans have emerged in the ontogeny of vocal communication in marmosets. Marmoset vocalizations manifest contextually correlated variations that might be analogous to morphemes or intonation contours in human speech. Additionally, pygmy marmoset perception of their own vocalizations occurs in a categorial manner analogous to human perception of speech sounds.

With respect to the ontogeny of vocal communication, we have identified four stages in the social development of pygmy marmosets which correspond to degree of dependence or independence from the parents [15]. In the first stage, which lasts about two weeks, the infants are constantly with their parents and there is no audible contact call. In stage 2 the parents initiate independence by leaving the infants alone for several minutes at a time. During this period the infants vocalize continuously until their parents retrieve them. Imbedded in a sequence of calls is a trill as shown at the top

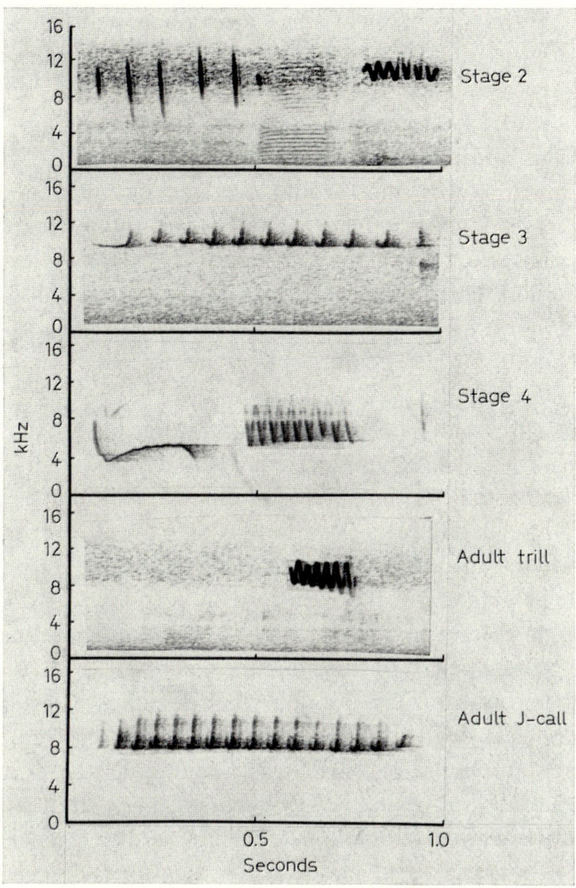

Fig. 1. Trill-like vocalizations of each developmental stage and adult forms. Stage 2: parental initiated independence; stage 3: parent and infant initiated independence; stage 4: complete independence.

of figure 1. In stage 3 both infants and parents initiate independence. As the infants move about on their own, rather than vocalize constantly they frequently emit what we have called a 'J-call' shown in the second row of figure 1. The trill-like form from stage 2 is not heard at all although the J-call seems to be based on a trill with the downsweep omitted. In stage 4 when the animals are generally four to five months old and almost completely independent, they emit both J-call and occasionally a trill embedded in a sequence of calls as shown in the third row of figure 1. The adults produced several types of trills (see below) but juveniles at stage 4 produced trills that were higher pitched and more irregular in form than the adults. In addition, adult trills were almost always observed in isolation from other calls, whereas the stage 4 juvenile trills were heard only within a sequence of calling.

Two of our animals showed a retarded social development. That is, they progressed at a much slower rate through the stages of independence and had not shown stage 4 independence at one year. Interestingly, the emergence of their vocalization paralleled the stages of social development rather than correlating with chronological age or physical maturation. The finding that vocalizations develop in a fixed sequence that correlates with social development strongly parallels results of phonemic and syntactic acquisition in human children [1, 11].

With respect to contextually correlated variation pygmy marmosets show five distinct variations in their contact trills [15] which are shown in figure 2. The closed mouth trill (A) is the most common trill used by adult marmosets; it is given in relatively calm situations as animals move from one place to another. Closed mouth trills usually elicit antiphonal trill responses. The open mouth trill (B) given with the mouth open is followed with high probability by some agonistic behavior (threat or retreat) and rarely elicits antiphonal responses. The open mouth trill differs acoustically from the closed mouth trill only in duration. The boundary between the longest closed mouth trill and the shortest open mouth trill was at 250 msec. The quiet trill (C) is given when two animals are quite close to each other (ca. 1 m) and has both a lower amplitude and a smaller frequency range relative to the closed mouth trill. The juvenile trill (D) has been described earlier. The J-call (E) is given both by stage 3 juveniles in all situations where adults would use trills and by adults when they are isolated from the rest of the colony. This isolation can be simply loss of visual contact or capture and restraint within a carrying cage. The form of the J-call has specific individual variations in structure allowing for individual recognition. These trill variations indicate that marmosets have a much larger and more subtle 'vocabulary' than has been previously recorded. The richness of information conveyed by these variations is analogous to the information conveyed by different words or different intonations in human speech.

Finally, one of the major phenomena of human speech perception literature is 'categorical perception' [8]. The acoustic parameters of a given speech sound can be varied over a broad range and still be perceived as that particular phoneme. However, there are sharp boundaries where a small change in an acoustic parameter yields perception of a different phoneme.

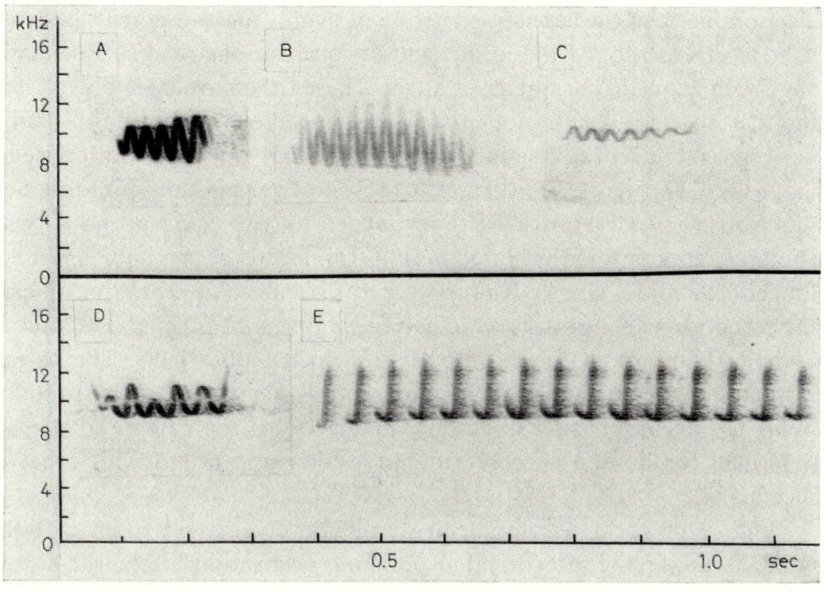

Fig. 2. Trill variations found in pygmy marmosets. A = Closed mouth trill; B = open mouth trill; C = quiet trill; D = juvenile trill; E = J-call.

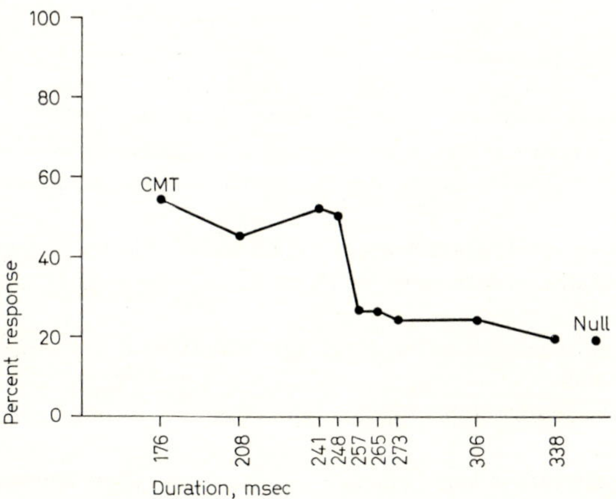

Fig. 3. Percent antiphonal responding by marmosets in response to playbacks of synthesized vocalizations varying in duration. CMT = Closed mouth trill; Null = no stimulus presented.

The simplicity of the acoustic structure of pygmy marmoset trills has enabled us to electronically synthesize marmoset vocalizations so as to systematically vary each acoustic parameter of a trill. These trills have been played back to the marmosets through hidden speakers, and their perception of a trill has been measured by the presence or absence of an antiphonal response to the playback [18]. Figure 3 illustrates the result of variation in duration. Several durations (up to 248 msec) were responded to with equal probability as the stimulus representing the closed mouth trill. However, a stimulus just 9 msec longer (257 msec) elicited a probability of response that did not differ from the response when no trills were played. It is significant that 250 msec is the dividing line between closed mouth and open mouth trills. The pattern of responding to several trill durations as equivalent with a sharp boundary point for discrimination is exactly analogous to the categorial perception described for human speech. Similar results were found with other parameters [18].

Thus, in vocal development, in the variety of trills used in different contexts, and in the categorical responding to variations in their trills, pygmy marmosets show analogues to human speech phenomena. The study of marmosets with their extensive vocal repertoires and frequent production of vocalizations shows great promise as a primate model of speech and language disorders.

References

1 Brown, R.: A first language (Harvard University Press, Cambridge 1973).

2 Dewson, J.H., III: Preliminary evidence of hemispheric asymmetry of auditory function; in Harnad Lateralization of the nervous system (Academic Press, New York 1976).

3 Dewson, J.H., III; Pribram, K.H., and Lynch, J.C.: Effects of ablation of temporal cortex upon speech sound discrimination in the monkey. Expl Neurol. *28:* 579–591 (1969).

4 Gardner, R.A. and Gardner, B.T.: Teaching sign language to a chimpanzee. Science *165:* 664–672 (1969).

5 Hamilton, C.R.: An assessment of hemispheric specialization in monkeys. Ann. N.Y. Acad. Sci. (in press).

6 Kellogg, W.N.: Communication and language in the home-raised chimpanzee. Science *162:* 423–427 (1968).

7 Lieberman, P.: On the origins of language (Macmillan, New York 1975).

8 Liberman, A.M.; Cooper, F.S.; Shankweiler, D.P., and Studdert-Kennedy, M.: Perception of the speech code. Psychol. Rev. *74:* 431–461 (1967).

9 Marler, P.: Birdsong and speech development: could there be parallels? Am. Scient. *58:* 669–673 (1970).

10 Maurus, M. and Ploog, D.: Social signals in squirrel monkeys. Analysis by cerebral brain stimulation. Expl Brain Res. *10:* 532–554 (1970).

11 Menyuk, P.: Acquisition and development of language (Prentice Hall, Englewood Cliffs 1971).

12 Nottebohm, F.: Ontogeny of bird song. Science *167:* 950–956 (1970).

13 Newman, J.D. and Symmes, D.: Vocal pathology in socially deprived monkeys. Devl Psychobiol. *7:* 351–358 (1974).

14 Newman, J.D. and Wollberg, Z.: Multiple coding of species specific vocalizations in the auditory cortex of squirrel monkeys. Brain Res. *54:* 287–304 (1973).

15 Pola, Y.V. and Snowdon, C.T.: The vocalizations of pygmy marmosets *(Cebuella pygmaea)*. Anim. Behav. *23:* 826–842 (1975).

16 Premack, D.: Language in chimpanzees? Science *172:* 808–822 (1971).

17 Rumbaugh, D.M. and Gill, T.V.: The mastery of language-type skills by the chimpanzee (Pan). Ann. N.Y. Acad. Sci. *280:* 562–578 (1976).

18 Snowdon, C.T. and Pola, Y.V.: Interspecific and intraspecific responses to synthesized pygmy marmoset vocalizations. Anim. Behav. (in press).

19 Sutton, D.; Larson, C.; Taylor, E.N., and Lindeman, R.C.: Vocalization in rhesus monkey. Conditionability. Brain Res. *52:* 225–231 (1973).

20 Talmage-Riggs, G.; Winter, P.; Ploog, D., and Mayer, W.: Effect of deafening on the vocal behavior of the squirrel monkey *(Saimiri sciureus)*. Folia primatol. *17:* 404–420 (1972).

21 Warren, J.M.: Handedness and cerebral dominance in monkeys; in Harnad Lateralization in the nervous system (Academic Press, New York 1976).

22 Wilson, W.A., jr.: Discriminative conditioning of vocalizations in *Lemur catta.* Anim. Behav. *23:* 432–436 (1975).

C.T. Snowdon, PhD, Department of Psychology, University of Wisconsin, *Madison, WI 53706* (USA)

Prim. Med., vol. 10, pp. 232–238 (Karger, Basel 1978)

Callithrix jacchus jacchus as a Subject for Behavioural Projects in Biomedical Research

T.B. POOLE, M.F. STEVENSON and A.G. SUTCLIFFE[1]

Department of Zoology, University College of Wales, Aberystwyth

Introduction

The common marmoset *(Callithrix jacchus jacchus)* is, in many ways, an ideal primate for behavioural studies in the laboratory. It is of small size, so that it is relatively cheap to maintain and also displays a wide behavioural repertoire in the relatively confined conditions of a laboratory cage. Unlike many species of primate it is possible to create groups in the laboratory with a natural age and sex structure, showing behavioural interactions which closely reflect those occurring under natural conditions.

There have, as yet, been few published quantitative studies of marmoset behaviour [1–4, 6, 8, 10, 11]. The main aim of this present paper is to point the way to the future use of the marmoset as a behavioural model for biomedical research.

Methodology

Common marmosets were kept in large observation cages on a mixed diet; details of their husbandry have been described by STEVENSON [9].

In behavioural studies it is essential that objective quantitative data should be collected and a brief outline will be given of the methods employed.

Two kinds of behavioural data have been collected, namely, ongoing behaviour and experimentally induced behaviour. In the case of ongoing behavioural data the animals

1 We should like to record our gratitude to the Science Research Council for financial assistance (Grant B/RG/44085).

were observed when behaving naturally in their home cage. Data of the second type have been recorded in experimental cages connected by flexible ducting to the home cage [5].

The following procedures were adopted for recording data. Firstly, behaviour patterns were identified and defined [11]. Secondly, data were collected using the three following methods:

(a) Check sheets were used to record relatively infrequent or long-lasting behaviour such as grooming, scent marking, copulation etc. The presence of these behaviour patterns in 30-sec intervals was scored for sessions of up to 30 min for a group.

(b) Data were also collected using tape-recorded commentaries which were subsequently transcribed (using a code number for each behaviour pattern) and processed through computer programmes. This method provided data on the frequency of occurrence and temporal associations of behaviour patterns. The precise methodology has been described elsewhere by Stevenson and Poole [11].

(c) Cine film (16 mm) or CCTV were used to record and analyze rapid sequences of behaviour. These methods provided the facility of frame by frame analysis which made it possible to identify the behaviour carried out by two or more individuals simultaneously. The precise durations of behaviour patterns could also be measured [7].

Results

Breeding Records

The Aberystwyth marmoset colony increased from eight individuals in May 1973 to 107 in March 1977. During this 46-month period two breeding adults died and four adults were introduced to the colony (May 1975). Of this stock, four animals were wild caught, all others having been laboratory bred.

Marmosets were kept in family groups and allowed to reproduce naturally until they contained eight to ten individuals (parents plus three or four successive sets of twin offspring). Although the older offspring in these groups were of adult age ($>$13 months) they did not breed; the dominant pair are believed to exert a suppressive effect on the breeding of other group members, but the mechanism involved is not understood.

Breeding data are shown in table I and it is apparent that we are more successful in rearing second generation offspring than we have been in rearing those of the first generation. The reason for this difference is that one of the laboratory bred founder females, which had been kept in social isolation from the age of four months, killed approximately half of her offspring at birth. All parents of second generation infants had experience of assisting in the rearing of younger sibs, in addition they invariably reared their first set of infants successfully.

Table I. Breeding results from Aberystwyth marmoset colony May 1973 to March 1977

Generation	Litter size, %			Total young		Reared, %	Success, %[1]
	2	3	4	born	reared		
F_1	39	57	4	74	44	59	70
F_2	26	63	11	77	54	70	90
Total	33	60	7	151	98	65	

Founder stock four pairs (V/73) + two pairs (V/75).

[1] Excluding triplet deaths (one triplet normally dies as families can only rear twins).

Behavioural Differences between Sexes

In the common marmoset the social behaviour of the two sexes was found to be similar. Both sexes care for young and no consistent sexual differences were found in the frequency of occurrence of major social behaviour patterns, although great individual differences existed. No differences were found between the sexes in the frequency of occurrence or duration of social play.

Agonistic Behaviour

Marmosets are known to be aggressive towards strangers [4]. Recent experimental studies of the Aberystwyth colony have shown that adults are aggressive towards adult or subadult members of their own sex; subadults (nonbreeding adults), however, are aggressive to other subadults but invariably submissive to adults. Neither adults, subadults, or juveniles show hostility towards unfamiliar juveniles. Aggression was characterized by the behaviour patterns 'attempted bite' (through transparent barrier), 'erh-erh vocalization', 'frown', and 'stare'; submission was inferred from 'gecker', 'avoid' and 'crouch'.

Intragroup aggression occurred most commonly in two contexts. In approximately half of the family groups in which younger sibs were present, five- to ten-month-old twins showed fighting which was sometimes prolonged and frequently resulted in a dominant/subordinate relationship which lasted from one to three days. Twin fights could often be seen to have occurred because two individuals were scarred and showed a dominant/subordinate relationship (table II).

Intragroup agonistic behaviour also may occur when family groups contain subadult individuals; one marmoset in the group attacks another

Table II. Twin fights recorded from the Aberystwyth marmoset colony

Sex of combatants	Family No.	No. in group	Duration of fight, min	Winner identifiable	Age of combatants, weeks
♂♂	I	6	–	+	22
	II	8	60	+	30
	II	10	–		30
	IV	6	30	+	24
	IX	6	25	+	26
	X	5	–	+	30
♂♀	II	6	15		28
	X	8	–		30
	V	6	–		24
♀♀	II	10	–		30
	I	8	30		25
	IV	8	–	+	44
	VI	6	7		38

– = Fight not observed.

Table III. Intrafamily aggressive behaviour in *C. j. jacchus* recorded from the Aberystwyth colony (excluding twin fights)

Family No.	No. in group	Attacker		Defender		Defender permanently peripheralized
		sex	age	sex	age	
IV	10	♀	*	♀	15	+
III	9	♀	*	♀	15	+
III	8	♀	13	♀	13	+
X	8	♂	13	♂	13	–
II	8	♂	15	♂	15	+
IV	6	♂	*	♂	15	+
IV	6	♂	11	♂	11	+
II	7	♂	15	♂	10	+
II	7	♂	15	♂	10	+
X	8	♂	16	♂	9	–
II	10	♀	18	♀	8	–

* = Breeding adult = (parent). Ages are given in months for non-breeding individuals (offspring).

which becomes submissive. In extreme cases the submissive animal becomes 'peripheralized'. Peripheralization may be defined as the situation in which an individual's social interactions with the rest of the group are significantly decreased as a result of attacks on it by one or more members of the group. In some cases of peripheralization the removal of a group member is essential but in other cases the attacked animal is accepted back into the group. It can be concluded from table III that attacks are confined to members of the same sex and that, in situations where the two individuals are of different status, it is invariably the younger animal which emerges as the loser.

Social Play

Common marmosets spend a great deal of time in play and this aspect of social behaviour has been analyzed in detail [11, STEVENSON and POOLE, in preparation]. Social play is a well-structured complex form of social behaviour which can readily be measured and analyzed. Juvenile marmosets show little consistent individual variation in the amount of time spent in play. The main forms of social play recorded could be divided into rough and tumble and approach/withdrawal. Play frequently occurs spontaneously in paired adults, but, after the birth of offspring, play is generally initiated by the young with all members of the family group subsequently participating. Breeding adults play less than other members of the group. When infant marmosets first begin to play they interact almost exclusively with their twin. They subsequently play with older members of the family, when the infants' physical development is more compatible with that of other members of the group.

General Conclusions

Before the behaviour of a species of animal can be manipulated to provide data of direct relevance to medicine, it is essential that the normal ongoing situation should have been thoroughly investigated. Such studies form an essential baseline from which experimental paradigms can be planned and also ensure that the methods used for assessing the influence of drugs or physiological manipulation are realistic in terms of the animal's known behavioural repertoire. The science of animal behaviour is too complex to favour the use of naive experimental methods and, although such methods have in the past been justified on the grounds that results may be obtained rapidly, the value of such an approach is minimal, for not only will apparently negative results conceal positive findings, but positive results may also acquire spurious validity.

The research on the common marmoset which has been briefly outlined shows that this species has many advantages as a behavioural model for experimental medicine. Its basic social structure is the family group, the parents, like human beings, form a long-lasting and close relationship and juvenile marmosets assist in taking care of infants.

The social behaviour of common marmosets can be defined and readily quantified by a variety of objective methods and behavioural units are stable and readily identifiable. Parental behaviour, play, pair formation and agonistic behaviour form readily identifiable complexes which offer excellent opportunities for experimental manipulation. The frequency of occurrence of different behaviour units and their sequential organization can readily be examined under different experimental conditions.

From a practical viewpoint, marmosets are adaptable, highly active animals with well-developed exploratory activity. Unlike many other small laboratory mammals, sight is a primary sense so that experiments involving visual discrimination are feasible. The fact that much of their communicatory system is visual and auditory also makes interpretation of their social behaviour relatively easy.

Marmosets do not show major sex-linked differences in behaviour and individual differences appear to be of much greater importance. Although individual variability may have the disadvantage of yielding data with high variances the situation is far more realistic in terms of human behaviour than one in which animals showing few individual differences are used.

It is clearly of importance that further baseline data should be acquired and that the development of experimental methods should proceed. Experimental studies are currently in progress in the Aberystwyth Animal Behaviour Group on the pair relationship and the influence of psychotropic drugs on marmoset behaviour.

References

1 Box, H.O.: Quantitative studies of behaviour within captive groups of marmoset monkeys *(Callithrix jacchus)*. Primates *16:* 155–174 (1975).

2 Epple, G.: Vergleichende Untersuchungen über Sexual- und Sozialverhalten der Krallenaffen (Hapalidae). Folia primatol. *7:* 37–65 (1967).

3 Epple, G.: Comparative studies on vocalization in marmoset monkeys (Hapalidae). Folia primatol. *8:* 1–40 (1968).

4 EPPLE, G.: Maintenance, breeding and development of marmoset monkeys (Callithricidae) in captivity. Folia primatol. *12:* 56–76 (1970).

5 HEARN, J.P.; LUNN, S.F.; BURDEN, F.J., and PILCHER, M.M.: Management of marmosets for biomedical research. Lab. Anim. *9:* 125–134 (1975).

6 INGRAM, J.C.: Parent-infant interactions in the common marmoset *(Callithrix jacchus)* and the development of young; thesis Bristol (1975).

7 POOLE, T.B.: Detailed analysis of fighting in polecats (Mustelidae) using cine film. J. Zool., Lond. *173:* 369–393 (1974).

8 ROTHE, H.: Some aspects of sexuality and reproduction in groups of captive marmosets *(Callithrix jacchus)*. Z. Tierpsychol. *37:* 255–273 (1975).

9 STEVENSON, M.F.: Maintenance and breeding of the common marmoset *Callithrix jacchus* with notes on hand rearing. Int. Zoo Yb. *16:* 110–116 (1976).

10 STEVENSON, M.F.: Birth and perinatal behaviour in family groups of the common marmoset *(Callithrix jacchus jacchus)*, compared to other Primates. J. hum. Evol. *5:* 365–381 (1976).

11 STEVENSON, M.F. and POOLE, T.B.: An ethogram of the common marmoset *(Callithrix jacchus jacchus):* general behavioural repertoire. Anim. Behav. *24:* 428–451 (1976).

T.B. POOLE, Department of Zoology, University College of Wales, *Aberystwyth, SY23 3DA* (UK)

Session V: Infectious Diseases
Chairman: F. DEINHARDT, München

Prim. Med., vol. 10, pp. 239–253 (Karger, Basel 1978)

Spontaneous Infectious Diseases of Marmosets[1]

R. D. HUNT, M. P. ANDERSON and L. V. CHALIFOUX

Harvard Medical School, New England Regional Primate Research Center, Southborough, Mass.

Introduction

Infectious diseases represent one of the major health problems of most species of non-human primates, the marmosets not excluded. Although our understanding of many of the agents and their associated diseases has greatly expanded in the past decade, our knowledge is pitifully small when compared with other laboratory animals, domestic animals or human beings. Progress will continue to be relatively slow when you consider how few laboratories are addressing problems of marmosets, and the relatively small number of animals actually used and available. The use of wild-caught marmosets and open colonies accentuates the occurrence of infectious diseases, but even as we move toward closed colonies we cannot expect infectious diseases to suddenly disappear. Our concern is threefold, the health of the animal, the potential of zoonoses, and the value of spontaneous infectious diseases as models for health problems of human beings. We are fortunate that our concerns can include the latter asset.

This introduction to the sessions on infectious diseases will review with intentional brevity those spontaneous infections which have been described in marmosets.

[1] Supported by grant number P40-RR00168 of the Division of Research Resources, National Institutes of Health.

Herpesvirus-T (H. tamarinus; H. platyrrhinae)

Herpesvirus-T infection has many similarities to *Herpesvirus simplex* and other herpesvirus diseases, however, the virus is distinct. Infection is limited to New World monkeys. The virus was first isolated independently by Holmes *et al.* [23] and Melnick *et al.* [39] in 1963 from white-lipped tamarins *(Saguinus nigricollis)* and cotton-topped marmosets *(Saguinus oedipus)*, in which fatal epizootics occurred.

The squirrel monkey *(Saimiri sciureus)* is believed to be the principal reservoir host for this agent. It is the only reservoir host in which clinical disease has been seen and from which virus has been isolated [33]. Significant serum antibody titers also occur in a high percentage of cinnamon ringtail monkeys *(Cebus albifrons)* and spider monkeys (*Ateles* spp.) which suggests that these species also serve as reservoir hosts [24]. Recorded lesions in the squirrel monkey consist of lingual, oral, and labial ulcers characterized by necrosis, multinucleated giant cells, and intranuclear inclusion bodies. Healing occurs spontaneously in 7–10 days. Visceral lesions have not been described. Aside from certain gaps in our knowledge, it appears that the relationships between the squirrel monkey and *Herpesvirus-T* is analogous to those between man and *H. simplex*.

Marmosets *(S. nigricollis, S. oedipus)* have a less fortunate relationship with *Herpesvirus-T*. In these species the virus causes a systemic infection with a high mortality rate. Owl monkeys *(Aotus trivirgatus)* are also susceptible to *Herpesvirus-T* in which the disease is essentially identical [27].

Following an incubation period of 7–10 days, a clinical syndrome develops, characterized by oral and labial vesicles and ulcers, ulcerative dermatitis, occasionally conjunctivitis, anorexia, lassitude, and hyperesthesia (as evidenced by intense scratching). After a duration of 2–3 days, the animal becomes moribund and dies. Ulcers are usually present throughout the oral cavity and on the lips, skin, esophagus, small intestine, cecum, and colon. Hemorrhage is present in most lymph nodes and the adrenal cortices. The capsular and cut surfaces of the liver contain innumerable, irregularly-sized grey to red foci of necrosis. Microscopically, lesions can be found in most organs. These lesions are characterized in all tissues by focal necrosis, multinucleated giant cells (in mucosal tissues), and intranuclear inclusion bodies and only vary from tissue to tissue due to each organ's unique cyto-architecture. Encephalitis can occur, but it is never extensive, and in most examples, absent.

Control of the disease can be partially effected by housing the fatally affected host separately from the reservoir hosts. This does not preclude exposure prior to their introduction into a colony, which stresses the importance of isolation or quarantine upon receipt. A live variant small plaque virus vaccine has been developed by DANIEL *et al.* [12] which has proved highly effective in preventing the infection in owl monkeys. The vaccine also affords protection to marmosets but a certain percentage of vaccinated animals develop a herpetic myelitis leading to paralysis and death.

Herpesvirus simplex (H. hominis)

Human beings are the reservoir host for *H. simplex*. The host-virus relationship is similar to *Herpesvirus-T* infection in squirrel monkeys, but the infection is better understood and has been associated with a wide variety of clinical and pathological conditions. The infection in man will not be reviewed here.

Among South American monkeys, *H. simplex* is known to cause a fatal systemic disease in owl monkeys that is essentially identical to *Herpesvirus-T* in this species or marmosets [28]. Marmosets have been shown to be experimentally susceptible to *H. simplex,* developing a systemic disease analogous to the disease in owl monkeys or *Herpesvirus-T* infection in marmosets.

NAHMIAS *et al.* [41] reported successful experimental infection of the cervix in *C. albifrons* with *H. simplex* type 2. A recurrent latent infection results which is not fatal.

As with *Herpesvirus-T* infection, we have shown that a small plaque variant of *H. simplex* provides excellent protection to owl monkeys, but in view of the rarity of the disease in marmosets it is not recommended.

Marmosets are also host to their own herpesvirus, termed *Herpesvirus saguinus,* but it has not been associated with any disease.

Herpesvirus saimiri

Herpesvirus saimiri is carried as a latent viral infection by squirrel monkeys *(S. sciureus)*. It has not been associated with any disease, spontaneous or experimental, in this species. The importance of this virus lies in the fact that it is the first herpesvirus shown to be oncogenic in mammals, the first herpesvirus proven to induce leukemia and malignant lymphoma in primates and the first virus of primate origin which is oncogenic in primates. *H. saimiri* induces leukemia or malignant lymphoma in several species of marmosets *(Saguinus fuscicollis, S. oedipus, S. nigricollis, Callithrix jacchus)*, owl monkeys *(A. trivirgatus)*, spider monkeys *(Ateles geoffroyi)* and rabbits *(Oryctolagus cuniculus)*.

At this writing, *H. saimiri* would not appear to be an important cause of spontaneous disease of any laboratory animal, however, the recent recovery of *H. saimiri* from a spontaneous case of leukemia in an owl monkey indicates it can cause a spontaneous malignancy, presumably contagious [26].

Herpesvirus ateles

Herpesvirus ateles is carried as a latent viral infection by spider monkeys *(A. geoffroyi)*. It has not been associated with spontaneous disease in any species but experimentally it induces malignant lymphoma with leukemia in cotton-topped marmosets *(S. oedipus)*, owl monkeys and rabbits [38].

Measles (Rubeola)

Measles is also known to be infectious for several species of primates including rhesus *(Macaca mulatta)*, cynomolgous *(M. fascicularis)*, Taiwan macaque *(M. cyclopis)*, baboons (*Papio* spp.), African Green *(Cercopithecus aethiops)*, marmosets (*Saguinus* spp.), squirrel monkeys *(S. sciureus)* and chimpanzees (*Pan* sp.).

In most species clinical disease is rarely recorded, although incidence of infection may be high.

An epizootic of measles in marmosets has been reported by LEVY and MIRCOVIC [36]. The infection resulted in death of 326 animals over a six-month period in a colony composed of *S. oedipus, S. fuscicollis* and *C. jacchus*. Clinical signs included lethargy, puffy upper eyelids, mucous nasal discharge and occasionally a maculopapular exanthema. Death occurred 8–18 h after the first signs were noted. There were no significant gross lesions. Microscopically the infection was characterized by an interstitial pneumonia with multinucleated giant cells in alveolar walls, alveolar ducts, bronchioles and free in the airways. Eosinophilic intranuclear inclusion bodies were present in the giant cells, but cytoplasmic inclusion bodies were not described. Warthin-Finkledey cells were present in lymph nodes, spleen and intestinal lymphoid tissues. These lesions are quite characteristic of but not pathognomonic of measles. Confirmation required virus isolation and identification.

To our knowledge this is the only record of measles in marmosets, which is surprising in view of the several colonies in existence and prevalence of the virus. No example has been seen at our center.

Rabies

Rabies is extremely uncommon in non-human primates. There is a single report of the disease in marmosets, in a *S. nigricollis* [1]. This monkey had received live, attenuated rabies vaccine shortly before developing clinical signs, which was the most probable source of infection. In a closed colony there is little likelihood of rabies, but it must be considered for any obscure neurologic disease in wild-caught animals.

Yellow Fever

Yellow fever is an important disease of human beings and non-human primates of Central and South America and the African continent. The disease has occurred in Europe and in the United States as recently as the twentieth century and its continual existence in a sylvatic cycle in tropical countries presents a potential danger for its return to the continents or its extension to other parts of the world.

The incubation period in man is three to ten days and the disease may vary from inapparent, to mild, to a fulminating disease with about 10% mortality. Clinical signs include fever, muscular aches, headache, nausea, jaundice, vomiting (sometimes with blood, hence 'black vomit' as a former name for the disease) and hemorrhages. Leukopenia and albuminuria are frequent. Pathological findings include icterus, and hemorrhages, but are dominated by degeneration of renal and hepatic parenchymal cells. There may be extensive necrosis of hepatocytes especially in the midzonal areas of lobules. Fatty change is seen in nonnecrotic cells. Necrotic hepatocytes undergo a peculiar hyalin change forming the characteristic Councilman body. Small irregular eosinophilic intranuclear inclusion bodies may develop, but they are probably not specific.

Most species of non-human primates are susceptible to yellow fever but there is marked species variation in the severity of the infection. In general,

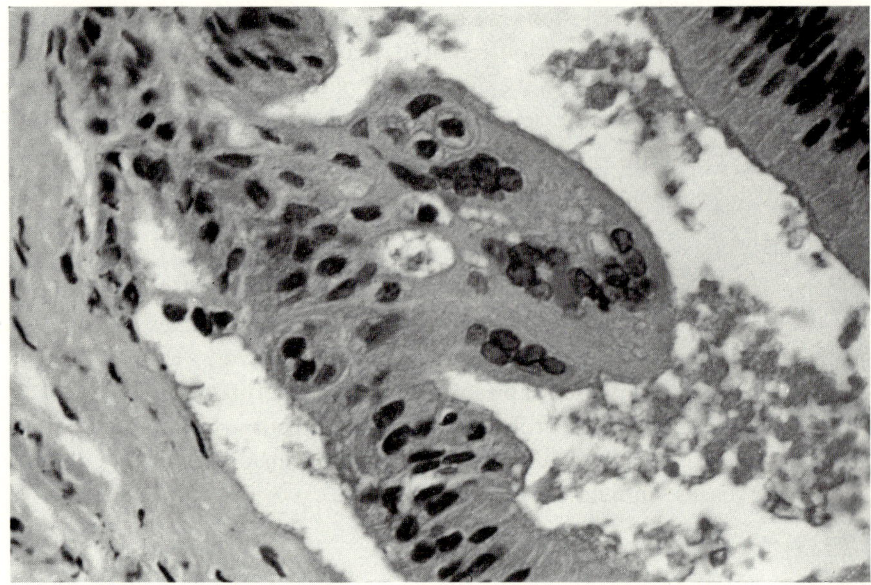

Fig. 1. Paramyxovirus cholangitis with syncytial cell formation in a marmoset. HE. ×400.

natural infection is rarely severe or fatal in African monkeys, whereas epizootics with relatively high mortalities have occurred in South America. *Alouatta* spp. are particularly susceptible, followed by *A. trivirgatus* and *Saguinus* spp., but clinical disease may also develop in *Ateles, Saimiri, Cebus,* and *Callicebus*. The pathological features of the disease in monkeys are similar to those found in man. In the healing state there is extensive regeneration of hepatocytes and a focal mononuclear cell infiltration. Although marmosets are susceptible, yellow fever is essentially unknown as a spontaneous disease in laboratory marmosets.

Paramyxovirus enterocolitis

A paramyxovirus, distinct from measles has been associated with a severe enterocolitis in marmosets which occurred in 1975 at the New England Regional Primate Research Center [5]. A similar disease and virus has also been recognized by DEINHARDT [13]. The classification of the agent is still incomplete as is the epizootiology and pathogenesis of the infection.

The disease occurred in *S. oedipus,* and *C. jacchus*. Clinically there was anorexia, diarrhea, and dehydration. Death usually occurred within 24h. Grossly there was congestion and hemorrhage of the mucosa of the stomach, cecum and colon and rarely in the small intestine. Peyer's patches and mesenteric lymph nodes were enlarged. Microscopically the mucosa of the stomach, cecum and colon was thin with foci of necrosis and hemorrhage, and associated infiltration of neutrophils. Epithelial syncytial cells containing intranuclear inclusion bodies were present in the mucosa. Syncytial cells were also present in bile ducts (fig. 1), hepatic cords, pancreatic ducts, renal tubules. There was necrosis of germinal centers in Peyer's patches and mesenteric lymph nodes with hyperplasia of lymphocytes and reticuloendothelial cells. The disease was reproduced in four *S. oedipus* inoculated via the intramuscular and intravenous routes. In addition to the lesions seen in the spontaneous disease, these animals also had an interstitial pneumonia. A survey for serum neutralizing antibody in *S. oedipus* and *C. jacchus* after the disease had abated revealed that 37% of 69 animals sampled had specific antibody. Specific antibody was also found in *A. trivirgatus* and *Cebus apella* but no other signs of the infection have been noted in these species.

Monkey Pox

There is a single report of monkey pox in *C. jacchus* during an outbreak at the Rotterdam Zoo [43].

Other Viruses

Serological surveys and virus isolations indicate that marmosets can be infected with a variety of other viruses but their importance as causes of spontaneous disease is unknown. Also marmosets can be experimentally infected with several viral agents that are not associated with spontaneous disease in these species, e.g. Epstein-Barr virus, simian sarcoma virus, Rous sarcoma virus, hepatitis viruses [15]. Some of these are discussed elsewhere in these proceedings.

Shigellosis

Clinical shigellosis is much more common in Old World monkeys than it is in New World monkeys, nevertheless, the disease is known to occur in several South American monkeys including marmosets. Recovery of the organism from fecal swabs is, however, more common than its association with dysentery. In marmosets, *Shigella sonnei* appears to be the most important pathogen [30]. The associated dysentery is not distinct from that described in rhesus monkeys. Clinically there is diarrhea often containing blood and dehydration. Grossly the lesions are mostly confined to the cecum and colon where the mucosa is swollen, ulcerated and hemorrhagic. Microscopically the lesions which are not pathognomonic are characterized by necrosis, hemorrhage and inflammatory cell infiltration. COOPER and NEEDHAM [8] reported controlling an outbreak in *S. nigricollis* and *C. jacchus* with neomycin sulfate added to drinking water.

Salmonellosis

Several species of *Salmonella* including *S. typhimurium, S. anatum, S. oranienburg* and *S. irumu* have been recovered from a variety of species of marmosets, but reports clearly describing and correlating infection with disease and deaths are lacking [30]. The latter probably does occur on occasion and most likely is characterized by necrotizing enterocolitis and septicemia as in other species.

Tuberculosis

There are few reports of tuberculosis in marmosets, as for that matter in New World monkeys, when compared with the occurrence of the disease in Old World species. This suggests that they are more resistant to tuberculosis, but supporting experimental evidence is lacking. MORELAND [40] in his review cites an example in *Callithrix penicillata* and *C. jacchus*. In any nonhuman primate, tuberculosis is a chronic granulomatous disease, most often of the lung, but may involve any organ or tissue.

Other Bacterial Diseases

Many known pathogenic bacteria have been recovered from healthy, ill or dead marmoset species, but their exact role or causes of specific diseases in these species is generally poorly defined. Several have been shown to be important pathogens to other New World monkeys and, therefore, deserve brief mention. These include: *Diplococcus pneumoniae* as a cause of pneumonia and/or meningitis; *Pasteurella multocida* as a cause of pneumonia;

Staphylococcus spp. and *Streptococcus* spp. as causes of purulent inflammation; *Escherichia coli* as a cause of enteritis; *Proteus* spp. as a cause of enteritis; *Yersinia enterocolitica* as a cause of pseudotuberculosis which has been reported by GIORGI *et al.* [21]; and *Klebsiella pneumoniae* as a cause of enteritis, pneumonia and septicemia. In our own experience, *K. pneumoniae* has been one of the most important bacterial pathogens.

Mycoses

The true incidence of mycotic diseases in marmosets in captivity and in their natural habitats is apparently unknown. As with other species of monkeys, captivity appears to predispose marmosets to a variety of mycoses and many of these are secondary invaders in lesions of uremia, viral and bacterial infections and in situations of antibiotic overuse.

TAKOS and ELTON [48] reported disseminated cryptococcosis in two juvenile marmosets *(Marikina (Saguinus) geoffroyi)* that were captive for several months in Panama. DEINHARDT *et al.* [16] reported an unidentified dermatomycosis associated with alopecia in captive marmosets. *Candida* spp. have been isolated by us and others from the stools and digestive tract of clinically healthy and of diseased marmosets [14, 16, 42]. This fungus is most frequently found around and in ulcers on the tongue, oral mucosa, esophagus and gastrointestinal tract.

The paucity of knowledge of mycoses in marmosets probably reflects the lack of investigation and/or lack of reporting of such mycoses. Marmosets appear to be susceptible to a variety of mycoses as indicated by the above reports and under appropriate conditions of exposure they would probably contract systemic and dermatophytic mycoses as is known in numerous other non-human primates and mammals [22, 46]. Some of these mycoses include histoplasmosis in squirrel monkeys *(S. sciureus)* [4], dermatophilosis in owl monkeys *(A. trivirgatus)* [32, 37] and woolly monkeys *(Lagothrix lagothricha)* [20], *Trichophyton* spp. in several Macaques and *Microsporum canis* in several New World species [46].

Parasitic Diseases

The majority of parasites that are known in marmosets appear to be innocuous relative to the general well being of the individual. This does not preclude that the parasitic burden of the marmoset should be taken for granted. Heavy infestations of certain 'non'-pathogenic parasites may produce disease or predispose to other diseases and some inocuous parasites may interfere with certain experimental manipulations.

The most important parasite of wild and captive marmosets is the acanthocephalan, *Prosthenorchis elegans*. This worm buries its proboscis in the wall of the ileum, cecum and colon resulting in ulceration and abscess formation. The all too frequent outcome, even with a light parasite burden, is mesenteric lymphadenitis, peritonitis and death from septicemia. *P. spirula* occasionally presents in a similar manner. There is no known effective and safe treatment to rid marmosets of acanthocephalans. Sometimes a marmoset will rid itself of a light burden due to attrition of the parasite, however, carriers are a threat to their colony mates in the presence of the intermediate host, cockroaches, which are frequently a problem in many animal facilities.

Helminths are relatively uncommon in marmosets but several occasionally occur in clinically significant numbers. The dicrocoelids *Athesmia foxi* and *Platynosomum amazonensis* are among the most commonly encountered trematodes and are in small and large bile ducts and in the gallbladder [10, 19]. The bile duct epithelium may have no morphologic evidence of damage or it may be hypertrophic and hyperplastic with pericholangial fibrosis and mixed inflammatory cell exudate. Less frequently encountered trematodes include *P. marmoseti* in bile ducts and gallbladder, and *Neodiplostomum tamarini* and *Phaneropsolus* sp. in the intestine [10, 19, 34].

Cestodes of interest in marmosets include spargana of *Diphyllobothrium erinacei* and *Spirometra reptens* in subcutaneous cysts, and the adult cyclophyllideans *Hymenolepis cebidarum, Raillietina* spp., *Bertiella mucronata, Atriotaenia (Oochoristica) megastoma* and *Paratriotaenia pedipomidatus* of the small intestine [10, 16, 17, 19, 34]. Unidentified cysticercae are frequently found in subcutaneous and muscular fascial planes in marmosets but these, as well as the adult cestodes mentioned above, are apparently of minor importance when present in small numbers.

A wide variety of nematodes are found in numerous locations of the body of marmosets. Filariasis is one of the most common parasitic infections encountered. Although microfilaria are abundant in the blood stream and vascular spaces of various organs, the adults are infrequently found except for *Tetrapetalonema marmosetae* which is situated in the fascia along the back and below the scapulae [18]. At least eight distinct microfilaria from marmosets have been identified on the basis of size, presence or absence of a sheath and the distribution of the acid phosphatase reaction [6]. The significance of these filarids with respect to animal health, their influence on experimental manipulations, or potential public health hazard is poorly defined.

Larvae and adult *Filaroides* spp. situated in alveoli and terminal portions of bronchioles caused a mild purulent pneumonia in marmosets [41]. Most other nematodes of marmosets localize in the digestive tract. *Spirura guianansis* may cause a mild inflammatory reaction in the esophagus and *Physaloptera dilatata* occasionally causes ulcerative gastritis [10, 16, 19]. The small intestine may contain several nonpathogenic or slightly pathogenic species including *Molineus vexillarius, Longistriata dubia, Subulura (Primasubulura) jacchi, Rictularia alphi* and *Strongyloides* spp. *Trichospirura leptostoma,* which is found in pancreatic ducts, *Trypanoxyuris minutus* in the colon and *T. tamarini* in the cecum and colon are also probably nonpathogenic in small numbers [9, 10, 47].

There are a wide variety of protozoans found in marmosets and only a few of these are of importance. *Trypanosoma cruzi* may produce a chronic mononuclear myocarditis which is frequently only recognized incidentally after necropsy [31]. The amastigote stages are found within myocardial fibers and only appear to incite an inflammatory response when the sarcoplasmic membranes rupture exposing the parasite and its products. In contrast to *T. cruzi, T. minasense* is a nonpathogenic species found as the trypomastigote form in the blood stream. At least two sarcodines must be differentiated in the intestinal tract of marmosets – *Entamoeba coli,* a nonpathogen and *Entemoeba histolytica* which may cause a mild to severe hemorrhagic ulcerative ileitis, typhlitis and colitis. Other intestinal protozoans of unknown significance include: *Isospora arctopitheci, Eimeria* spp., *Trichomonas* spp. and *Giardia.* The latter two may frequently be seen in relative abundance situated along the intestinal mucosa in the absence of any significant inflammatory reaction.

Toxoplasma gondi is occasionally associated with a severe fatal hepatitis and encephalitis in marmosets [3, 31]. Lesions are typically necrotizing and with or without intense mononuclear inflammatory infiltrates. Lung, spleen, lymph nodes, adrenal, heart, and intestine may also contain typical lesions.

Malaria does not occur in marmosets under natural conditions but the simian plasmodia *Plasmodium brasilianum* and *P. simium* and the human *P. vivax* and *P. falciparum* can be experimentally maintained in several species of marmosets [2, 49].

Sarcocystis sp. and *Pneumocystis carinii* are found in marmosets under circumstances similar to other species of mammals [7, 29, 31, 44].

Larval forms of two pentastomids, *Linguatula serrata* and *Porocephalus clavatus,* are frequently found in viscera and under the serosa of the thoracic and abdominal cavities [11, 31]. They are generally insignificant but may create problems when in locations such as meninges where they occasionally migrate.

The only other parasites of significance in marmosets are mites and lice. The mites, *Fonsecalges saimiri, Listrocarpus cosgrovei, L. hapalei, Dunnalges lambrechti,* and *Rhyncoptes anastosi,* may cause a mild to severe dermatitis [19, 34, 35]. *Mortelmansia duboisi* is a rare mite of the nasal cavity and is of unknown pathogenicity [34]. Two species of sucking lice, *Harrisonia uncinata* and *Gliricola pinto,* were reported in *Callithrix* spp. and *Saguinus* spp. [25].

Summary

The various species of marmosets are susceptible to a wide variety of infectious agents of which only a few have been fully characterized. Little is known concerning spontaneous disease in their natural habitat, and often deaths in the laboratory go unexplained. In captivity, *Herpesvirus-T* infection appears to be the most important viral infection, but serious disease may also follow infection with measles virus (rubeola) and an unidentified paramyxovirus. Bacterial diseases are multiple, but rarely occur as epizootics. Various species of *Salmonella, Yersinia, Klebsiella,* and *Diplococcus* are among the more frequent pathogens. Mycoses and parasitic infections are also numerous, but most do not result in major losses.

References

1 AARON, E.; KAMEII, I.; BAYER, E.V.; EMMONS, R., and CHIN, J.: Probable vaccine-induced rabies in a pet marmoset-California; in Primate Zoonoses Surveillance, Report No. 12 (Center for Disease Control, Atlanta 1975).

2 BAERG, D.C. and ROSSAN, R.N.: *Plasmodium vivax* tissue stage in *Saguinus geoffroyi.* Trans. R. Soc. trop. Med. Hyg. *70:* 167–168 (1976).

3 BENIRSCHKE, K. and RICHART, R.: Spontaneous acute toxoplasmosis in a marmoset monkey. Am. J. trop. Med. Hyg. *9:* 269–273 (1960).

4 BERGELAND, M.E.; BARNES, D.M., and KAPLAN, W.: Spontaneous histoplasmosis in a squirrel monkey; in Primate Zoonoses Surveillance, Report No. 1 (Center for Disease Control, Atlanta 1970).

5 CHALIFOUX, L.; SEHGAL, P.K.; HUNT, R.D.; FRASER, C.E.O., and KING, N.W.: Paramyxovirus enterocolitis in marmosets *(Saguinus oedipus* and *Callithrix jacchus)* (in press).

6 CHALIFOUX, L.V.; HUNT, R.D.; GARCIA, F.G.; SEHGAL, P.K., and COMISKEY, J.R.: Filariasis in New World monkeys: histochemical differentiation of circulating microfilariae. Lab. Anim. Sci. *23:* 211–220 (1973).

7 CHANDLER, F.W.; MCCLURE, H.M.; CAMPBELL, W.G., jr., and WATTS, J.C.: Pul-

monary pneumocystosis in nonhuman primates. Archs Path. Lab. Med. *100:* 163–167 (1976).

8 COOPER, J.E. and NEEDHAM, J.R.: An outbreak of shigellosis in laboratory marmosets and tamarins (Family Callithricidae). J. Hyg. *76:* 415–424 (1976).

9 COSGRAVE, G.E.; HUMASON, G., and LUSBAUGH, C.G.: *Trichospirura leptostoma,* a nematode of the pancreatic ducts of marmosets (*Saguinus* spp.). J. Am. vet. med. Ass. *157:* 696–698 (1970).

10 COSGROVE, G.E.; NELSON, B., and GENGOZIAN, N.: Helminth parasites of the tamarin, *Saguinus fuscicollis.* Lab. Anim. Care *18:* 654 (1968).

11 COSGROVE, G.E.; NELSON, B.M., and SELF, J.T.: The pathology of pentastomid infection in primates. Lab. Anim. Care *20:* 354–360 (1970).

12 DANIEL, M.D.; BARAHONA, H.; MELENDEZ, L.V.; HUNT, R.D.; SEHGAL, P.; MARSHALL, B.; INGALLS, J., and FORBES, M.: Prevention of fatal herpes infections in owl and marmoset monkeys by vaccination. Presented 6th Cong. Int. Primatol. Soc., Cambridge 1976.

13 DEINHARDT, F.: Personal commun. (1977).

14 DEINHARDT, J.B.; DEVINE, J.; PASSOVOY, M.; POHLMAN, R., and DEINHARDT, F.: Marmosets as laboratory animals. I. Care of marmosets in the laboratory, pathology and outline of statistical evaluation of data. Lab. Anim. Care *17:* 11–29 (1967).

15 DEINHARDT, F.; HOLMES, A.W.; CAPPS, R.B., and POPPER, H.: Studies on the transmission of human viral hepatitis to marmoset monkeys. J. exp. Med. *125:* 673–688 (1967).

16 DEINHARDT, F.; HOLMES, A.W.; DEVINE, J., and DEINHARDT, J.: Marmosets as laboratory animals. IV. The microbiology of laboratory kept marmosets. Lab. Anim. Care *17:* 48–70 (1967).

17 DUNN, F.F.: Acathocephalans and cestodes of South American monkeys and marmosets. J. Parasit. *49:* 717–722 (1963).

18 DUNN, F.L. and LAMBRECHT, F.L.: On some filarial parasites of South American primates with a description of *Tetrapetalonema tamarinae* n. sp. from the Peruvian tamarin marmoset, *Tamarinus nigricollis* (Spix 1823). J. Helminth. *37:* 261–286 (1963).

19 FLYNN, R.J.: Nematodes; in FLYNN Parasites of laboratory animals (Iowa State University Press, Ames 1973).

20 FRASER, C.E.O. and GARCIA, F.G.: Cutaneous streptothricosis (dermatophilosis) in a wooly monkey; in Primate Zoonoses Surveillance, Report No. 5 (Center for Disease Control, Atlanta 1971).

21 GIORGI, W.; MATERA, A., y MOLLARET, H.H.: Isolamento de *Yersinia enterocolitica* de abscesos de saguis *(Callithrix penicillata* e *Callithrix jacchus).* Arq. Inst. Biol. *36:* 123–127 (1969).

22 HESSLER, J.R.; WOODARD, J.C.; BEATTIE, R.J., and MORELAND, A.F.: Mucormycosis in a rhesus monkey. J. Am. vet. med. Ass. *151:* 909–913 (1967).

23 HOLMES, A.W.; CALDWELL, R.G.; DEDMON, R.E., and DEINHARDT, F.: Isolation and characterization of a new herpes virus. J. Immun. *92:* 602–610 (1964).

24 HOLMES, A.W.; DEVINE, J.A.; NOWAKOWSKI, E., and DEINHARDT, F.: The epidemiology of a herpes virus infection of New World monkeys. J. immun. *96:* 668–671 (1966).

25 HOPKINS, G.H.E.: The host associations of the lice of mammals. Proc. zool. Soc. Lond. *119:* 387 (1949).

26 HUNT, R.D.; GARCIA, F.G.; BARAHONA, H.H.; KING, N.W.; FRASER, C.E.O., and MELENDEZ, L.V.: Spontaneous *Herpesvirus saimiri* lymphoma in an owl monkey. J. infect. Dis. *127:* 723–725 (1973).

27 HUNT, R.D. and MELENDEZ, L.V.: Spontaneous *Herpes T* infection in the owl monkey *(Aotus trivirgatus)*. Pathol. vet. *3:* 1–26 (1966).

28 HUNT, R.D. and MELENDEZ, L.V.: Herpes virus infections of non-human primates. A review. Lab. Anim. Care *19:* 221–234 (1969).

29 KARR, S.L., jr. and WONG, M.M.: A survey of sarcocystis in nonhuman primates. Lab, Anim. Sci. *25:* 641–645 (1975).

30 KAUFMANN, A.F.; MORRIS, G.; RICHARDSON, J.H.; HEALY, G., and KAPLAN, W.: A survey of newly arrived South American monkeys for potential human pathogens; in Primate Zoonoses Surveillance, Report No. 1 (Center for Disease Control, Atlanta 1970).

31 KING, N.W.: Synopsis of the pathology of New World monkeys. 1st Inter-Am. Conf. on Conservation and Utilization of American Nonhuman Primates in Biomedical Research. Scientific Publication No. 317, pp. 169–198 (PAHO, WHO, Washington 1976).

32 KING, N.W.; FRASER, C.E.O.; GARCIA, F.G.; WOLF, L.A., and WILLIAMSON, M.E.: Cutaneous streptothricosis (dermatophiliasis) in owl monkeys. Lab. Anim. Sci. *21:* 67–74 (1971).

33 KING, N.W.; HUNT, R.D.; DANIEL, M.D., and MELENDEZ, L.V.: Overt Herpes-T infection in squirrel monkeys *(Saimiri sciureus)*. Lab. Anim. Care *17:* 413–423 (1967).

34 KINTZ, R.E. and MYERS, B.J.: Parasites in South American primates; in LUCAS and DUPLAIX-HALL International zoo yearbook, vol. 12 (Biological Society, London 1972).

35 LAVOIPIERRE, M.M.J.: A new family of acarines belonging to the suborder Sarcoptiformes parasitic in the hair follicles of primates. Ann. Natal Mus. *16:* 1–18 (1964).

36 LEVY, B.M. and MIRCOVIC, R.R.: An epizootic of measles in a marmoset colony. Lab. Anim. Care *21:* 33–39 (1971).

37 MCCLURE, H.M.; KAPLAN, W.; BONNER, W.B., and KEELING, M.E.: Dermatophilosis in owl monkeys. Sabouraudia *9:* 185–190 (1971).

38 MELENDEZ, L.V.; HUNT, R.D.; DANIEL, M.D.; FRASER, C.E.O.; BARAHONA, H.H.; KING, N.W., and GARCIA, F.G.: *Herpesviruses saimiri* and *ateles* – their role in malignant lymphoma of monkeys. Fed. Proc. Fed. Am. Socs exp. Biol. *31:* 1643–1650 (1972).

39 MELNICK, J.L.; MIDULLA, M.; WIMBERLY, I.; BARRERA-ORO, J.G., and LEVY, B.M.: A new member of the herpes v rus groups solated from South American marmosets. J. Immun. *92:* 596–601 (1964).

40 MORELAND, A.F.: Tuberculosis in New World primates. Lab. Anim. Care *20:* 262–264 (1970).

41 NAHMIAS, A.J.; LONDON, W.T.; CATALANO, L.W.; FUCCILLO, D.A.; SEVER, J.L., and GRAHAM, C.: Genital *herpesvirus hominis* type 2 infection. An experimental model in Cebus monkeys. Science *171:* 297–298 (1971).

42 NELSON, B.; COSGROVE, G.E., and GENGOZIAN, N.: Diseases of an imported primate *Tamarinus nigricollis*. Lab. Anim. Care *16:* 255–275 (1966).

43 PETERS, J.C.: Eine 'Monkey-Pox'-Enzootie im Affenhaus des Tiergartens 'Blijdorp'. Kleintier-Prax. *11:* 65–71 (1966).

44 Poelma, F.G.: *Pneumocystis carinii* infections in zoo animals. Z. ParasitKde *46:* 61–68 (1975).

45 Richardson, J.H. and Humphrey, G.L.: Rabies in imported nonhuman primates. Lab. Anim. Sci. *21:* 1082–1083 (1971).

46 Ruch, T.C.: Diseases of laboratory primates (Saunders, Philadelphia 1959).

47 Smith, W.M. and Chitwood, M.B.: *Trichospirura leptostoma* gen et sp n (Nematoda: Thelaziodea) from the pancreatic ducts of the white-eared marmoset, *Callithrix jacchus*. J. Parasit. *53:* 1270–1272 (1967).

48 Takos, M.J. and Elton, N.W.: Spontaneous cryptococcosis of marmoset monkeys in Panama. Archs Path. *55:* 403–407 (1953).

49 Young, M.D.: Natural and induced malarias in Western Hemisphere monkeys. Lab. Anim. Care *20:* 361–367 (1970).

R.D. Hunt, DVM, Harvard Medical School, New England Regional Primate Research Center, One Pine Hill Drive, *Southborough, MA 01772* (USA)

Prim. Med., vol. 10, pp. 254–260 (Karger, Basel 1978)

Human Spongiform Encephalopathies in Marmoset Monkeys (*Saguinus* sp.)[1]

D.A. Peterson, L.G. Wolfe and F.W. Deinhardt

Departments of Microbiology, Rush-Presbyterian-St. Luke's and University of Illinois Medical Centers, Chicago, Ill.

Introduction

During the past 16 years 'slow virus infections' have been found associated with a variety of chronic degenerative diseases of the central nervous system (CNS) of man. From the list of 'candidate' diseases kuru has been experimentally transmitted to chimpanzees [1, 4], five species of New World monkeys [1, 4, 8], and three species of Old World monkeys [1, 4], and Creutzfeldt-Jakob disease (CJD) has been experimentally transmitted to chimpanzees [1, 4], six species of New World monkeys [1, 4, 8], nine species of Old World monkeys [1, 4], domestic cats [1] and guinea pigs [7]. Incubation periods for primary passage from human to animals ranged from 26–105 months. Gajdusek and Gibbs [2] have reported also the transmission of subacute spongiform encephalopathy to nonhuman primates from a patient with papulosis atrophicans maligna of Degos, a patient with familial Alzheimer's disease, and a patient with progressive supranuclear palsy (Steele-Richardson syndrome). In 1969, we initiated studies of kuru and CJD and subsequently reported primary transmission of kuru and CJD to marmosets [8]. The objectives of this study were (a) to refine the model system for kuru and CJD in marmosets, (b) to examine the pathogenesis of the experimentally induced disease and (c) to identify and characterize the infectious agent(s).

1 These studies were supported in part by research grant No. 5 R01 AI 09628096, National Institute of Allergy and Infectious Disease, National Institutes of Health, US Public Health Service and research grants No. 415–13 and No. 727–22, Department of Mental Health and Developmental Disabilities, State of Illinois. We thank the Board of Health, City of Chicago, for providing space for housing many of our experimental animals.

Materials and Methods

Inocula. Five or 10% brain homogenates in phosphate-buffered saline (PBS) were prepared from seven kuru and four CJD patients for the initial inoculations (table I). These were kindly supplied by Drs. D.C. GAJDUSEK and C.J. GIBBS, jr., National Institute of

Table I. Inoculation of human brain homogenates into marmosets

Inoculum	Inoculation			Results		
	species	age	number	SE	NS	MPI
Kuru						
Eiro	WL	9 months	1		1	46
	CT	6 months	1	1[1]		26
Enage	WL	4 days	1		1	6
	CT	9 days	1		1	8
	CT	8 months	1	1[1]		36
Igierakaba	WL	9 months	1		1	64
	WL	adult	1		1	34
Kabuinampa	WL	4 days	1		1	6
	WL	adult	2		2	3, 20
	CT	9 days	1		1	8
	CT	8 months	1	1[1]		31
Kigea	WL	9 months	2		2	40, 42
	WL	6 months	1	1[1]		94
	WL	adult	1		1	40
Kwase	WL	6 months	1		1	9
	WL	adult	1		1	39
Moba	WL	1 month	2		2	20, 38
	WL	adult	2	1[1]		76
					1	14
Creutzfeldt-Jakob disease						
Burgess	WL	3 days	2		2	3, 38
Reeson	WL	6 months	1	1[1]		43
	WL	adult	1		1	15
Tate	WL	4 days	1		1	13
	WL	adult	1	1[1]		54
Woolfe	WL	9 months	1		1	37
	CT	7 months	1		1	25

SE = Spongioform encephalopathy; NS = nonspecific deaths, unrelated to inoculation; MPI = months postinoculation.

[1] Identification codes of positive animals: Eiro, 71–FL–1; Enage, IS–1; Kabuinampa, IS–2; Kigea, JK–1; Moba, HF–1; Reeson, JM–1; Tate, 69–DH–2.

Table II. Serial passage of experimental kuru and Creutzfeldt-Jakob disease in marmosets[1]

Inoculum		Inoculation			Dead				Alive
passage No.	animal No.	species	age days	No.	animal No.	symptoms days[2]	pathology	MPI	MPI
Kuru									
2	IS-1	WL	10	2	AE-1	80	SE	20	
					AE-2	70	SE	24	
		WL	adult	1	AQ-2	10	SE	28	
		CT	adult	1	FH-2	112	SE	25	
	IS-2	WL	2–8	4	FN-2	2	SE	9	
					HD-1	18	SE	9.5	
							NS	1.5	
					GG-1	40	SE	11	
					DU-1	12	SE	9	
		WL	13	2	HT-1	6	SE	1.5	
					HT-2	3	SE	2	
3	FN-2	WL	1	2	DM-1	1	SE	6	
					DM-2	3	SE	7.5	
		WL	2	1	DK-1	28	SE	10	
	HT-1	WL	7, 8	4	Z-1	6	SE	1.5	62
					Z-2	1	SE	2	
					AB-2	42	SE	34	
		WL	4	2			NS	5, 9	
	HT-2	WL	adult	1			NS	2	
		CT	adult	1	V-1	43	SE	20	
4	Z-1	WL	10	2			NS	8, 9	
	Z-2	WL	2, 5	2	DG-2	1	SE	10	
							NS	25	
		WL	14	2			NS	12, 31	
	DM-2	WL	3–10	3	Q-1	57	SE	10	
					Q-2	47	SE	10	
					T-2	30	SE	9	
		WL	4, 5	4	BD-1	35	SE	8	
					BD-2	3	SE	6	
					BE-1	55	SE	8	
					BE-2	70	SE	8	
5	DG-2	WL	3	2			NS	39	45
Creutzfeldt-Jakob disease									
2	JM-1	WL	2, 5	2	FA-2	60	SE	18	
							NS	7	
		WL	13	2	KF-1	17	SE	34	
							NS	1.2	

[1] See table I and text for abbreviations; results as of 3/10/77. Mild spongiosus in animals that died 1.5 and 2 MPI; moderate to severe spongiosus in remainder.

[2] Length of time symptoms of disease were observed.

Neurological and Communicative Disorders and Stroke, NIH. For serial passage 10% marmoset brain homogenates in PBS were prepared from histopathologically positive animals (table II). Afterwards 0.1 ml of brain homogenates of kuru and CJD patients were inoculated intracerebrally (i.c.), intraperitoneally (i.p.) and/or intravenously (i.v.) into 22 and eight marmosets, respectively. For serial passage neonatal (one to 14 day old) and adult marmosets were inoculated i.c. with 0.1 ml and i.p. with 0.5 ml of marmoset brain homogenates prepared from marmosets with experimentally induced kuru or CJD [8].

Animals. Cotton-topped (CT) marmosets *(Saguinus oedipus oedipus)* and white-lipped (WL) marmosets *(Sanguinus fuscicollis* and *Saguinus nigricollis)*, both wild caught and hand-reared in captivity, were inoculated and maintained as described previously [8, 9].

Brain explant cultures. Brain tissue obtained at necropsy was washed (three times with cold HBSS), minced into small fragments (0.5–1 mm cubes), five to ten pieces placed into a 30-ml screw cap plastic flask containing 7 ml of media (Dulbecco's [Modified Eagle's] Media, BioLab, Inc. Northbrook, Ill.; containing 10% fetal calf serum, 1.5% sodium bicarbonate, 100 U penicillin/ml, 100 μg kanamycin/ml) and incubated at 37°C. Cultures were examined at two-day intervals and monitored by standard virological techniques.

Results

Of 23 marmosets inoculated with brain homogenates from seven human kuru patients, 18 died from nonspecific intercurrent disease without signs of CNS disease and five marmosets (23%) developed rapidly progressive CNS disease 26, 31, 36, 76 and 94 months after inoculation (table I). Thus five of the seven (71%) kuru patients (Eiro, Enage, Kabuinampa, Kigea and Moba) induced kuru in marmosets (three CT and two WL). Second, third and fourth serial marmoset passages of kuru were accomplished by i.c. inoculation (0.1 ml) of neonatal marmosets (one to 14 days of age) with brain homogenates (10% w/v) from marmosets with experimentally induced kuru. The incubation periods ranged from 1.5 to 11 months (average 7.3) in 19 animals, 20–25 months in five animals, and greater than 26 months in two animals (table II). Marmosets with kuru exhibited symptomatic periods of one to 112 days (average 20 days) marked by tremors, incoordination, lack of tactile response, increased aggression, ataxia, visual impairment and a progressive course to death. Pathologic findings in the 30 positive kuru animals were compatible with experimentally induced spongiform encephalopathy in other nonhuman primates [6]. Characteristic microscopic lesions included spongiform polioencephalopathy with neuronal loss in the cerebral cortex and basal ganglia, spongiosus in the molecular layer with moderate loss of Purkinje cells and loss of granule cells in the cerebellum, and spongioform polioencephalopathy in the spinal cord. The severity of the lesions was variable but in general reflected the length of the post-inoculation interval. Mild astrogliosis accompanied the spongiosus and neuronal loss; perivascular cuffing was not observed. Similar neuropathologic findings have not been observed in more than 2,500 control (uninoculated animals or animals inoculated by various routes with tumor or hepatitis viruses) or 'sham-inoculated' marmosets (inoculated with normal marmoset brain).

CJD has been induced by the i.c. inoculation of marmosets with brain homogenates from two human CJD patients (Reeson and Tate) with incubation periods of 43 and 54 months (table I). Second serial passage of CJD was achieved in neonatal marmosets with incubation periods of 18 and 34 months (table II). As with kuru the symptomatology and histopathology are compatible with CJD in man and nonhuman primates [6].

Marmosets inoculated with brain homogenates from two human focal epilepsy patients and two human multiple sclerosis patients are under observation 29–91 months postinoculation and have not developed any signs of CNS disease. Several animals died of nonspecific, spontaneous diseases 29–68 months postinoculation.

Explant cultures of brain (with or without cocultivation with various normal marmoset cell lines) from six marmosets with experimental kuru and one with experimental CJD were harvested at various tissue culture passages and inoculated i.c. into neonatal marmosets. One animal, FK-1, inoculated with CJD marmoset brain explant cells (3.2×10^6 cells, tissue culture passage 3, 33 days in culture, passage split ratio 1:2) developed CJD at 34 months postinoculation. Of the remaining 12 animals inoculated with tissue culture preparations, five died of intercurrent disease six to 18 months postinoculation and the remaining seven animals are alive at 33–59 months postinoculation.

Brains were removed from 54 marmosets inoculated with either kuru or CJD material and from six control animals, and explant cultures ($\geq$passage 5) were established from 44 diseased and three normal brains. All attempts at virus isolation have failed although many cultures have been maintained for over one year. However, cell alterations and morphological changes in kuru and CJD experimental marmoset brain cultures, similar to cell alterations described by others, have been observed [3, 5]. In addition, the cultures from kuru or CJD affected brains grew with more vigor, reaching confluency in one half to one third the time required for 'normal' brain cell cultures.

Discussion

The latent periods of 2.3–7.8 years for primary transmission of kuru (five of seven patients) and CJD (two of four patients) from man to marmosets do not differ significantly from the latent periods (2.2–8.7 years) reported in other nonhuman primates [1, 2, 4]. In addition, only 23% (five of 22) of kuru and 25% (two of eight) of CJD animals inoculated developed disease. Even though more than 50% of the human kuru and CJD brain tissue induced disease in marmosets the combined effect of long incubation period and normal attrition rates (58.2% at one year; 78.4% at three years; 84% at five years) offer no advantages over other susceptible nonhuman primates for primary transmission experiments.

Serial passage of kuru in neonatal marmosets has significantly reduced the latent period and increased the incidence of experimental disease. Of 45 neonatal marmosets inoculated with brain material of the 1st, 2nd, 3rd or 4th marmoset passage of kuru, 13 (29%) died within 30 days and were omitted from table II since the attrition rate for uninoculated neonatal marmosets for the first 30 days of life is 24–26% in our colony. Of the remaining 32 neonatal marmosets two (6%) are still alive, eight (25%) died of nonspecific causes and 22 (69%) developed kuru with latent periods of 1.5–11 months (mean 7.3) in 19 (86%) and 20–34 months (mean 26) in three (14%).

The altered morphology and increased growth rate of kuru and CJD explant cultures and the demonstration of infectivity in one of these cultures provide a tool for further studies *in vitro* aimed at identifying the agent(s) and examining the immune response of experimental hosts.

Neonatal marmosets offer a reliable indicator system – experimental disease – for further studies of the etiological agent(s) of kuru and CJD and to identify the host immune response(s) to these agents. Marmosets should be used in attempts to transmit other chronic degenerative, neurologic, renal and connective tissue disorders of man to experimental animals.

Summary

Brain homogenates (10% w/v) from five of seven kuru patients inoculated intracerebrally (i.c., 0.1 ml) into marmosets (*Saguinus* sp.) induced a rapidly progressive CNS disease 26, 31, 36, 76 and 94 months postinoculation. Serial marmoset passages of kuru were accomplished by i.c. inoculation of neonatal marmosets with brain homogenates from marmosets with experimentally induced kuru. The incubation periods ranged from

1.5 to 11 months (average 7.3) in 19 animals, 20–25 months in four animals and greater than 26 months in two animals. Symptoms and brain lesions of the induced disease were compatible with kuru as observed in humans and other nonhuman primates. Creutzfeldt-Jakob disease (CJD) has been induced by i.c. inoculation of marmosets with brain homogenates from two of four human CJD patients with incubation periods of 43 and 54 months and is in serial passage.

References

1 FUCCILLO, D.A.; KURENT, J.E., and SEVER, J.L.: Slow virus diseases. A. Rev. Microbiol. *28:* 231–264 (1974).

2 GAJDUSEK, D.C. and GIBBS, C.J., jr.: Familial and sporadic chronic neurological degenerative disorders transmitted from man to primates. Adv. Neurol. *10:* 291–317 (1975).

3 GAJDUSEK, D.C.; GIBBS, C.J., jr.; ROGERS, N.G.; BASNIGHT, M., and HOOKS, J.: Persistence of viruses of kuru and Creutzfeldt-Jakob disease in tissue cultures of brain cells. Nature, Lond. *235:* 104–105 (1972).

4 GIBBS, C.J., jr. and GAJDUSEK, D.C.: Experimental subacute spongiform virus encephalopathies in primates and other animals. Science *182:* 67–68 (1973).

5 HOOKS, J.; GIBBS, C.J., jr.; CHOPRA, M.L., and GAJDUSEK, D.C.: Spontaneous transformation of human brain cells grown *in vitro* and description of associated virus particles. Science *176:* 1420–1422 (1972).

6 LAMPERT, P.W.; GAJDUSEK, D.C., and GIBBS, C.J., jr.: Subacute spongiform virus encephalopathies. Scrapie, kuru and Creutzfeldt-Jakob disease. A review. Am. J. Path. *68:* 626–646 (1972).

7 MANUELIDIS, E.E.; KIM, J.; ANGELO, J.N., and MANUELIDIS, L.: Serial propagation of Creutzfeldt-Jakob disease in guinea pigs. Proc. natn. Acad. Sci. USA *73:* 223–227 (1976).

8 PETERSON, D.A.; WOLFE, L.G.; DEINHARDT, F.; GAJDUSEK, D.C., and GIBBS, C.J., jr.: Transmission of kuru and Creutzfeldt-Jakob disease to marmoset monkeys. Intervirology *2:* 14–19 (1973/74).

9 WOLFE, L.G.; DEINHARDT, F.; OGDEN, J.D.; ADAMS, M.R., and FISHER, L.E.: Reproduction of wild-caught and laboratory-born marmoset species used in biomedical research *(Saguinus* sp., *Callithrix jacchus)*. Lab. Anim. Sci. *25:* 802–813 (1975).

D.A. PETERSON, PhD, Department of Microbiology, Rush-Presbyterian-St. Luke's Medical Center, 1753 W. Congress Parkway, *Chicago, IL 60612* (USA)

Prim. Med., vol. 10, pp. 261–270 (Karger, Basel 1978)

A Paramyxovirus Causing Fatal Gastroenterocolitis in Marmoset Monkeys[1]

C.E.O. Fraser, L. Chalifoux, P. Sehgal, R.D. Hunt and N.W. King
New England Regional Primate Research Center, Harvard Medical School, Southborough, Mass.

Introduction

Recently an outbreak of disease occurred in the marmosets housed at the New England Regional Primate Research Center. The disease was characterized clinically by severe diarrhea, dehydration and death. Nonspecific diarrhea and dehydration are common in monkeys, however, the relatively high death rate of this outbreak and pathological features suggested the occurrence of a new disease. This study was undertaken to describe the pathology of the disease and the isolation and characterization of a paramyxovirus which was established as the causative agent of the disease. The name *Paramyxovirus saguinus* is suggested for this agent.

Materials and Methods

Cell cultures. Continuous cell cultures were maintained and prepared by standard procedures. The owl monkey kidney cells (OMK 637) are a cell line developed in these laboratories. Vero cells (African green monkey kidney) were obtained from Dr. H. Rabin (Frederick Cancer Center, Frederick, Md.). Cells were grown in plastic flasks (Falcon) with Eagle's minimum essential medium containing 10% fetal calf serum (MEM-10) with penicillin 250 U/ml and streptomycin 250 μg/ml and cells were transferred to Linbro trays for virus assays. For immunofluorescence cells were grown on glass slides in 60-mm plastic dishes.

1 Supported by a grant from the National Institutes of Health, NIH-USPHS grant No. RR00168–15.

Viruses: isolation of virus. The original virus was obtained from a cotton-top marmoset *(Saguinus oedipus oedipus)* which had died with a severe enterocolitis. The spleen was minced and the spleen cells were cocultivated with a monolayer of owl monkey kidney cell line (OMK 637) with RPMI 1640 and 10% fetal calf (IgG free) serum. Isolations from other organs and tissues were done by mincing the tissue and inoculating it onto a monolayer of owl monkey cells.

Measles virus (Edmondson strain) was obtained from Dr. R. RUSTIGIAN, Veterans Administration Hospital, Brockton.

Virus assay. The infectivity was assayed in cell cultures grown in Linbro trays. The 50% infective dose ($TCID_{50}$) was calculated by the method of REED and MUENCH [6].

Neutralization. The method of constant (1:5) serum and varying virus dilution was used. The diluted serum was mixed with an equal volume of the appropriate dilutions of virus. The mixture was incubated at 4 °C for 1 h and then inoculated onto OMK and Vero cell cultures. The neutralization indices were calculated by the method of REED and MUENCH [6].

Experimental infection. Four antibody negative *S. oedipus* were inoculated intravenously and orally with 0.5 ml, respectively, of stock virus E877N containing 100 $TCID_{50}$/ml.

Immunofluorescence. The antigen was made by establishing monolayers of owl monkey kidney cells on slides (ten per slide). The monolayers are separated from each other by spraying the slides with fluoroglide. The monolayers were infected with 10^2 $TCID_{50}$ of the virus and fixed in acetone at room temperature for 10 min when adequate cytopathic effect (CPE) had developed. Procedures for performing indirect immunofluorescence have been described [2].

Staining procedures. Infected monolayer cultures and tissue sections were fixed and stained by standard methods with hematoxylin and eosin. The procedures used for *electron microscopy* have been described in detail previously [1].

Antisera. A marmoset antiserum to *P. saguinus* was obtained from marmoset No. 223–75 which was one of the animals involved in the original disease outbreaks. Goat antiserum to *P. saguinus* was prepared by an initial inoculation of 10^2 $TCID_{50}$ intradermally followed by booster inoculations given intravenously on days 34 and 41 and at biweekly intervals thereafter. Antiserum against measles (Edmondson strain) was prepared in an African green monkey by Dr. R. RUSTIGIAN.

Results

The disease was characterized clinically by loss of appetite, diarrhea, dehydration and death. Gross examination showed congestion and some hemorrhage in the mucosa of the stomach, cecum and colon and occasionally also in the small intestines. Peyer's patches were sometimes congested and prominent, mesenteric lymph nodes and spleen were often enlarged.

The most characteristic microscopic lesions were colitis and typhlitis. There was thin mucosa with sloughing of the surface epithelium, large syncytial cells were present on the surface and in the crypts. Intranuclear inclusions were seen in the giant cells. There was often karyorrhexis of the remaining epithelium and many mitoses were present. Within the lamina propria there was necrosis and regeneration of cells. There was also infiltration of the lamina propria and the submucosa by mononuclear cells. Aggregations of polymorphonuclear leukocytes were often seen in the crypts which were often dilated and on occasion were entirely obliterated. In some severe cases the mucosa was necrotic and hemorrhagic with foci of ulceration. Gastritis was commonly seen and could be severe. Intestinal lesions were less common and when they did occur the duodenum near the pancreas was the region most likely to be affected.

Cholangitis was often present, characterized by the formation of syncytial cells in the bile duct epithelium. These were the only lesions in the spontaneous disease in which both intracytoplasmic and intranuclear inclusions were seen.

Syncytial cells were also seen in pancreatic duct epithelium, pancreatic acini, hepatic cords, kidney tubules and in one case in endometrial epithelium.

Lymph nodules in lymph nodes, spleen and Peyer's patches were hyperplastic with diffuse or focal necrosis of germinal centers. No lesions were seen in esophagus, brain or respiratory epithelium.

Several attempts to isolate an agent from enteric contents, stomach, intestines and colon were unsuccessful. Marmoset No. 90–74 died suddenly and was found to have enterocolitis, peritonitis, hepatitis and myocarditis. The spleen and mesenteric lymph nodes were enlarged and congested. Leukocytes from spleen were harvested and cocultivated with a monolayer of owl monkey kidney cells (OMK 637). A definite CPE developed by the fifth day and fluid and cells were collected and stored at –86 °C (stock E756M). This isolate is referred to as *P. saguinus*.

Cultural Characteristics

Growth in owl monkey cells was slow. Although CPE first developed in five days, it required 14 days for full expression of CPE. Monolayers stained by hematoxylin and eosin revealed numerous polykaryocytes and eosinophylic intracytoplasmic and intranuclear inclusions.

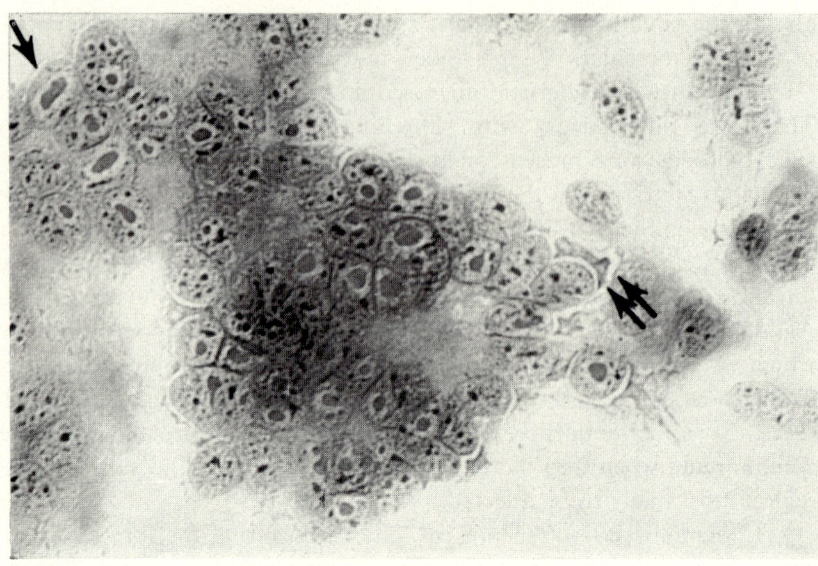

Fig 1. Typical polykaryocyte with eosinophilic intranuclear (↓) and intracytoplasmic inclusions (↓↓) in owl monkey cells infected with *P. saguinus.* ×580.

The CPE was very similar to that of measles virus (fig. 1). The virus grows well in Vero cells and develops a higher titer than in owl monkey cells, e.g. an increase of 10^2 $TCID_{50}$/ml.

Chemical Characteristics

The virus was titrated in cells treated with media containing 40 mg/ml 5-bromodeoxyuridine (5-BDUR) and in untreated control cells. There was no significant reduction in titer in the BUDR-treated cells, suggesting that this is probably an RNA virus.

Sensitivity to liquid solvents was demonstrated by adding chloroform (20% final dilution) to undiluted aliquots of the virus for 10 min at 4°C. Virus infectivity was destroyed.

Physical Characteristics

Virus infectivity was completely destroyed by submerging undiluted aliquots sealed in ampoules, in a waterbath at 56°C for 5 min. The virus passed through a filter of 450 nm but was retained by one of 220 nm.

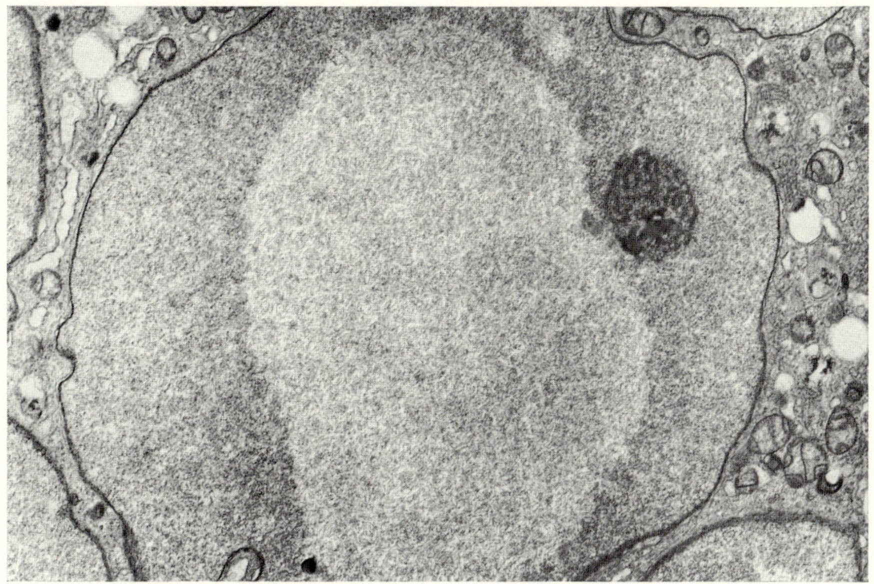

Fig. 2. Portion of a nucleus of a polykaryocyte containing a centrally placed incluosio body composed of numerous twisted, tubular paramyxovirus nucleocapsids. The chromtinn and nucleolus are displaced peripherally. ×6,200.

Electron Microscopy

Infected OMK cells formed multinucleate cells whose nuclei contained peripherally displaced heterochromatin and a central lucid zone containing myriads of elongate, tangled tubular structures (fig. 2). The diameter of the individual tubules was 16–17 nm but their length was impossible to determine because it passed in and out of the planes of section (fig. 3). The cytoplasm of infected cells contained large, irregular dense masses that appeared to be composed of tubular structures that were heavily coated with a granular material (fig. 4) making the tubules much less distinct than those in the nuclei. The morphologic features of the tubular nucleocapsids in the nuclei and the cytoplasmic inclusion bodies are consistent with those of a paramyxovirus.

Experimental Infection

There were four experimentally infected monkeys and four uninfected controls. All the animals in this experiment were shown to have no measurable levels of antibody by immunofluorescence prior to the start of the ex-

Table I. Antibodies to *P. saguinus*

Species	Number positive	Number tested	Positive, %
New World			
Saguinus	26	69	37
Aotus	2	45	4
Saimiri	0	6	0
Cebus	1	6	17
Ateles	0	6	0
Old World			
African green	12	12	80
M. mulatta	11	18	61
M. fascicularis	0	4	0
M. arctoides	1	6	17
M. cyclopis	8	68	12

Table II. Reciprocal cross-immunofluorescence of measles and *P. saguinus*

Antisera	Virus	
	measles	*P. saguinus*
Measles (African green monkey)	80[1]	160
P. saguinus (owl monkey)	0	40
P. saguinus (goat)	80	640

[1] Reciprocal of highest serum dilution giving definite staining.

periment. Of the four infected animals, one died eight days after infection and one was killed on the tenth day and two on the 14th day after infection in a moribund condition.

The lesions seen in these animals were similar to those described in the spontaneous disease, except that they were more acute with more necrosis and hemorrhage. In addition interstitial pneumonia was evident in the experimentally infected animals and inclusions were seen in the bladder of one animal.

In the two animals which survived for 14 days it was possible to demonstrate a leukopenia and a rise in antibody titers from zero to 1:20 by immunofluorescence.

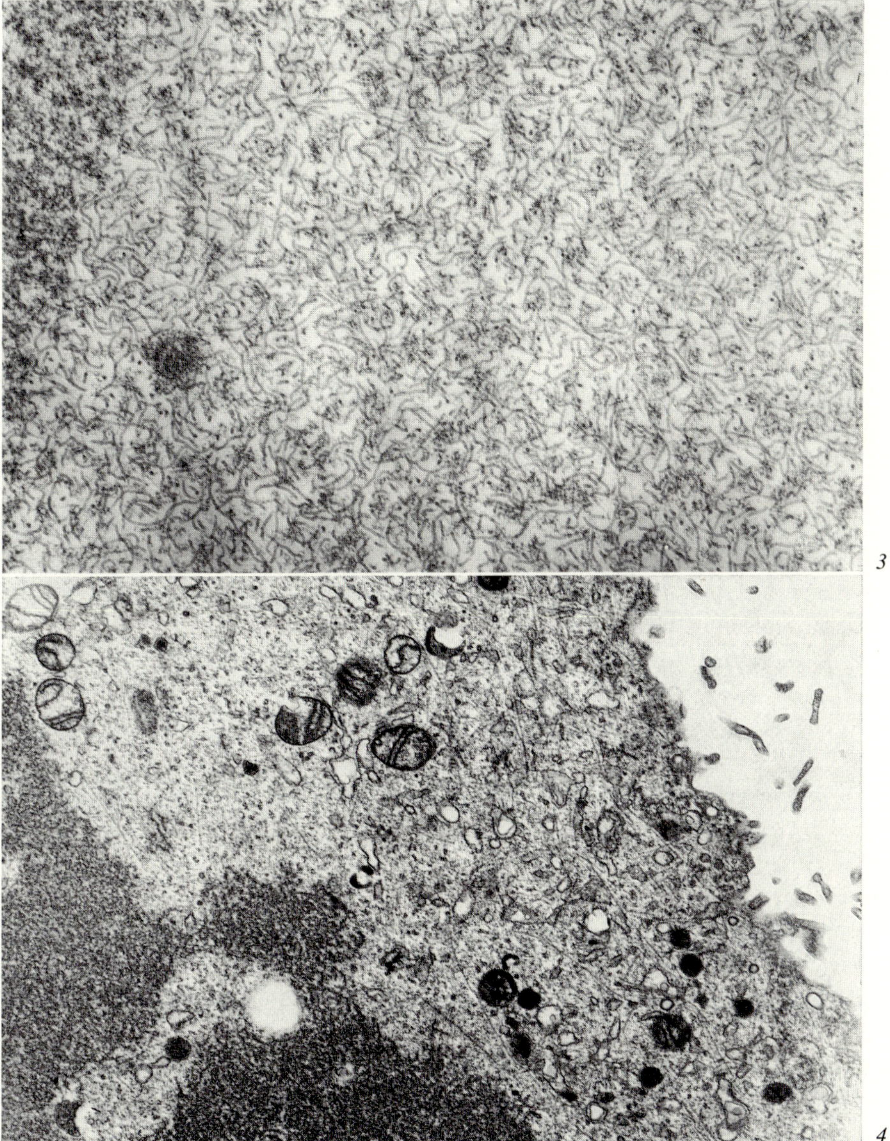

Fig. 3. Portion of an intranuclear inclusion body in which the twisted, tubular nature of the viral nucleocapsids is shown. ×22,000.

Fig. 4. Portion of the cytoplasm of an infected OMK cell containing irregular cytoplasmic inclusions (lower left) The inclusion is composed of granular coated tubular structures. ×14,200.

The virus was reisolated from all four infected animals. The virus was isolated not only from spleen but also from mesenteric lymph nodes, stomach, ileum, cecum, colon and in one case from heparinized blood. The reisolated virus was found to be indistinguishable from the original isolate.

In view of the fact that this disease had not been recognized before in marmosets, even though many necropsies had been done in this and other laboratories, it was suspected that marmosets are not the endemic host for this virus and that they were infected through contact with another species. It can be seen from table I that 37% of captive Saguinus had antibodies to the virus. Very little activity was found in the sera of other New World species studied.

On the other hand a high incidence of antibodies, which could be due however to a previous exposure to either *P. saguinus* or measles virus, was found in the Old World species *C. aethiops* and *M. mulatta* (80% and 61%, respectively).

In a comparison of *P. saguinus* with measles virus by cross-immunofluorescence (table II) it was found that whereas antisera to measles reacted with both viruses to about the same titer, antisera to *P. saguinus* reacted with the homologous virus severalfold higher than with measles.

Discussion

A severe gastroenterocolitis in marmosets has been described. This newly recognized entity resulted in the death of about 10% of the colony. The disease is characterized by syncytial cell formation and necrosis in several organs but the most significant lesions were seen in the intestinal tract. There was no exanthema which is so characteristic of measles virus. A virus was isolated from spleen leukocytes from an infected animal. It was identified as a paramyxovirus on the basis of its cytopathology, ultrastructure and its physiochemical characteristics. In tissue culture the virus elicits polykaryocytes and eosinophilic inclusions in the nucleus and cytoplasm which are indistinguishable from those of measles [4]. Like measles the virus is closely associated with the leukocytes.

Measles is a highly infectious agent which often infects captive Old World monkeys. In Old World monkeys the disease is milder than in man but is also characterized by an exanthema in a high percentage of cases [5]. Measles would appear to be rare in marmosets. The only report is that of Levy and Mirkovic [3] who described an epizootic with high mortality due to a classical measles giant cell pneumonia. In contrast, *P. saguinus* causes a disease which is characterized by gastroenterocolitis. In preliminary studies *P. saguinus* virus gave an eightfold higher titer with homologous sera than with measles antiserum. In a limited retrospective study, sera from marmosets involved in the outbreak were titered by immunofluorescence against the two viruses simultaneously. Although 80% had at least a fourfold higher titer to the homologous virus than to measles there was enough variation to suggest that it would be difficult to differentiate infection caused by *P. saguinus* virus from that caused by measles solely on the basis of retrospective serology. The neutralization test has also demonstrated very extensive cross-neutralization. This is not surprising in view of the fact that great cross-reactivity has been recognized among the parainfluenza viruses and in the measles-distemper-rinderpest complex [7]. This virus, like distemper and rinderpest, shares extensive cross-reacting antigens with measles but elicits a different disease. So far no hemagglutination has been demonstrated with guinea pig, chicken, Rhesus and African green monkey red blood cells. It is clear that more work is required to clarify the relationship of this virus to measles and other closely related paramyxoviruses.

Summary

A paramyxovirus has been isolated, identified and characterized. The virus causes a disease characterized clinically by severe diarrhea, dehydration and death. Microscopically the disease is characterized as a proliferative gastroenterocolitis with focal necrosis of the mucosa, by giant syncytial cells and eosinophilic intranuclear and intracytoplasmic inclusions. The virus is closely related to measles virus antigenically.

Acknowledgement

The authors wish to thank Mr. Douglas Jackman and Mrs. Terry Simeone for their excellent technical help.

References

1 Barahona, H.H.; Melendez, L.V.; King, N.W.; Daniel, M.D.; Fraser, C.E.O., and Preville, A.C.: *Herpesvirus aotus* type 2. A new viral agent from owl monkeys *(Aotus trivirgatus)*. J. infect. Dis. *127:* 171–178 (1973).

2 Fraser, C.E.O.; Melendez, L.V.; Barahona, H.H., and Daniel, M.D.: The effect

of poly I:C on the fluorescent and neutralizing antibodies to *Herpesvirus saimiri* in goats. Int. J. Cancer *7:* 397–404 (1971).

3 LEVY, B.M. and MIRKOVIC, R.R.: An epizootic of measles in a marmoset colony. Lab. Anim. Sci. *21:* 33–39 (1971).

4 MARCY, S.M. and KIBRICK, S.: Infectious diseases; 1st ed., pp. 865–880 (Harper & Row, Baltimore 1972).

5 PEEBLES, T.C.; MCCARTHY, K.; ENDERS, J.F., and HALLOWAY, A.: Behavior of monkeys after inoculation of virus derived from patients with measles and propagated in tissue culture together with observations of spontaneous infections of these animals by an agent exhibiting similar antigenic properties. J. Immun. *78:* 63–74 (1957).

6 REED, L.J. and MUENCH, H.: A simple method of estimating fifty per cent end points. Am. J. Hyg. *27:* 493–497 (1938).

7 WARREN, J.: The relationships of the viruses of measles, canine distemper and rinderpest; in Advances in virus research, vol. 7, pp. 727–760 (Academic Press, New York 1960).

Dr. C.E.O. FRASER, New England Regional Primate Center, Harvard Medical School, 1 Pine Hill Drive, *Southborough, MA 01772* (USA)

Prim. Med., vol. 10, pp. 271–276 (Karger, Basel 1978)

Transfer Factor and Its Protective Effect against Herpesvirus Infections of Marmosets[1]

J. W. Eichberg, R. L. Heberling, S. S. Kalter, R. W. Steele and W. T. Kniker

Southwest Foundation for Research and Education; Brook Army Medical Center, and The University of Texas Health Science Center, San Antonio, Tex.

Introduction

Using criteria developed for human and nonhuman primates [3, 6, 7], marmosets demonstrated deficiencies in both their humoral and cell-mediated immunocompetence. This may be the underlying reason for their susceptibility to a variety of oncogenic and nononcogenic viruses [1, 4, 5, 8].

The present study was undertaken to evaluate the ability to change the cellular immune responsiveness of marmoset monkeys *in vivo* and *in vitro* through administration of transfer factor (TF) from either human or baboon donors. In addition, the efficacy of TF to protect marmosets from fatal *Herpesvirus hominis* type 1 (HVH-1) disease was evaluated.

Materials and Methods

Animals

Adult cotton-top marmoset monkeys *(Saguinus oedipus)* were used for the following studies: three for the evaluation of transfer of cellular reactivity by human dialyzable TF (TF_d), two for the evaluation of transfer of cellular reactivity by baboon TF, 29 for the evaluation of efficacy of human TF_d to protect from fatal HVH-1 disease, and two for the evaluation of efficacy of marmoset TF to protect from fatal HVH-1 disease.

1 This work was supported in part by NIH Grants CA15842, RR05519, and RR00361, Contracts N01–CP–43214 and N01–CB–64055 within the Virus Cancer Program of the National Cancer Institute and Clinical Investigation Grant C–19–75, Brooke Army Medical Center.

Table I. Conversion of skin test (ST) and blastogenic (B) responses in marmosets following administration of human and baboon transfer factor

Antigen	Human TF		Baboon TF	
	ST[1]	B[2]	ST	B
TF positive				
Monilia	3/3[3]	1/1	2/2	2/2
Tetanus	2/3	2/3	0/2	1/2
Mumps	1/3	1/3	0/2	1/2
Cocci	0/3	1/3	ND	ND
HVH-1	2/3	2/3	0/2	0/2
NDCB	ND	ND	0/2	0/2
All TF positive antigens	8/15	7/13	2/10	4/10
	(53%)	(54%)	(20%)	(40%)
TF negative				
Histo	0/3	0/3	ND	ND
PPD	0/3	0/3	0/2	0/2
HVH-2	ND	0/3	ND	ND
All TF negative antigens	0/6	0/9	0/2	0/2
	(0%)	(0%)	(0%)	(0%)

ND = Not done.

1 Positive ST: ≥ 5 mm of induration.

2 Positive B: BI$\geq$2.5 where BI = cpm of [^{3}H]thymidine uptake for lymphocytes incubated with antigens or infected cells divided by uptake after incubation with media or uninfected control cells, respectively.

3 Expressed as number of animals converting to positive responses divided by number of animals negative prior to transfer factor administration.

TF Donors

For the evaluation of transfer of cellular reactivity and for the protection study, a human donor was selected who had high skin test and *in vitro* blastogenic responses to a variety of antigens, particularly to HVH-1 [9]. For the evaluation of transfer of cellular immunity, four adult baboons were sensitized to five antigens (table I). Those procedures have been described in detail elsewhere [9].

Two marmosets which had been pretreated with human TF_d and survived the subsequent HVH-1 challenge served as donors to evaluate the efficacy of protection provided by marmoset TF.

Administration of TF and HVH-1

Transfer of cellular reactivity by human and baboon TF. TF was administered in a dose equivalent to 10^8 lymphocytes per 7 kg of body weight. Since a marmoset weighs approximately 0.5 kg, the actual dose was 1×10^7 in a volume of 0.1 ml per animal administered subcutaneously. Three days after TF injection, skin tests were repeated and blood was again drawn for *in vitro* studies.

Table II. Fourteen days survival of HVH-infected marmosets treated with human TF (10^8 cells/7 kg)

	Animals infected	Survivors	Survival %		HVH isolated
–3 weeks	4	3	75		
–2 weeks	2	1	50	8/16	15/16
–1 weeks	4	2	50	50%	
–3 days	6	2	33		
±0 day	2	0	0		
+1 day	2	0	0		
+3 days	2	0	0	0/13	13/13
+7 days	2	0	0	0%	
Controls	5	0	0		

Protection study. Initial TF_d injections, in a dose equivalent to that described above, were given as listed in table II. Control animals did not receive any TF_d. Animals which received TF_d at –3 weeks, –2 weeks, or –1 week were also treated with TF_d at –3 days. After day 0, TF doses were repeated twice weekly until death or 14 days. On day 0, 0.1 ml of HVH-1 with a titer of $10^{6.3}$ $TCID_{50}$/0.2 ml was instilled into each nostril. TF prepared from marmosets was administered at –3 days prior to challenge with HVH-1.

Test Performed

Skin tests were performed in the usual manner by injecting 0.1 ml of the antigenic material intradermally into the volar aspect of the forearm and read at 24 and 48 h. A skin test was considered positive if the amount of induration was 5 mm or greater.

The lymphocyte blastogenesis assay employed in this study has been described [9]. A blastogenic index (BI) was derived by dividing counts per minute (cpm) of lymphocyte cultures containing specific antigens by cpm of lymphocyte cultures containing medium or control cells. A BI of ≥ 2.5 was considered a positive response.

Results

Table I summarizes skin tests and blastogenic responses of marmosets following administration of human TF_d and baboon TF. Cell-mediated immunity reaction was transferred to more than half of the human TF_d positive antigens evaluated either by skin tests and/or by blastogenic responses. Transfer was most successful with the antigens monilia and tetanus, and HVH-1. No animal became positive to the TF_d negative antigens. Efficacy of baboon TF was considerably lower than that of human

TF_d. Only reactivity to monilia was transferred when skin tests were applied. In addition to monilia, the lymphocyte transformation assay revealed transfer of reactivity to tetanus and mumps. No animal converted its response to PPD, the negative control antigen.

A summary of the protection provided by human TF_d is presented in table II. Only treatment with TF_d prior to inoculation with HVH-1 provided protection. When treatment was begun at –3 weeks 75% survived; at –2 weeks and –1 week, 50%; and at –3days, 33%. This resulted in a total of eight out of 16 or 50% survival. In contrast, none of the animals survived which were not treated, or TF_d treatment was begun at time of infection or thereafter. Although TF_d did provide protection in 50% of all cases, it did not prevent infection. This was supported by the fact that HVH-1 was isolated on one or several occasions from 28 of 29 marmosets from either throat swabs taken postinfection (twice weekly) or organs from dead animals at necropsy.

In addition, when TF prepared from marmosets which had been treated with TF_d and had survived HVH-1 infection, was given to two untreated marmosets both animals survived a subsequent HVH-1 challenge.

Discussion

These studies demonstrate conversion of *in vitro* and *in vivo* cellular reactivity in marmosets following an *in vivo* administration of human or baboon TF. The cumulative conversion ratios of 53% (skin test) and 54% (blastogenesis) after administration of human TF_d and 20% and 40% of baboon TF, respectively, was in a similar range as for baboon and cebus monkey TF recipients [9]. The relative immune deficient marmosets [3, 6, 7] were therefore able to respond to xenogeneic TF application as were more immune competent nonhuman primate species like the baboon and cebus monkey. However, examination of more animals may have revealed more subtle species differences.

The efficacy of transfer of cellular reactivity by human TF_d was superior to the baboon TF. However, this difference could be accounted for by the fact that the human donor was selected on the basis of very high skin test and blastogenic responses to several antigens and the baboon donors had been immunized one year prior to TF preparation and had varying degrees of reactivity (TF pooled).

Specificity of transfer was maintained by the fact that positive responses were only detected with TF positive antigens and no positive responses could be detected with TF negative antigens.

Several statements pertaining to the protection study can be made. Pretreatment of marmosets was required for protection to occur, since no animals survived the HVH-1 challenge when treatment was begun at or postinfection. Although TF did confer 50% protection from disease induced by HVH-1, it did not prevent infection. This was evidenced by HVH-1 isolation from 28 of 29 animals at one or several occasions from throat swabs taken postinfection or organs from dead animals. In addition, specificity of TF_d was supported also *in vivo* by the failure of this TF preparation specific for HVH-1 to protect three pretreated marmosets against *Herpesvirus saimiri*-induced malignant lymphoma.

The marmoset with its relative functional immunocompetency [3, 6, 7] offers a unique opportunity to correlate cellular immune reactivity to acute and neoplastic viral diseases. In the present HVH-1 infection model, TF_d appeared to be of significant benefit.

Acknowledgements

The authors wish to thank Janice Morrison, Isidoro Chapa, and David Lawlor for their technical assistance.

Summary

Using skin test response and lymphocyte blastogenesis as indicators of cell-mediated immunity, we have been able to demonstrate *in vivo* transfer of cell-mediated immunity to marmosets both with dialyzable transfer factor (TF_d) prepared from a human donor and transfer factor (TF) from baboon whole cell lysates. We were able to protect marmosets with TF_d from fatal *Herpesvirus hominis* type 1 (HVH-1) disease. When TF_d was administered prior to challenge with HVH-1, 50% of the marmosets survived. Of the untreated control animals and those first treated on day 0 or postinfection, 100% succumbed to disseminated HVH-1 disease. In addition, when TF prepared from TF_d-treated marmosets which had survived HVH-1 disease, was given to other marmosets, it conferred protection from subsequent HVH-1 challenge.

References

1 Deinhardt, F.; Wolfe, L.G.; Theilen, G.H., and Snyder, S.P.: ST-feline fibrosarcoma virus. Induction of tumors in marmoset monkeys. Science, N.Y. *167:* 881 (1970).

2 HARRINGTON, J.T., jr.: Macrophage migration from an agarose droplet. A micromethod for assay of delayed hypersensitivity in the mouse. Cell. Immunol. *12:* 476–480 (1974).

3 HARVEY, J.S., jr.; FELSBURG, P.J.; HEBERLING, R.L.; KNIKER, W.T., and KALTER, S.S.: Immunological competence in non-human primates. Differences observed in four species. Clin. exp. Immunol. *16:* 267–277 (1974).

4 HOLMES, A.W.; WOLFE, L.; ROSENBLATE, H., and DEINHARDT, F.: Hepatitis in marmosets. Induction of disease with coded specimens from a human volunteer study. Science, N.Y. *165:* 816–817 (1969).

5 KALTER, S.S.; FELSBURG, P.J.; HEBERLING, R.L.; NAHMIAS, A.J., and BRACK, M.: Experimental *Herpesvirus hominis* type 2 infection in nonhuman primates. Proc. Soc. exp. Biol. Med. *139:* 964–968 (1972).

6 KALTER, S.S.; HEBERLING, R.L.; FELSBURG, P.J.; MCCULLOUGH, B.; EICHBERG, J. W.; KNIKER, W.T.; HARVEY, J.S., jr.; MACIAS, E.G.; ELLER, J.J.; ROMMEL, F.A., and STEELE, R.W.: Immunological aspects of herpesvirus infections in primates; in DE-THÉ, EPSTEIN and ZUR HAUSEN Oncogenesis and herpesviruses, vol. II, pp. 133–136, IARC Publ. No. 11 (International Agency for Research on Cancer, Lyon 1975).

7 KNIKER, W.T.; MACIAS, E.G.; STEELE, R.W.; HEBERLING, R.L., and ELLER, J.J.: Relative cellular immunity in three nonhuman primate species. Fed. Proc. Fed. Am. Socs exp. Biol. *34:* 824 (1975).

8 MELENDEZ, L.V.; HUNT, R.D.; DANIEL, M.D.; GARCIA, F.G., and FRASER, C.E.O.: *Herpesvirus saimiri.* II. Experimentally induced malignant lymphoma in primates. Lab. Anim. Care *19:* 378–386 (1969).

9 STEELE, R.W.; EICHBERG, J.W.; HEBERLING, R.L.; ELLER, J.J.; KALTER, S.S., and KNIKER, W.T.: *In vivo* transfer of cellular immunity to primates with transfer factor prepared from human or primate leucocytes. Cell. Immunol. *22:* 110–120 (1976).

J.W. EICHBERG, DVM, PhD, Department of Microbiology and Infectious Diseases, Southwest Foundation for Research and Education, PO Box 28147, *San Antonio, TX 78284* (USA)

Prim. Med., vol. 10, pp. 277–279 (Karger, Basel 1978)

Hepatitis in Marmosets: An Introduction

F. Deinhardt

Departments of Microbiology, Rush-Presbyterian-St. Luke's and University of Illinois Medical Centers, Chicago, Ill.

Viral hepatitis remains a disease of major public health importance; in the USA it ranks fourth in incidence, behind gonorrhea, chicken pox and mumps, on the list of notifiable infectious diseases. Specific control measures (chemotherapy and vaccination) are available for gonorrhea and mumps and the third disease, chicken pox, is generally mild and of no major public health importance, making hepatitis the single, most common, notifiable and serious infectious disease for which no specific treatment exists. About 50,000–60,000 cases of the three recognized forms of hepatitis, hepatitis A, B and 'non-A–non-B' hepatitis, are reported annually in the USA; this figure probably should be multiplied at least by two to obtain a more accurate incidence of clinical disease and by an even greater factor for an estimate of the total number of infections. On a worldwide basis, the World Health Organization reports about one million cases annually. As with the figures from the USA, this number can only represent a fraction of actual infections and the gap between reported and actual cases is probably far greater. These figures illustrate the urgent need for the development of control measures but despite strenuous efforts by many scientists advances in understanding the epidemiology and pathogenesis of human hepatitis came slowly, largely because attempts to isolate the causative agents of all three forms of hepatitis were unsuccessful for so long. During a period when the causative agents of most of the common virus diseases of man were isolated and characterized in great detail, all efforts to isolate the hepatitis viruses in cell culture systems or to transmit the disease to laboratory animals failed.

Transmission of viral hepatitis from man to some species of marmosets (*Saguinus* spp.) and the regular passage of the virus from animal to animal in the 1960s [1, 2] provided the first model system (other than human volunteers) for further investigation of hepatitis. A number of laboratories confirmed the early reports [for a general review and individual references, see DEINHARDT, 3] and although the first isolate in marmosets, the so-called GB strain, proved not to be classical hepatitis A, transmission of hepatitis A strains already well-characterized in human volunteers (MS-1 strain of Willowbrook) and isolations from various outbreaks of typical hepatitis A from North and South America and from Western Germany soon followed. It can now be said that some species of marmosets are regularly susceptible to hepatitis A viruses originating in various parts of the world and that all these viruses seem to be antigenically identical or at least closely related and Dr. HILLEMAN discusses in detail the differences in susceptibility of various *Saguinus* spp.

The original isolate GB and an isolate made by KÖHLER *et al.*, the so-called Berlin strain [for review see DEINHARDT, 3] appear to be antigenically identical and to differ from hepatitis A; they are probably related to human non-A–non-B hepatitis. Several attempts to transmit hepatitis B to marmosets failed and this primate family is probably of little importance for the study of this form of hepatitis. Studies of hepatitis A and of 'GB-hepatitis' (non-A–non-B) have furthered the characterization of various aspects of the pathogenesis of hepatitis and the identification of the actual hepatitis A virus, viremic periods, virus excretion, changes in lymphocyte populations during the disease and antibody development, as discussed by Dr. PETERSON, have been studied in experimentally infected marmosets, and apart from acute phase human feces, infected marmoset livers remain until now the only source of hepatitis A virus antigen for the development of diagnostic tests.

The final identification of the GB strain still remains unsettled but it is hoped that marmosets will provide, as they have for hepatitis A, at least the antigens for a definite identification of this disease as one form of non-A–non-B hepatitis, thus enabling the development of diagnostic tests for this third form of human viral hepatitis. Marmosets will continue to play a major role in our efforts to prevent or eliminate viral hepatitis through public health measures and the development of viral vaccines. Nevertheless the pressing need for their use in this field of biomedical research must not conflict with the need to conserve those species whose survival in the wild is threatened. Laboratory use of primates must be limited to those experiments for which no alternative model exists, and the essential needs of hepatitis and other biomedical research for experimental animals must be met more and more by the establishment of marmoset breeding colonies. On the other hand, an earnest plea must be made to the legislative bodies which control the conservation of primates: efforts to safeguard the wide range of primate species in the wild must be increased by controls over the ever-encroaching demands of civilization on the natural habitats of nonhuman primates, a far greater danger than the demands made by biomedical research, and in the interests of us all, the need for certain numbers of feral animals, to sustain the efforts to eliminate serious and worldwide diseases, until enough breeding colonies have been established must be recognized and met.

References

1 DEINHARDT, F. and HOLMES, A.W.: Epidemiology and etiology of viral hepatitis; in POPPER and SCHAFFNER Progress in liver diseases, vol. II, pp. 373–394 (Grune & Stratton, New York 1965).

2 DEINHARDT, F.; HOLMES, A.W.; CAPPS, R.B., and POPPER, H.: Studies on the transmission of human viral hepatitis to marmoset monkeys. I. Transmission of disease, serial passage and description of liver lesions. J. exp. Med. *125:* 673–689 (1967).

3 DEINHARDT, F.: Hepatitis in primates. Adv. Virus Res. *20:* 113–157 (1976).

Dr. F. DEINHARDT, Max v. Pettenkofer Institut, Pettenkoferstrasse 9a, *D–8000 München 2* (FRG)

Prim. Med., vol. 10, pp. 280–287 (Karger, Basel 1978)

Virus Excretion, Antibody Development and Changes in Rosette Formation of Lymphocytes during Non-B Hepatitis in Marmosets[1]

D. A. Peterson, F. W. Deinhardt, L. G. Wolfe, D. R. Johnson, W. T. Hall and G. G. Froesner

Department of Microbiology, Rush-Presbyterian-St. Luke's and University of Illinois Medical Centers, Chicago, Ill.; Electro-Nucleonics Laboratories, Inc., Bethesda, Md., and Max v. Pettenkofer Institute, University of Munich, Munich

Introduction

Although intensely studied for the last 55 years, viral hepatitis remains the last major public health problem for which no specific treatment exists. With the discovery of the so-called Australia antigen or hepatitis B surface antigen (HBsAg) [2] major inroads have been made in understanding hepatitis B which currently accounts for only 20–40% of the reported viral hepatitis [4]. In the last five to six years advances in hepatitis A (HA) have been possible because of (1) human volunteer studies which provided clinical characterization of the disease and specimens for future evaluation [3, 14], (2) development of the marmoset experimental model [7, 8], (3) demonstration of the etiological relationship of the 27-nm particle in feces with HA [12], (4) development of techniques to detect HA virus (HAV) and antibody to HAV [12, 16, 17] and (5) development of the chimpanzee experimental model [10]. Thus the model systems and techniques have provided a better understanding of HA and have enabled us to study the following aspects of the disease: (a) the interrelationships between acute and inapparent (subclinical) disease, (b) occurrence and duration of HAV excretion, (c) development of antibody to HAV (anti-HAV), (d) effect on peripheral T lymphocyte function and (e) are plasma factors produced which affect T lymphocytes?

1 This work was supported in part by research contract 74–130 Food and Drug Administration, by the Richard B. Capps Liver Research Fund of Rush-Presbyterian-St. Luke's Medical Center and by a grant from the Deutsche Forschungsgemeinschaft (Fr 400/3). We thank the City of Chicago for housing most of our experimental animals.

Materials and Methods

Animals and inocula. Twenty adult white-lipped marmosets *(Saguinus fuscicollis, Saguinus nigricollis)* were divided into four groups: (A) six inoculated intramuscularly (i.m.) with 0.5 ml of a 6.6% (w/v) human fecal suspension containing HAV (GBG isolate), (B) six inoculated i.m. with 0.5 ml of a 4% (w/v) human fecal suspension containing HAV (F. day 38 isolate) which had previously induced subclinical disease and seroconversion in marmosets; (C) four inoculated i.m. with 0.5 ml of non-A–non-B (1:50 dilution, acute phase serum, marmoset passage 11, GB strain) human hepatitis, (D) four uninoculated controls. The marmosets were evaluated for transaminase levels (SGOT/SGPT) weekly, and liver biopsies were performed at two week intervals [6, 7].

Detection of HAV. Preinoculation and postinoculation serial stool specimens were frozen at –20 °C after collection from animals inoculated with HAV. Tests for HAV or associated antigens were performed on 2% stool filtrates by the solid phase radioimmunoassay (RIA) system of Ling and Overby [15] using human anti-HAV serum.

Detection of anti-HAV. Serum and/or plasma specimens were tested for anti-HAV by immune adherence hemagglutination [17] and by a RIA blocking assay.

T lymphocyte E-rosette formation. Marmoset lymphocytes were separated by Ficoll-Hypaque sedimentation and tested for E-rosette formation with sheep erythrocytes by the technique of Jondal *et al.* [13] at weekly intervals. A minimum of 200 lymphocytes were examined in duplicate samples and the percentage of lymphocytes with ≥ 3 erythrocytes attached was determined. Average values of E-rosette forming peripheral lymphocytes in normal adult marmosets was 65% with a range of 54–72%.

Rosette inhibitory factor (RIF). Tests for a serum factor, which might inhibit E-rosette formation by lymphocytes *in vitro* were performed as described by Chisari *et al.* [6]. Lymphocytes taken from a marmoset more than four weeks after recovery from infection with GB strain hepatitis were incubated before evaluation with weekly samples of autologous plasma obtained after infection with GB virus.

Results

The results of experimental infection of marmosets are shown in table I. In group A, four of six marmosets inoculated with HAV (GBG isolate) developed acute hepatitis 29–43 days postinoculation (DPI) which persisted for one to three weeks. HAV was detectable by RIA in the feces from eight to 48 DPI, and the excretion period of ten to 20 days generally preceded maximal SGPT elevations by six to 21 days; anti-HAV was detectable one to two weeks after peak SGPT levels (results obtained from one animal are illustrated in figure 1). Two animals which did not develop abnormal trans-

Table I. Non-B hepatitis in marmoset monkeys

Inoculum	No. hepatitis / No. inoculated	No. seroconv.[1] / No. tested	Mean incubation period, days
Hepatitis A:			
GBG; 6.6% fecal susp. 0.5 ml, i.m.	4/6	6/6	34
F, d38; 4% fecal susp. 0.5 ml, i.m.	0/6	4/6	–
GB strain, p. 11; 1:50 dil. serum; 0.5 ml.	4/4	0/4	17
Uninoculated controls	0/4	0/4	–

[1] Preinoculation and convalescent sera tested for anti-HAV.

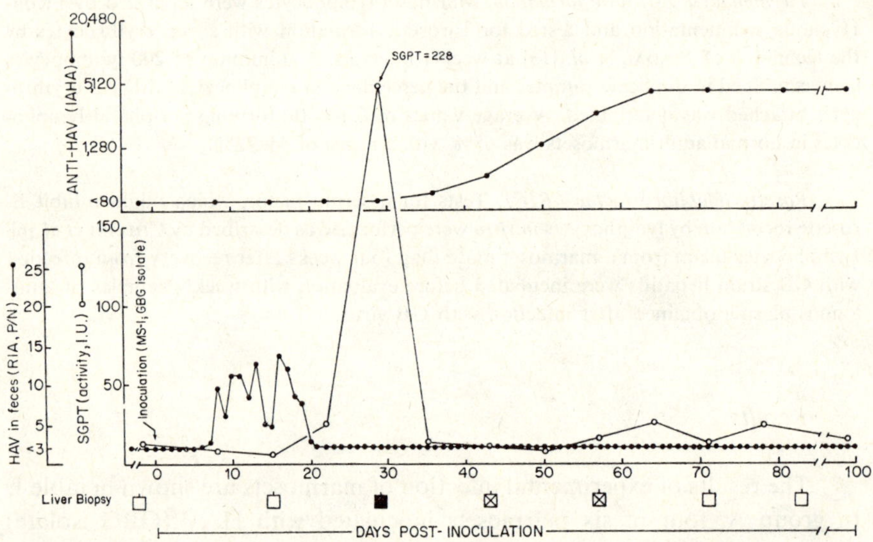

Fig. 1. Acute hepatitis in a marmoset experimentally infected with HAV, serial determinations of liver histology (normal histology, □; acute hepatitis, ■; resolving lesions, ☒; serum alanine aminotransferase (O) activity (SGPT, IU); HAV excretion (●) in stools (solid phase RIA, positive/negative ratio of counts per minute, P/N); and antibody to HAV (anti-HAV determined by immune adherence hemagglutination, IAHA, expressed as inverse titer).

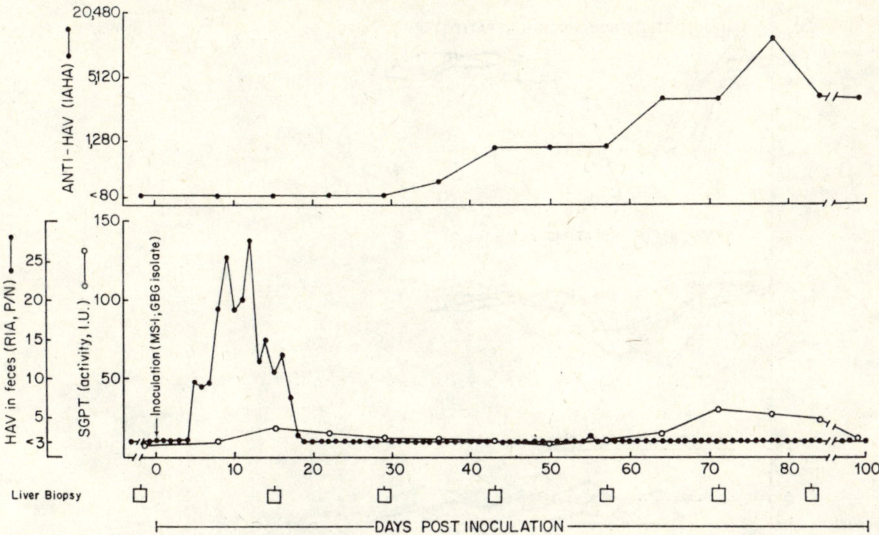

Fig. 2. Inapparent hepatitis in a marmoset experimentally infected with HAV, serial determination of liver histology, SGPT activity, HAV excretion in stool and anti-HAV. See figure 1 for abbreviations and symbols.

aminase levels or morphologic liver changes had inapparent HAV infections as demonstrated by HAV fecal excretion (five to 19 DPI) and development of anti-HAV by 36 DPI (one animal illustrated in figure 2). Selected fecal specimens from the HAV excretion periods in animals with acute or inapparent disease were examined by immune electron microscopy and revealed typical 27-nm HAV particles.

Group B marmosets inoculated with HAV (F d38 isolate) developed slightly increased transaminase levels, with no morphological evidence of hepatitis, and four of six developed anti-HAV (table I). All animals in group C, inoculated with non-A–non-B hepatitis virus (GB strain) developed enzymatic and morphologic evidence of hepatitis 14–21 DPI which persisted for two to three weeks (table I). Figure 3 shows the percentages of circulating marmoset T lymphocytes forming E-rosettes during the course of infection with human hepatitis and compares them to values in uninoculated control animals. Marmosets inoculated with the GB strain (fig. 3, top panel) showed a significant decrease ($p<0.001$) of E-rosette formation slightly before and during the acute phase of the disease (14–28 DPI); E-rosette-forming cells decreased from 60–65% in preinoculation samples to 35–40% during the

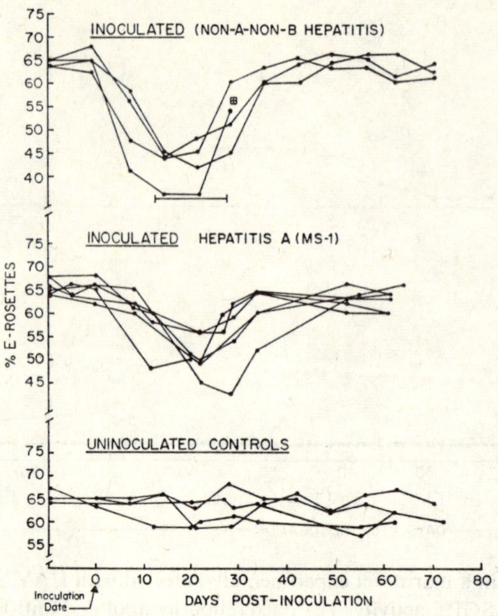

Fig. 3. Peripheral E-rosette-forming lymphocytes of marmosets infected with human viral hepatitis. Lower panel: uninoculated control animals; middle panel; six animals inoculated with HAV (MS-1 prototype); top panel; four animals (⊞ died at 30 days) inoculated with non-A–non-B human hepatitis (GB strain); bar indicates period of elevated plasma transaminase levels.

acute phase of the disease and then returned to baseline values by five weeks postinoculation. Marmosets inoculated with HAV (fig. 3, middle panel) developed less severe, but significant decreases ($p<0.001$) of E-rosette-forming cells during a period which correlated with slight transaminase elevations. CHISARI *et al.* [5, 6] have reported the presence of RIF, a serum lipoprotein, in human hepatitis B patients which inhibits lymphocyte E-rosette formation *in vitro*. Plasma from a marmoset infected with the GB strain of hepatitis was examined for the presence of RIF (fig. 4). The decrease (65–35%) in circulating E-rosette-forming cells (filled circles) observed during the acute phase of the disease (solid bar at 14–28 DPI correlated with enhanced activity of a plasma factor that inhibited E-rosette formation by normal marmoset lymphocytes (open circles). This effect was reversible, as

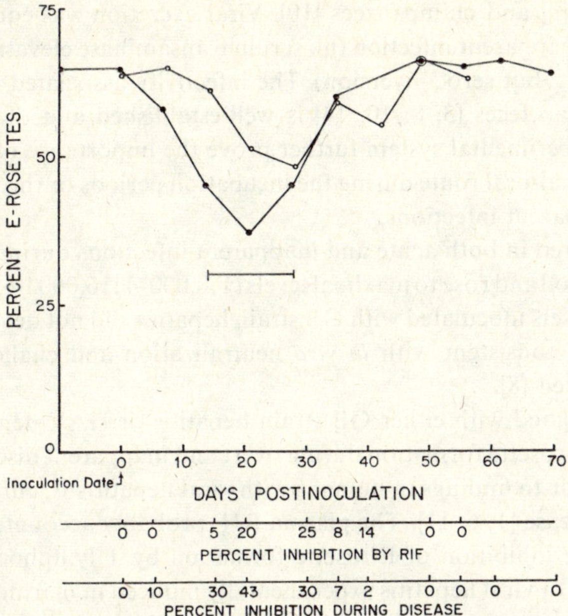

Fig. 4. RIF in plasma of a marmoset with non-A–non-B human hepatitis (GB strain). Peripheral rosette-forming cells (●) during the course of experimental hepatitis, bar indicates period of elevated plasma transaminase levels; rosette-forming cells (O) after incubation *in vitro* of 'normal lymphocytes' (obtained four weeks after transaminase levels and percentage of rosette-forming cells had returned to normal levels) with weekly plasma samples from the same animal during the course of hepatitis.

RIF-treated lymphocytes formed E-rosettes normally, when they were washed, incubated overnight at 37 °C and again washed before the E-rosette test.

The number of circulating E-rosette-forming cells (fig. 3, bottom panel), transaminase levels (table I) and liver histology (table I) in control animals remained normal throughout the observation period.

Discussion

The HAV fecal excretion period of ten to 20 days, which preceded elevation of transaminase levels by six to 21 days, falls within the range found for acute HA in man [9] and chimpanzees [10]. Viral excretion was equally high in animals with inapparent infection (no serum transaminase elevations, no liver histopathology, but seroconversion). The infectivity associated with HAV-containing human feces [3, 8, 10, 14] is well-established and the excretion data in this experimental system further prove the importance of the spread of HA by the anal-oral route during the incubation periods or through individuals with inapparent infections.

Anti-HAV appeared in both acute and inapparent infections during the early convalescent period and rose to maximal levels (1:8,000–1:16,000 IAHA) in 20–30 days. Marmosets inoculated with GB strain hepatitis did not develop anti-HAV and this is consistent with *in vivo* neutralization and challenge data previously reported [8].

Marmosets inoculated with either GB strain hepatitis or HAV demonstrated a decrease in E-rosette formation during overt and inapparent disease. These results are similar to findings in humans with viral hepatitis B, chronic and alcoholic liver disease [1, 6, 11]. The plasma RIF probably accounts for part but not all of the inhibition of E-rosette formation by T lymphocytes during the acute phase of viral hepatitis experimentally induced in marmosets or naturally occurring in man. The decrease of E-rosette formation may be an indication of impaired T lymphocyte functions which probably are important in the control of viral replication and the development of hepatic cytopathology.

Summary

Four of six marmosets infected with HAV developed acute disease; HAV excretion in feces occurred six to 21 days (duration ten to 20 days) before maximal transaminase elevation. Two animals with inapparent disease exhibited a similar excretion of HAV and all six animals developed antibodies to HAV. Significant decrease in percentages of circulating T lymphocytes forming E-rosettes occurred during the course of non-B human hepatitis in marmosets.

References

1 Berstein, I.M.; Webster, K.H.; Williams, R.C., and Strickland, R.G.: Reduction in circulating T lymphocytes in alcoholic liver disease. Lancet *ii:* 488–490 (1974).

2 Blumberg, B.S.; Alter, H.J., and Visnich, S.: A 'new' antigen in leukemia sera. J. Am. med. Ass. *191:* 541–546 (1965).

3 BOGGS, J.D.; MELNICK, J.L.; CONRAD, M.E., and FELSHER, F.: Viral hepatitis: clinical and tissue culture studies. J. Am. med. Ass. *214:* 1041–1046 (1970).

4 BRYAN, J.A. and GREGG, M.B.: Viral hepatitis in the United States 1970–1973: an analysis of morbidity trends and the impact of HBsAg testing on surveillance and epidemiology. Am. J. med. Sci. *270:* 271–282 (1975).

5 CHISARI, F.V. and EDGINGTON, T.S.: Lymphocyte E-rosette inhibitory factor: a regulatory serum lipoprotein. J. exp. Med. *142:* 1092–1107 (1975).

6 CHISARI, F.V.; ROUTENBERG, J.A., and EDGINGTON, T.S.: Mechanisms responsible for defective human T lymphocyte E-rosette function associated with hepatitis B virus infections. J. clin. Invest. *57:* 1227–1238 (1976).

7 DEINHARDT, F.; HOLMES, A.W.; CAPPS, R.B., and POPPER, H.: Studies on the transmission of human viral hepatitis to marmoset monkeys. I. Transmission of disease, serial passage and description of liver lesions. J. exp. Med. *125:* 673–689 (1967).

8 DEINHARDT, F.; PETERSON, D.; CROSS, G.; WOLFE, L., and HOLMES, A.W.: Hepatitis in marmosets. Am. J. med. Sci. *270:* 73–80 (1975).

9 DIENSTAG, J.L.; FEINSTONE, S.M.; KAPIKIAN, A.Z., and PURCELL, R.H.: Fecal shedding of hepatitis-A antigen. Lancet *i:* 765–767 (1975).

10 DIENSTAG, J.L.; FEINSTONE, S.M.; PURCELL, R.H.; HOOFNAGLE, J.H.; BARKER, L.F.; LONDON, W.T.; POPPER, H.; PETERSON, J.M., and KAPIKIAN, A.Z.: Experimental infection of chimpanzees with hepatitis A virus. J. infect. Dis. *132:* 532–545 (1975).

11 EDGINGTON, T.S. and CHISARI, F.V.: Immunological aspects of hepatitis B virus infection. Am. J. med. Sci. *270:* 213–227 (1975).

12 FEINSTONE, S.M.; KAPIKIAN, A.Z., and PURCELL, R.H.: Hepatitis A: detection by immune electron microscopy of a virus-like antigen associated with acute illness. Science *182:* 1026–1028 (1973).

13 JONDAL, M.; HOLM, G., and WIGZELL, H.: Surface markers on human T and B lymphocytes. I. A large population of lymphocytes forming nonimmune rosettes with sheep red blood cells. J. exp. Med. *136:* 207–215 (1972).

14 KRUGMAN, S.; WARD, R., and GILES, J.P.: The natural history of infectious hepatitis. Am. J. Med. *32:* 717–728 (1962).

15 LING, C.M. and OVERBY, L.R.: Prevalence of hepatitis B virus antigen as revealed by direct radioimmune assay with ^{125}I-antibody. J. Immun. *109:* 834–841 (1972).

16 MILLER, W.J.; PROVOST, P.J.; MCALEER, W.J.; ITTENSOHN, O.I.; VILLAREJOS, V.M., and HILLEMAN, M.R.: Specific immune adherence assay for human hepatitis A antibody. Application to diagnostic and epidemiologic investigations (38783). Proc. Soc. exp. Biol. Med. *149:* 254–261 (1975).

17 MORITSUGU, Y.; DIENSTAG, J.L.; VALDESUSO, J.; WONG, D.C.; WAGNER, J.; ROUTENBERG, J.A., and PURCELL, R.H.: Purification of hepatitis A antigen from feces and detection of antigen and antibody by immune adherence hemagglutination. Infec. Immunity *13:* 898–908 (1976).

D.A. PETERSON, PhD, Department of Microbiology, Rush-Presbyterian-St.-Luke's Medical Center, 1753 West Congress Parkway, *Chicago, IL 60612* (USA)

Prim. Med., vol. 10, pp. 288–294 (Karger, Basel 1978)

Tests in *Rufiventer* and other Marmosets of Susceptibility to Human Hepatitis A Virus

P. J. PROVOST, V. M. VILLAREJOS and M. R. HILLEMAN

Division of Virus and Cell Biology Research, Merck Institute for Therapeutic Research, West Point, Pa., and Louisiana State University, International Center for Medical Research and Training, San José

Introduction

White moustached marmosets, *Saguinus mystax,* have been the most suitable species for studies of human hepatitis A virus [1–3] and for preparing hepatitis A antigens for serologic tests for antibody [4, 5]. Embargoes placed on export of these animals from the Amazon basin in 1974 prevented their further use at that time and a suitable substitute was sought. Tests were carried out to measure and compare susceptibilities of six different marmoset species to CR326 strain human hepatitis A virus and to hepatitis A virus in plasmas from human hepatitis A cases. The marmoset species included *S. mystax, S. weddelli, S. o. oedipus, Callithrix jacchus,* and *C. argentata* as described by NAPIER and NAPIER [6] and an additional marmoset of imprecise definition that is simply referred to as the *rufiventer* marmoset. Marmosets similar to but perhaps not identical with the *rufiventer* marmoset have been designated by HERSHKOVITZ [7] as *Marikina labiata (Jacchus rufiventer)* and by NAPIER and NAPIER [6] as *S. labiatus.* All animals used in our studies were legally wild-caught and were purchased through licensed importers.

Material and Methods

In testing for susceptibility, liver extract of *S. mystax* marmosets infected with *S. mystax* marmoset-adapted CR326 (fourth passage) human hepatitis A virus was used. Primary isolation of hepatitis A virus from human specimens was made using serum and clot extract (1:5 dilution) from Costa Rican patient 033–03 and using serum (1:12) from volunteer K. who had been infected with MS-1 strain human hepatitis A virus [8]. The latter was kindly furnished by Dr. M. CONRAD. Serum glutamic oxaloacetic transaminase

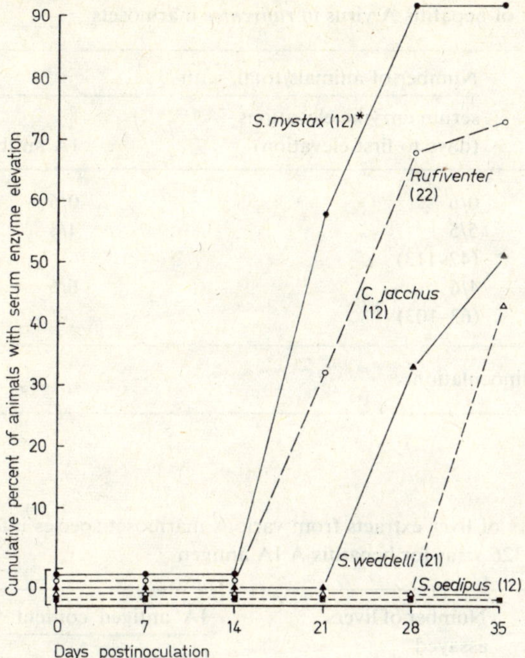

Fig. 1. Comparison according to time and percent of serum enzyme elevations in five species of marmosets injected with CR326 strain human hepatitis A virus. In parentheses number of animals.

(SGOT) and serum isocitric dehydrogenase (SICD) enzyme assays were performed on the plasmas obtained from the marmosets at weekly intervals. Tests for hepatitis A antibody were by the immune adherence (IA) assay. To test for hepatitis A antigen, livers were harvested from the infected marmosets at the time of the first pronounced serum enzyme elevation. The livers were processed by cesium chloride density gradient separation and tested for antigen content in the IA assay. The best antigen preparations had titers of 1:4 or higher ($\geq$4 units). Details of the procedures employed have already appeared in previous publications [1–5, 9].

Results

Comparison of susceptibility of marmoset species to mystax-adapted CR326 hepatitis A virus. Figure 1 shows the percent of animals with serum enzyme elevations, according to time, following injection with CR326 hepatitis A mormoset liver extract. Greatest susceptibility was indicated by

Table I. Primary isolation of hepatitis A virus in *rufiventer* marmosets

Inoculum	Number of animals/total, with serum enzyme elevations (days to first elevation)	IA antibody[1]
Saline solution	0/6	0/6
Patient 033–03	5/5 (42–113)	4/5
K. day 30 (MS–1)	4/6 (63–103)	6/6

[1] Through 21 weeks postinoculation.

Table II. Results of assays of liver extracts from various marmoset species infected with *mystax*-adapted CR326 virus for hepatitis A IA antigen

Type of marmoset	Number of livers assayed	IA antigen content, units		
		0	2	4
S. mystax	12	3	2	7
Rufiventer	14	9	5	0
C. jacchus	8	8	0	0
C. argentata	9	9	0	0
S. weddelli	5	5	0	0

Table III. Evidence for adaptation to *rufiventer* marmosets of *mystax*-adapted CR326 hepatitis A virus

Passage level in *rufiventer* marmoset	Inoculum (liver dilution)	Numbers infected total (%)	Days to first enzyme elevation	Number of livers with ≥ 4 units IA antigen/total (%)
1	1:1500 (mystax)	18/22 (82)	21	0/14 (0)
2	1:100	20/24 (80)	14	1/12 (8)
3	1:100	23/24 (96)	7	3/18 (17)
4	1:100	23/23 (100)	7	7/16 (44)
5	1:100	12/12 (100)	7	6/12 (50)

the shortest incubation period and the largest percentage of animals with enzyme elevation. *S. mystax* and *rufiventer* marmosets were the most susceptible species. *C. jacchus* and *S. weddelli* were less susceptible, and *S. o. oedipus* gave no evidence of infection by this method of detection. *C. argentata* marmosets, not shown in the figure, behaved similarly to *C. jacchus* animals. Importantly, even though 50% of *C. jacchus* animals showed enzyme elevations through day 35, these were generally at low level and less than the 200 and 2,000 SGOT and SICD values, respectively, considered indicative of infection in *S. mystax* marmosets [1].

Hepatitis A virus in human specimens. Table I shows that *rufiventer* marmosets were very susceptible to hepatitis A virus in human specimens. Nine of 11 animals showed enzyme elevations within 42–113 days and all but one of the animals (033–03) developed hepatitis A antibody. Importantly, patient 033–03 was the same individual from whom the CR326 strain of hepatitis A virus was originally isolated [1].

Hepatitis A antigen titers in livers of marmosets infected with mystax-*adapted CR326 virus.* Livers of infected marmosets from the experiments shown in figure 1 and from *C. argentata* were perfused and harvested at the time of first pronounced serum enzyme elevation. Most livers were harvested within the 35-day period shown in figure 1. Table II shows that antigen detectable by IA was present only in infected livers from *S. mystax* and *rufiventer* marmosets. The antigen was most commonly found and was present in highest titer ($\geq$1:4) in *S. mystax* liver.

Adaptation of CR326 virus to rufiventer *marmosets.* The *mystax*-adapted CR326 virus was passed serially for five passages in *rufiventer* marmosets and the IA antigen titer was measured in individual marmoset livers at each passage level. As shown in table III, the *mystax*-adapted CR326 strain hepatitis A virus was readily adapted to *rufiventer* marmosets on serial passage. This was evidenced by (a) increase in the percent of animals infected, (b) progressive shortening of the incubation period for disease (time of first enzyme elevation) and (c) progressive increase in the percent of liver extracts with an IA titer $\geq$1:4.

Use of rufiventer-*adapted CR326 virus in diagnostic tests.* Serum samples from hepatitis A and B cases and from normal persons were tested for hepatitis IA antibody content using *mystax*- or *rufiventer*-adapted CR326

Table IV. Hepatitis A antibody titers of human sera tested using *mystax*-adapted and *rufiventer*-adapted CR326 hepatitis A virus antigen

Subjects	Case	Time of specimen, days	IA antibody titer	
			mystax antigen[1]	*rufiventer* antigen[2]
Hepatitis A cases	033–02	− 21	<5	<5
		+ 36	20,480	≥2,560
		+102	40,960	≥2,560
	056–08	− 57	<5	<5
		− 28	<5	<5
		+185	10,240	≥2,560
	343–09	− 6	<5	<5
		+ 3	800	160
		+104	12,800	≥2,560
Hepatitis B cases	039–01	−108	<5	<5
		+ 10	5	<5
		+149	<5	<5
	039–03	− 27	2,560	≥2,560
		+ 7	5,120	≥2,560
		+192	2,560	≥2,560
Normal controls	2	–	1,280	640
	16	–	<5	<5
	20	–	<5	<5

[1] Tests performed September 1974.
[2] Tests performed March 1976.

virus as antigen. Table IV shows that essentially identical results were obtained, indicating that the virus retained its original antigen specificity and hepatitis A antibody-measuring capacity on serial passage in *rufiventer* marmosets.

Discussion

The present findings show that the *rufiventer* marmoset is a suitable substitute species for *S. mystax* in studies of human hepatitis A both for initial virus recovery from human specimens and for production of antigen for use in the IA assay. Additional studies, not recorded in the text, showed the presence in infected *rufiventer* marmoset liver of typical 27-nm particles previously demonstrated by us in infected *mystax* marmosets [3] and by FEINSTONE *et al.* [10] in human stools. Further, the density of the *rufiventer*-adapted CR326 virus was found to be in the 1.32–1.36 g/cm^3 range based on cesium chloride density gradient centrifugation, being identical with our previous findings with *mystax* liver-derived hepatitis A antigen [3, 5].

The early appearance of serum enzyme elevation in *rufiventer* marmosets, within seven days following inoculation at passage levels three, four and five, was earlier than seen with *mystax* marmosets in our previous studies [1–3]. However, this is probably due to the use of more potent inocula in the present work. Livers from such 'early onset' animals in a number of cases gave excellent yields of hepatitis A IA antigen (titer $\geq$1:4). These findings, in the marmoset at least, remove hepatitis A from the category of 'slow-growing' virus and indicate that hepatitis A can be the result of an acute liver infection by the virus probably without need for a secondary factor, such as an autoimmune response [11] to liver tissue antigens. A number of additional passages of CR326 virus have been made in *rufiventer* marmosets with no important changes noted either with respect to time of onset of disease or content of antigen in the liver. Such passages are continuing to be made.

Patient 033–03 from whom hepatitis A virus was isolated in *rufiventer* marmosets in the present study was the same individual from whom the original CR326 virus isolation was made in *mystax* marmosets. Thus, two CR326 strains have been derived, one by original passage in *mystax* marmosets and the other by original passage in *rufiventer* marmosets.

Though none of the five additional marmoset species tested other than *rufiventer* was a satisfactory replacement for *S. mystax,* it should be recorded that all species presented evidence of infection with human hepatitis A. Even though serum enzyme elevations were low or absent, a portion of animals of all species developed hepatitis A IA antibody.

Summary

The *rufiventer* marmoset proved equally satisfactory to *S. mystax* for studies of human hepatitis A virus. *C. jacchus*, *C. argentata*, *S. weddelli*, and *S. oedipomidas oedipus* were not satisfactory. Livers of *rufiventer* marmosets produced satisfactory CR326 strain hepatitis A antigen for immune adherence tests both in amount and specificity. *Rufiventer* marmosets infected with human hepatitis A virus showed enzyme elevations and high titers of viral antigen in their livers as early as seven days after viral inoculation, indicating that a primary viral infection can cause hepatitis without need for a secondary autoimmune response to liver tissue.

Acknowledgements

We gratefully acknowledge the expert technical assistance of F.S. BANKER, Dr. P.A. CONTI, W.P.M. FISHER, J.R. GEHRET, P.A. GIESA, B.J. HARDER, Dr. O.L. ITTENSOHN, M. JOHNSTON, and R.A. MACHLOWITZ.

References

1 MASCOLI, C.C.; ITTENSOHN, O.L.; VILLAREJOS, V.M.; ARGUEDAS G., J.A.; PROVOST, P.J., and HILLEMAN, M.R.: Recovery of hepatitis agents in the marmoset from human cases occurring in Costa Rica. Proc. Soc. exp. Biol. Med. *142:* 276–282 (1973).

2 PROVOST, P.J.; ITTENSOHN, O.L.; VILLAREJOS, V.M.; ARGUEDAS G., J.A., and HILLEMAN, M.R.: Etiologic relationship of marmoset-propagated CR326 hepatitis A virus to hepatitis in man. Proc. Soc. exp. Biol. Med. *142:* 1257–1267 (1973).

3 PROVOST, P.J.; WOLANSKI, B.S.; MILLER, W.J.; ITTENSOHN, O.L.; MCALEER, W.J., and HILLEMAN, M.R.: Physical, chemical and morphologic dimensions of human hepatitis A virus strain CR326. Proc. Soc. exp. Biol. Med. *148:* 532–539 (1975).

4 PROVOST, P.J.; ITTENSOHN, O.L.; VILLAREJOS, V.M., and HILLEMAN, M.R.: A specific complement-fixation test for human hepatitis A employing CR326 virus antigen. Diagnosis and epidemiology. Proc. Soc. exp. Biol. Med. *148:* 962–969 (1975).

5 MILLER, W.J.; PROVOST, P.J.; MCALEER, W.J.; ITTENSOHN, O.L.; VILLAREJOS, V.M., and HILLEMAN, M.R.: Specific immune adherence assay for human hepatitis A antibody. Application to diagnostic and epidemiologic investigations. Proc. Soc. exp. Biol. Med. *149:* 254–261 (1975).

6 NAPIER, J.R. and NAPIER, P.H.: A handbook of living primates (Academic Press, London 1967).

7 HERSHKOVITZ, P.: Mammals of northern Colombia. Preliminary report No. 4: Monkeys (Primates), with taxonomic revisions of some forms. Proc. US natn. Mus. *98:* 323–427 (1949).

8 BOGGS, J.D.; MELNICK, J.L.; CONRAD, M.E., and FELSHER, B.F.: Viral hepatitis. Clinical and tissue culture studies. J. Am. med. Ass. *214:* 1041–1046 (1970).

9 VILLAREJOS, V.M.; PROVOST, P.J.; ITTENSOHN, O.L.; MCLEAN, A.A., and HILLEMAN, M.R.: Seroepidemiologic investigations of human hepatitis caused by A, B, and a possible third virus. Proc. Soc. exp. Biol. Med. *152:* 524–528 (1976).

10 FEINSTONE, S.M.; KAPIKIAN, A.Z., and PURCELL, R.H.: Hepatitis A. Detection by immune electron microscopy of a viruslike antigen associated with acute illness. Science *182:* 1026–1028 (1973).

11 Editorial: What about type-A hepatitis? Lancet *ii:* 1007–1008 (1973).

P.J. PROVOST, PhD, Division of Virus and Cell Biology Research, Merck Institute for Therapeutic Research, *West Point, PA 19486* (USA)

Prim. Med., vol. 10, pp. 295–299 (Karger, Basel 1978)

Experimental Infection of Marmosets with Hepatitis A Virus

J. W. Ebert, J. E. Maynard, D. W. Bradley, D. Lorenz and D. H. Krushak

Phoenix Laboratories Division[1], Bureau of Epidemiology, Center for Disease Control, Phoenix, Ariz. and Division of Blood and Blood Products, Bureau of Biologics, Food and Drug Administration, Bethesda, Md.

Introduction

The marmoset has become an accepted model of human disease for the study of hepatitis A [3–8]. Experimental infection of these animals with hepatitis A virus (HAV) is accompanied by elevated SGPT and SICD activities (reflecting hepatic damage), histopathologic changes in liver, and the acquisition of specific antibody to HAV. This report briefly describes three studies in which marmosets were intravenously infected with HAV-positive acute-illness phase serum or stool materials.

Materials and Methods

Wild-caught *Saguinus mystax*, *S. nigricollis*, and *S. oedipus oedipus* received from commercial suppliers were conditioned for a period of six weeks prior to being inoculated with HAV. They were fed a mixture of commercially available, high-protein monkey biscuits, applesauce, and vitamin supplements. Water was supplied *ad libitum*. Animals were bled weekly or twice weekly by femoral venipuncture, and liver biopsies were performed biweekly with a 1.2-mm Menghini needle. The liver tissue was fixed in 10% buffered formalin and liver sections were stained with hematoxylin and eosin. Activities of alanine aminotransferase (SGPT) and isocitrate dehydrogenase (SICD) were determined by standard Sigma colorimetric methods.

Some animals were killed immediately following the first elevation of SGPT or SICD activity. Homogenates of their livers were examined by immune electron microscopy (IEM) for the presence of HAV particles.

1 World Health Organization Collaborating Centre for Reference and Research on Viral Hepatitis.

In study I (passage 1), ten *S. mystax* were each given 0.25 ml of MS-1 human acute-illness phase hepatitis A serum supplied by Dr. S. KRUGMAN. In study I (passage 2), acute phase marmoset serum from passage 1 (collected between seven and 14 days prior to elevated enzyme activities) was pooled and inoculated into animals in passage number 2. Animals in succeeding passages (numbers 3 and 4) were each inoculated with 0.25 ml of a serum pool derived (as described above) from the preceeding passage. In study II, acute-illness phase human plasma from a patient associated with an outbreak of foodborne hepatitis A in Phoenix, Ariz. [2], was inoculated into six *S. mystax*. In study III, cesium chloride (CsCl) density gradient fractions containing HAV were inoculated into a total of seven marmosets. An acute phase chimpanzee stool preparation was isopycnically banded on a CsCl gradient. Two major buoyant density species of HAV were detected by IEM in this gradient. A primary HAV peak was found at 1.32 g/cm^3, while a secondary peak was found at 1.41 g/cm^3 [1, 2]. Aliquots of 0.05 ml of the 1.32 and 1.41 g/cm^3 gradient fractions were inoculated into four and three animals, respectively. One hundred and thirty-three days following inoculation, three animals inoculated with the 1.32 g/cm^3 gradient fraction (one *S. mystax* and two *S. nigricollis*) were challenged with a 1.39 g/cm^3, HAV-positive CsCl gradient fraction derived from the banding of another acute phase chimpanzee stool.

Results

Study I. Of the experimentally inoculated animals, 42–88% showed enzymatic evidence of infection by HAV (table I). Most striking was the finding that serial passage of HAV in marmosets led to decreased incubation periods, the shortest being 25–34 days for animals in the fourth passage. Of particular interest was the finding that early acute phase livers from two of the fourth passage animals contained HAV.

Study II. All six animals showed enzymatic evidence of infection with HAV. In two of these animals, liver homogenates examined by IEM, were found to contain HAV particles. Two other animals tested had antibody at 96 days after inoculation.

Study III. Four animals inoculated with the 1.32 g/cm^3 CsCl gradient fraction, and three inoculated with the 1.41 g/cm^3 gradient fraction showed signs of infection 21–49 days following inoculation. Three of the animals originally inoculated with the 1.32 g/cm^3 CsCl gradient fraction showed no signs of infection when challenged with the 1.39 g/cm^3 gradient fraction described above.

Table I. Studies of hepatitis A in marmosets

Study Passage	Species	n	Inoculum[1]	Incubation[2]	Enzyme elevations[3]	Histo-pathology	Virus in liver by IEM	Antibody by IEM
I/1	*S. mystax*	10	MS-1	49–63	5/10	5/10	NT	6/10
I/1	*S. mystax*	8	control		0/8	0/8	NT	NT
I/2	*S. mystax*	19	M-1 (MS-1)	35–45	8/19	7/19	NT	4/16
I/2	*S. mystax*	5	control		0/5	0/5	NT	NT
I/3	*S. mystax*	7	M-2 (MS-1)	28–38	3/7	3/7	NT	3/7
I/4	*S. mystax*	8	M-3 (MS-1)	25–34	7/8	6/8	2/3	2/2
II	*S. mystax*	6	human plasma	27–41	6/6	3/6	2/2	2/2
III	*S. nigricollis*	2	1.32	21, 28	2/2	1/2	NT	2/2
	S. mystax	2	1.32	21, 49	2/2	1/2	NT	2/2
	S. oedipus	1	1.41	31	1/1	1/1	NT	1/1
	S. mystax	2	1.41	21, 24	2/2	2/2	NT	2/2

[1] MS-1 = human acute phase hepatitis serum; M-1 to M-3 = marmoset acute phase hepatitis serum; 1.32, 1.42 = CsCl density gradient fractions, g/cm³.

[2] Days postinoculation to first enzyme elevation.

[3] Number of animals with SGPT activity greater than 50 Sigma units or SICD activity greater than 1,500 Sigma units.

Discussion

The shortened incubation periods seen in study I may reflect either an increase in the titer of HAV in serum from animals in subsequent passages, or an adaptation of the virus to the marmoset. Differences in the titer of the inoculum, or in the susceptibility of marmosets to different strains of HAV may account for the variation in the percentage of animals showing evidence of disease. The correlation between serum enzyme elevations and abnormal liver histopathology was highest when biopsies were obtained during the peak of SGPT activity. The occasional failure of histopathologic examination to reveal evidence of viral hepatitis may be related to the relatively long (two week) intervals between liver biopsies. The animal with the 49-day incubation period (study III) may have secondarily acquired an HAV infection from its cage-mate. All other animals in this study demonstrated incubation periods ranging between 21 and 31 days.

Marmosets have been shown to be susceptible to infection with HAV under experimental conditions. In the absence of a tissue culture system for the propagation of HAV, the marmoset appears to be particularly well-suited for studies of viral pathogenesis (replication), and for titration of HAV-positive materials. Livers obtained from marmosets during the acute phase of disease may also provide HAV in quantities suitable for diagnostic tests for anti-HAV, including radioimmunoassay and immune adherence hemagglutination. Breeding of marmosets in captivity should produce a reliable, and readily available, biomedical model for the study of viral hepatitis A.

Summary

Saguinus mystax marmosets were experimentally infected with two strains of human hepatitis A virus. One of these strains of HAV was successfully subpassaged in this species of marmosets. In another experiment, the 1.32 and 1.41 g/cm^3 buoyant density species of HAV derived from an infected chimpanzee stool were shown to be infectious in three species of marmosets. The value of the marmoset as an experimental model for hepatitis A infection was demonstrated by these studies.

Acknowledgement

The authors gratefully acknowledge the services of Dr. S. FEINSTONE, Dr. L. WOLFE, Mr. C. GRAVELLE, and Mr. C. HORNBECK.

References

1 BRADLEY, D.W.; HORNBECK, C.L.; GRAVELLE, C.R.; COOK, E.H., and MAYNARD, J.E.: CsCl banding of hepatitis A-associated virus-like particles. J. infect, Dis. *131:* 304–306 (1975).

2 GRAVELLE, C.R.; HORNBECK, C.L.; MAYNARD, J.E.; SCHABLE, C.A.; COOK, E.H., and BRADLEY, D.W.: Hepatitis A. Report of a common source outbreak with recovery of a possible etiologic agent. II. Laboratory studies. J. infect. Dis. *131:* 167–171 (1975).

3 HOLMES, A.W.; DEINHARDT, F.; WOLFE, L.; FROESNER, G.; PETERSON, D., and CASTRO, B.: Specific neutralization of human hepatitis type A in marmoset monkeys. Nature, Lond. *243:* 419–420 (1973).

4 HOLMES, A.W.; WOLFE, L., and DEINHARDT, F.: Infectious hepatitis in marmosets; in Pathology of simian primates, part II, pp. 684–701 (Karger, Basel 1972).

5 KRUSHAK, D.H.; MAYNARD, J.E.; BRADLEY, D.W., and HORNBECK, C.L.: Hepatitis A in marmosets. J. med. Primatol. *4:* 341 (1975).

6 Lorenz, D.; Barker, L.; Stevens, D.; Peterson, M., and Kirschstein, R.: Hepatitis in the marmoset, *Saguinus mystax*. Proc. Soc. exp. Biol. Med. *135:* 348–354 (1970).
7 Maynard, J.E.; Lorenz, D.; Bradley, D.W.; Feinstone, S.M.; Krushak, D.H.; Barker, L.F., and Purcell, R.H.: Review of infectivity studies in nonhuman primates with virus-like particles associated with MS-1 hepatitis. Am. J. med. Sci. *270:* 81–85 (1975).
8 Provost, P.J.; Ittensohn, O.L.; Villarejos, V.M.; Arguedas G., J.A., and Hilleman, M.R.: Etiologic relationship of marmoset propagated CR326 hepatitis A virus to hepatitis in man. Proc. Soc. exp. Biol. Med. *142:* 1257–1267 (1973).

J.W. Ebert, VMD, Phoenix Laboratories Division, Bureau of Epidemiology, Center for Disease Control, 4402 North Seventh Street, *Phoenix, AZ 85014* (USA)

Author Index

Subject Index[1]

[1] Submitted by Mary S. Hart, January 1978.

U.C.W. ABERYSTWYTH
LLYFRGELL